Robots for Shearing Sheep

Robots for Shearing Sheep

Shear Magic

James P. Trevelyan
Technical Consultant
Automated Sheep Shearing Project
University of Western Australia

Oxford New York Melbourne Tokyo
OXFORD UNIVERSITY PRESS
1992

Oxford University Press, Walton Street, Oxford OX2 6DP

Oxford New York Toronto
Delhi Bombay Calcutta Madras Karachi
Petaling Jaya Singapore Hong Kong Tokyo
Nairobi Dar es Salaam Cape Town
Melbourne Auckland

and associated companies in
Berlin Ibadan

Oxford is a trade mark of Oxford University Press

Published in the United States
by Oxford University Press, New York

A catalogue record for this book is available from the British Library

Library of Congress Cataloging in Publication Data
Trevelyan, James P.
Robots for shearing sheep: shear magic/James P. Trevelyan.
Includes bibliographical references and index.
1. Sheep-shearing. 2. Robotics. I. Title
SF379.T74 1992 636.3' 145–dc20 91-34936

ISBN 0-19-856252-7

Set by
Pentacor PLC, High Wycombe, Bucks
Printed by St. Edmundsbury Press, Bury St. Edmunds, Suffolk

Preface

This is a book about an extraordinary adventure in robotics. It is an adventure that has captivated the interest of many people, myself included. Robots shearing sheep, no less. I thought the idea ridiculous when it was first suggested to me; technically feasible, yes, but who could afford to build one? I underestimated the determination of the wool industry—they have supported this amazing research project through all the technical difficulties and sometimes frustrating delays.

I am writing this book to present a coherent account of the sheep-shearing project, right from its earliest origins. I am writing it for the robotics and artificial intelligence research community, because their early achievements inspired the imaginations of many farmers who thought there had to be a better way of taking wool from the backs of sheep.

The story begins in 1972 and is not finished yet—a long time for one project. Yet, at a time when we often read about 'the ever increasing rate of technological change' it might be reassuring. Technology is changing—there is no doubt about that, but the rapid rate of change is an illusion, a convenient one which helps to sell computers. The real rate of technological change is much slower than we are led to believe, and certainly is no more dramatic and has no greater impact on our lives than the technological changes which happened a century ago. My eldest child has nearly finished his schooling, yet the *last* time men landed on the moon was two years before his birth!

While this is a book about robots, it is also about the people who built them. Few people realize the extent to which personalities dominate technical decision making. Engineering is an applied science: this is the illusion of an 'exact' discipline where design is the logical outcome of the laws of physics and mathematical equations. Real engineering, both in research and application, is a human activity, and like every human activity it is complex, unpredictable and can only be rationalized with slanted hindsight.

Sheep-shearing robots are a uniquely Australian achievement. Ridiculed by 'international experts' a decade ago, they are now hailed by the same experts as an example of just how clever robots can be.

Yet this too is an illusion. 'Intelligent robots', like 'intelligent computers', are illusions. If it were not for their apparent utility, the programmers and designers could be seen as magicians; just putting on a show.

To me, the greatest irony of research in artificial intelligence is that it shows how shallow our concepts of mind and intelligence really are. When computers beat grand masters at chess, we all said 'wow'. We all agreed that you have to be really intelligent to play chess that well. Yet, we now realize we all share perception and thinking skills which we haven't even begun to understand. All we have learned is that the thinking which *we associate* with intelligence is the easiest part to do with computers.

Robots and computers are useful, and will become more useful as we learn more. Yet not as replacements for human workers. The machines of the industrial revolution extended our physical abilities. Computers and robots also extend our control, thinking and perceptual abilities but only in a limited sense. People are an essential part of the picture. If we fail to understand the human contribution, we head for disaster.

The sheep-shearing project is an attempt to replicate a manual skill in machinery. It has been an immense technical challenge and a thrilling one to meet. Yet the outcome is a greater understanding and respect for the human skill of shearing.

To me, the most rewarding part of this challenge has been to discover simplicity within the technical complexities emerging from our workshops and laboratory. Part of my job has been to present our work to visitors, farmers, colleagues, shearers and students. Many people instinctively react with the phrase 'it must be so complicated!' Yet I found that the effort of explaining the technical side in simple, concise, non-technical language has been rewarded by the discovery of simplicity. An idea is only simple if it can be communicated simply, yet it may be much harder to communicate an idea simply than to think of it in the first place. One of the maxims for successful engineering is expressed by the KISS acronym—'Keep it simple, SIMPLE'. To be useful, robots must not only be seen to be simple, they must also be simple. Our elegant solutions have often emerged from this process of explaining, and this book is a part of that process.

I believe that the skill of explaining technical concepts in simple terms has been an important contribution to the success of the sheep-shearing project. Robots can only be built by teams of specialists cooperating with each other. And not all the specialists have engineering skills—public relations, accounting, shearing and sewing have been just as important. The only common language they share is English, free of technical jargon. Successful cooperation relies on good communication skills; a common language and the ability to use it.

I also believe that the explanations we have used are as valuable as the technical ideas which have emerged within them. I have found them valuable for teaching, and other robotics specialists may find them useful in other contexts.

Robots will be more useful if more people understand how they work and what they can and cannot do. As long as we robot researchers hide behind a veil of arcane language and complex technicalities, we will keep robots to ourselves. Jargon provides some protection too. Researchers often accidentally 'reinvent wheels' at great cost to their sponsors. However, if the reinvented wheel is called a 'high transverse mobility load bearing element' or 'HTMLB element' for convenience, the sponsor usually feels much better.

Has it all been worthwhile?

Shearers' backs still break with the sheer physical stress and strain, day after day, week after week, more demanding than any professional sport. On average, men stop shearing after ten years, many with cumulative injuries for which there is little compensation. Perhaps in respect of the myth we all like to perpetuate, only a handful of the eligible few line up to claim their payouts. The legendary Australian shearer—the iron man of the outback who works hard, drinks hard and plays hard—is a myth built on broken backs and arthritic joints.

Now robots provide an alternative. The new challenge is to take them out of our laboratory and consign the hard labour of hand shearing to the myths and legends where it belongs. That challenge is now one of economics and politics and is still to be accepted.

Yet no machine can work effectively without the human skills and experience of the shearers who will look after it. The machine serves only as an extension of human activity, not as a replacement.

The farmers who thought that rows of obedient robots would replace the truculant, independent and often rebellious shearers may be in for a surprise. The new shearers will be articulate, highly trained and experienced. Cooperation will be the same path to success that it always has been. The confrontations which led to the great shearers' strikes last century, and gave birth to the Australian union movement, may be as obsolete as shearing blades.

Acknowledgements

The research described in this book was inspired and paid for by Australian woolgrowers through contributions to the Wool Research and Development Fund, managed by the Australian Wool Corporation.

I would like to thank all the people who have contributed to this project and for the support they have given me. A full list appears on p. 361: I apologize for accidental omissions.

While writing this book has been an independent project, it would not have been possible without the encouragement of the Corporation.

I would like to specially thank my colleagues in the research team, for their loyalty, trust, tolerance, and for contributions to this book. David Elford, Peter Kovesi, Jan Baranski, Steve Ridout, and Graham Walker read draft material and contributed useful suggestions. Lynette Lynn provided secretarial assistance and support. Professor D. Allen-Williams proof-read the final draft.

Most of all I would like to thank my wife Jo and children Charles, Nicholas, and Clare for their patient support, love, and understanding.

A note to readers

I have written this book for a technical readership, representing the many specialist backgrounds which contribute to robotics. I have attempted to strike a balance in the text between presenting a readable account and providing useful technical details. Much of the book is devoted to the major technical results, but this is presented within a framework which recounts some of the history too. Again, I have tried to balance the need for a chronological account with the need for coherent descriptions of our techniques. A chronology has been included at the end of the book. A few important events are described in detail to convey the nervous tension and excitement or humour that accompanied them. To do otherwise would create a false impression of orderly relaxed progression from one stage to another.

We have often needed to communicate technical concepts to farmers, shearers and other people without a technical background. Explanations which we have found useful for this are presented in boxes, or boxed subsections to keep them separate from the main account. A glossary of shearing terms has been included for reference.

Mathematical notation helps to save space and is used to present useful results. Readers can safely skip mathematical sections until they need to make use of these results.

If further technical information is needed readers are invited to write to the author. However, permission will have to be obtained from the Australian Wool Corporation before release of unpublished information.

Notation

I have attempted to be conventional with notation. Scalar variables are referred to by Greek or italicized roman letters, e.g. ϕ *x t dist error*. Vectors are denoted in lower case bold, e.g. **i**, **a**, **frm**. Arrays and matrices, usually used for coordinate frames, are denoted by upper case bold, e.g. **As**, **J**. Where it has been essential to distinguish between the text and computer code, I have used plain text for the codes, e.g. **movetowards**. |**a**| denotes magnitude of **a**, **x** denotes a vector cross product and · denotes a dot product. Several operations are denoted by function names—these are defined in the text.

Contents

1
First steps

First encounters

It was a hot summer day in late January when I had my first experience of mechanized shearing. In fact, I find it hard to remember if I had a close encounter with sheep before.

Muresk Agricultural College lies about 80 km inland from Perth. A dry wind carried the brown dust and smell of grass baked grey by the sun across the gravel tracks at the bottom end of the campus. The glare of sunlight and radiant heat off the corrugated walls of the shed at the end of the track prepared us for the oppressive heat within. In a small pen beside the shed, three sheep stood still and listless in the rippling waves of heat. Once inside, the dry wind cooled perspiring necks, arms and faces in the relative darkness.

A dozen or so expectant people stood quietly waiting, chatting in groups of two or three. Farmers mostly, I thought, and faces which I was to come to know well during the years which followed.

Standing on the wooden floor at one end of the shed was an almost alien machine, far removed from the machines I associate with farming. A two metre high white frame carried a complex array of machined aluminium, bristling with precision slides, exotic sensors and bundles of neatly tied coloured wiring. Below this a shearing handpiece protruded from an articulated mechanism with two shiny wheels to measure the shape of the sheep. Lower down, a bright blue canvas belt draped loosely over two long rollers. Grass seeds and tufts of wool lying in the hollow showed that sheep had been lying on it. White clamps on revolving mountings carried straps to hold legs at each end of the belt.

At the end closer to the doorway was an electronic controller—about the size of a two drawer filing cabinet, with the rear door wide open and two fans blowing warm air into it. A thermometer lay on top of the cabinet indicating 45°C to a worried Jim Brown, a tall English engineer, his pale face more accustomed to the chill of winter fogs at Cambridge. A few more degrees and a hundred or so delicate integrated circuit modules inside would start cooking themselves to oblivion.

A familiar face emerged from under the machine—Roy Leslie, a young graduate student specializing in automatic process control who had been co-opted to help Jim set up the machine. Roy was an asset, not only for his physical size, but also for his farming background. He took me round the machine pointing out some of the modifications which he and Jim had made just days before. Burnt paint on the pristine white frame showed where a steel section had been

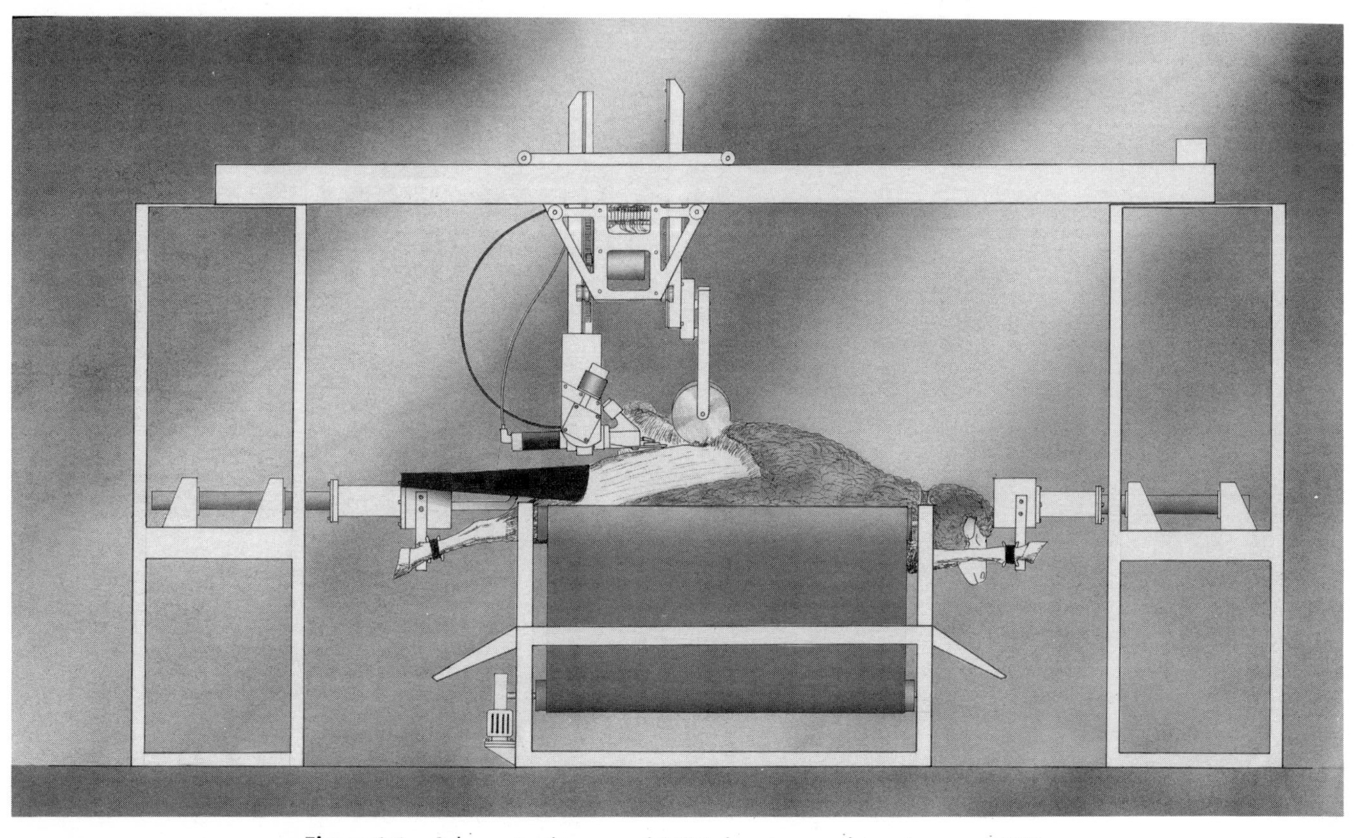

Figure 1.1 Schematic diagram of PATS shearing machine, January 1977.

cut out to accommodate the longer necks of Australian sheep. Every now and then the machine would come to life as motors hummed and Jim and Roy performed last minute adjustments and checks.

I was introduced to Norman Lewis, nephew of Essington Lewis who was a powerful Australian steel industry figure in the early days. Norman was the power behind the push to build an automatic shearing machine. Marjorie Lewis, the power behind Norman, was with him as she was to be on every occasion a critical decision was to be made. Marjorie thought sheep, felt for sheep, had a perfect affinity with sheep. She knew each of the several hundred ewes and lambs in her nucleus flock by sight. With her gentle wrinkly face and thick curly white hair it was difficult not to sense her feelings for the animal which was being lifted onto the cradle.

This was the moment of truth for them both. In 1972 they had witnessed the collapse of woolgrowing communities in the Wyoming pastoral districts of the USA. Caught in a relentless grip between falling wool prices and rising shearing costs, many had been forced to leave their ranches since the climate and land was too harsh for fat lamb production. Norman and Marjorie saw this as the inevitable fate for the Australian sheep industry too, once the safety margin provided by the higher value of the Australian Merino[1] fleece had been eroded away.

On their way back to Australia the Lewises had visited the world famous Stanford Research Institute (SRI), world leader in computers and the new science of robotics. Could the Institute build a robot to shear sheep? The reply had been definitely affirmative, provided enough money could be found. The asking price was very, very high but the Lewises were not easily deterred.

The foresight and vision of the Lewises can only be appreciated by realizing that the very first factory robot, the Unimate, had only just become a commercial proposition. The now familiar lines of nodding orange robots welding cars together in the huge factories of Detroit and Tokyo were just a dream in the mind of the Unimate's inventors, Joe Engelberger and George Devol.[2] Their robot was a clumsy and crude device which had none of the dexterity or sensitivity needed for shearing sheep.

Richard Paul, a young American graduate student brought up in Sydney, had just finished his Ph.D. thesis at Stanford University, a short stroll away from SRI. He and his colleague Victor Scheinman had developed a more precise robot with a far superior control system to the Unimate. It could follow lines in space, or circles, a clear advance on the crude point-to-point jerks of the Unimate but a far cry from the complex curves and heaving of a living breathing sheep.

[1] Merino—the name of the most common sheep breed in Australia. Merino sheep were originally imported from Spain, and have been refined in Australia to produce heavy fleeces of fine wool sought after for clothing.

[2] *Robotics* published by Time Life books provides an excellent and clearly illustrated introduction to the mechanics of industrial robots, and some historical details. See Chapter 2, pp. 35–38 for an account of the Unimate robot.

The Basic Industrial Robot

An explanation for general audiences

A typical industrial robot can be thought of as a positioning machine which carries a tool appropriate for the job it is used for. Thus a welding robot in a car factory carries a welding tool. The robot moves the tool to each position where a weld is needed, and holds the tool there while the weld is made.

A typical robot consists of a mechanical arm with several sliding or hinged joints. The position or angle of each joint is adjusted by an electronically controlled electric motor or hydraulic piston.

The robot is 'programmed' or 'taught' by a technician who uses buttons or switches to adjust each joint so that the tool on the end of the arm is at the right position. The buttons are usually mounted on a portable box at the end of a cable so the technician can work from a good viewing position—the box is called a teach pendant. At each position, the joint angles are memorized by the controller, and the technician adjusts the arm to the next position, and so on until an entire sequence has been memorized.

Once this has been done, the robot performs its task by repeating the memorized sequence of movements endlessly. Nearly all of the robots now in use in factories around the world are at most minor elaborations of this basic machine.

These simple robots work provided they can repeat the taught movements accurately enough, and provided the parts which they make or use are the same, time after time.

Even at Stanford, the first mobile robot was still just a curiosity. It could respond to very simple commands like 'go through the door'. However, every aspect of the robot's environment had to be accurately defined in its program. How could a woolly sheep be accurately defined?

On their return to Western Australia (WA), the Lewises started to collect support for their dream. They were no strangers to long and difficult research projects. They had pioneered the use of clover to boost the capacity of their land from 0.1 sheep per acre to 1 sheep per acre and then had observed the disastrous effects of some clover varieties on pregnant ewes. They found the blood group factor responsible and are still improving fertility levels.

The Lewises formed 'The Southern Districts Sheep Research Council' (SDSRC) with friends and neighbours and collected research funds. 'Ten cents a head' was their slogan. They raised $40,000 and started to look for a firm to carry out the engineering work. Their fellow members of the SDSRC executive were not so fiercely independent, and insisted that pressure be applied to the Australian Wool Corporation and ministers for agriculture in the state and federal governments. They saw that the SDSRC funds would not be enough. Their first formal application was rejected by the Corporation in November 1973. The Corporation favoured the idea of chemical shearing as a long term solution and mechanization of sheep handling in shearing sheds as an intermediate step. Shearing robots were considered 'way out' and impractical.

The Corporation had just set up its own attack on the problem—the 'Australian Wool Harvesting Programme'.

CSIRO,[3] with long experience of research in animal physiology and biochemistry, had suggested biological defleecing. A pill or injection could cause the sheep to stop growing wool for a short time; each wool fibre would have a break or weakened zone so the wool could be pulled off after a few weeks had passed. This occurs naturally with some sheep, and often follows severe stress caused by illness. But they realized that it could take as long as five to ten years to find a safe drug to use.

A second effort was aimed at coordinating and building on the efforts of inventive farmers who had devised machines to help with sheep handling. The Hamilton brothers had built their own production line[4] shearing system on their farm in Victoria and had shorn their flock of 10,000 sheep with it. Tragically, the shed was burned to the ground during the night after an open day. Other farmers, particularly in Western Australia, had been inventing sheep-shearing cradles, sheep-handling machines to reduce the need for pushing the sheep through races. Farmers all round the country were cooperating and pooling their resources, trying each others' equipment and suggesting new ideas. There was an expectant atmosphere and a 'breakthrough' was thought to be imminent.

The Lewises and their friends were sceptical of chemical and biological defleecing. Marjorie, with her long experience and deep knowledge of fertility and reproduction, knew that a ewe with a broken fleece would not produce a lamb. How could a drug powerful enough to disrupt wool growth not have the same effect? Mechanized shearing still relied on human shearers leaving farmers as vulnerable to union action as they had been since the shearers' strikes of the 1890's heralded the union movement in Australia. But Norman also had a more distant vision. He had seen how harvesting machines had revolutionized the wheat industry in Australia and around the world, and was convinced that wool-harvesting machines could have the same impact. He saw the involvement of people in shearing as the major obstacle to expansion of the wool industry.

Events in Sydney were soon to prove him right—in part. Justice Mary Gaudron of the Australian Conciliation and Arbitration Commission ruled that the rate for shearing sheep be raised to 43.5 cents, a rise of nearly 50% on the

[3] Commonwealth Scientific and Industrial Research Organisation—a government agency with over 100 research divisions in all aspects of science and technology.

[4] This was known at the time as 'chain shearing'. Sheep were suspended from moving cradles on an overhead track, hanging upside down by their legs. A number of shearers worked either side of a 'production line', each shearing a particular part of the sheep, and pushing the animals on to the next shearer. It was claimed that less skill was required to obtain a higher tally per shearer, and that there was less physical stress.

rate for the previous season.[5] Far from sheep and shearing sheds, with just the stroke of a pen, she provided the single most pressing stimulus for robot shearing. The executive of the SDSRC met the next day in Wagin and decided to follow Norman's advice and commit the funds they had raised to go it alone without waiting for the Corporation to act. PA Technical and Science (PATS) Centre in Cambridge, England, was commissioned to start work on Norman's machine.

The Corporation reacted to this by forming a working group to look at the idea of automating sheep shearing. They sought the views of inventive farmers, engineers, and others, including the Lewises and Lance Lines,[6] an engineer turned farmer in South Australia. Lance had his own ideas for a shearing machine, but was keeping them to himself while collaborating with an independent West Australian inventor, Denton Roberts. They urged the Corporation to support some preliminary experiments with a single automated shearing blow along the back of a sheep.

Finally, in November 1975 with labour costs rising at record rates and wool prices still falling, the SDSRC persuaded the Corporation to back the PATS Centre work with the $250,000 needed to complete the experimental machine. They managed more than that—they retained all rights to the machine. Years were to pass before the Corporation discovered their embarrassing mistake.

The Corporation did not put all their eggs in one basket. CSIRO physicist, David Henshaw had become convinced by his experiences on the Corporation's working group that automatic shearing was possible, and in his laboratory in Geelong he had already started working on a machine to shear a single blow along the back of a sheep. John Clegg, also on the working group, had suggested that David provide the machine with a means of sensing the skin of the sheep at the points of the comb.[7] He had discovered that a small electrical current could pass between the comb and the sheep skin—as the comb was pressed harder on to the skin, the electrical resistance to the current was reduced. He and David coined the term 'resistance sensing'—see p. 146 for details.

David had realized that it would not be necessary to measure the profile of a sheep if a machine could sense the skin. Feedback control could be used to adjust the position of the shearing handpiece automatically. This would have the advantage of responding automatically to animal movements such as breathing or struggling.

The Corporation provided CSIRO with funds and the race was on to build the world's first sheep-shearing machine. Within weeks, by mid 1975, David had

[5] Australian Conciliation and Arbitration Commission Reports for 21st June 1974 show a rise from $29.90 piecework rate to $31.44. On 9th September, this was raised to $43.50. Comparable rises were awarded to wool classers and shed hands, and shearing costs were further raised by increases in allowances to cover 4 weeks leave with a 17.5% pay loading and a week's sick leave to bring them into line with other workers. Thus over two shearing seasons, with high subsequent cost of living adjustments, the cost of shearing appeared to nearly double.

[6] Lance Lines went on to start Merino Wool Harvesting Pty Ltd—see p. 190.

[7] 'A note on sensors', 18th June 1975.

Feedback Control

The essence of feedback control is sensing. A sensor is used to measure the state of the process we wish to control and some input to the process is adjusted on the basis of the measurement. This happens continuously, keeping the process automatically controlled.

For example, a sensor in an air-conditioner measures the temperature of the room air. If the air is too hot, the cooler is turned on. If the temperature is too low, then a heater is turned on. If the temperature is about right, then just the fan is left running to circulate the air in the room.

Feedback control is very effective and is used for nearly all automatically controlled machines.

It relies on two most fundamental requirements, often called the 'Laws of Control'. First, the process must be observable—that is, there must be a means of sensing the state of the process we wish to control. Second, the process must be controllable—that is, there must be some means by which the process can be adjusted to alter its state.

the first machine ready. The front of the handpiece could be driven up and down by one electric motor and the back end by another. The motors were driven by the outputs of powerful amplifiers operating on the electric current measurements.

The sheep was strapped on to a wooden beam mounted on a trolley which could be moved under an archway. The handpiece hung from the middle of the arch on a long lever which was moved up and down by the motors. When the sheep emerged through the archway, a neat strip of wool had been shorn along its backbone. David was elated and pressed on with a more complex machine which would be able to shear the side of a sheep.

Norman's idea was quite different. Instead of using electronic sensors to detect the skin in contact with the comb, he decided to use a pair of profile measuring wheels which would measure the shape of the sheep, rolling through the wool just ahead of the cutter. The controller stored the profile in its memory and the shearing handpiece was moved by electric motors to follow the stored profile. Within months of starting, the principles had been demonstrated, again with a single successful blow along the back of a sheep.

Until now, neither of these developments had penetrated the peace and tranquillity of the beautifully landscaped campus of The University of Western Australia at Crawley, nestling between the bushland of Kings Park and the Swan River. However, at the end of 1975 the head of the Department of Mechanical Engineering, Professor David Allen-Williams, met the local Corporation manager who told him about the developments in wool harvesting. Prof. A-W (as we called him) was always on the look-out for industrially relevant research programmes, particularly ones which could attract financial

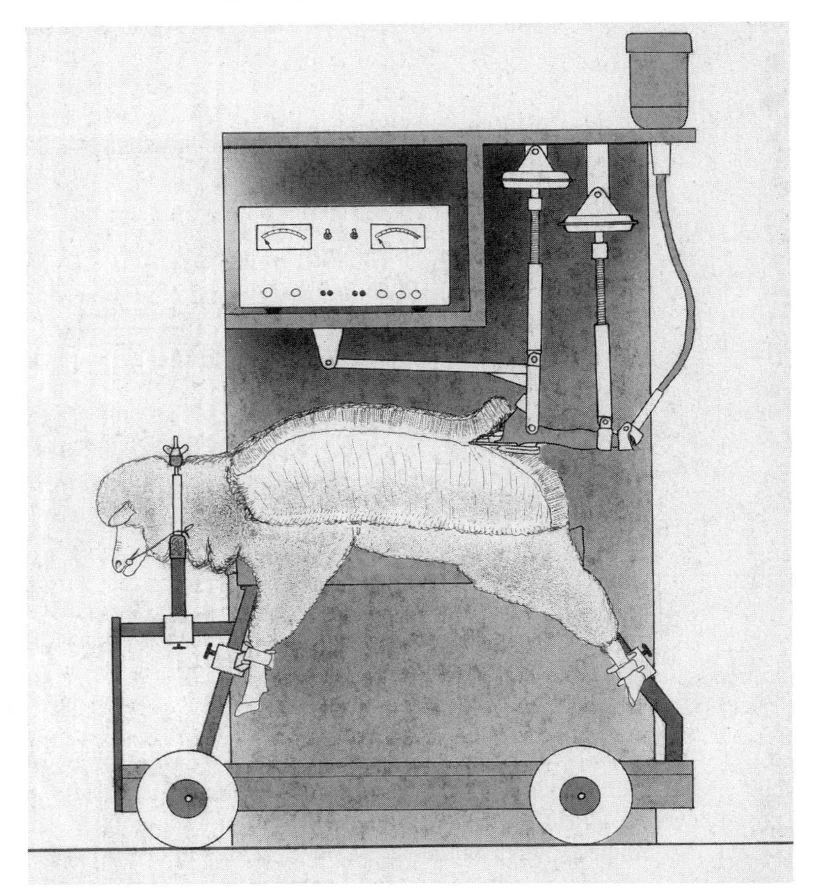

Figure 1.2 Schematic diagram of CSIRO shearing machine for single blow, June 1975.

support. When he learned of the enormous sums being expended on the wool harvesting research programme, he was particularly keen to meet the Corporation's development engineer, Alan Richardson. A meeting was arranged in Sydney, and Alan was persuaded to present a seminar during his next visit to Perth.

I still have a vivid recollection of the light hearted afternoon tea conversation before the seminar on March 26th. I was both amused and incredulous when Prof. A-W mentioned the idea of farmers spending tens of thousands of dollars on a robot. But Roy Leslie quickly pointed out that wheat farmers were prepared to pay hundreds of thousands of dollars on a harvester to use for just a few days per year. I thought again and went along to the seminar.

Alan Richardson presented an amazing story. He showed photographs of experimental production line shearing sheds with sheep hung upside down for shearing. Other sheep stood naked with pink wrinkly skin, their fleece having fallen out through the action of chemicals. Vague drawings of the two shearing machines being built were flashed in front of us too quickly to take it all in. Yet the one fact that remained in all of our minds was the amount of money being spent to find a solution. Hundreds of thousands of dollars, even millions were being spent. For us, a ten thousand dollar research grant seemed like a king's ransom, yet here was a quarter of a million being spent overseas in the belief that there was no expertise in Australia. We knew we could do better at a fraction of the cost. Alan was hard pressed to answer the questions and we were left with an invitation to join in the race for a solution.

Meanwhile, Jim Blair, who was in charge of research in control and computer applications in the department, had written a proposal to commence research on a sheep-shearing system. To avoid the appearance of outright competition with the CSIRO and SDSRC projects, the proposal focused on linking the components of a shearing system together by control computers, rather than the more narrow focus of the other projects on the shearing process itself.

By September, Roy Leslie had started with modest funding which was sufficient for a study tour to see sheep-handling devices which had already been invented in WA and the eastern states.

By the end of 1976, the first act was nearing its climax. PATS had commissioned their machine and reported good results with initial shearing tests at Cambridge, but wanted to develop additional refinements before more testing was undertaken. The Corporation was concerned about the relevance of these tests, and was keen to see the machine tried with Australian sheep. English sheep are round, with open fleeces and firm flesh—ideal subjects for the machine to work with. The Corporation decided to mount a review of the PATS machine in January 1977 at Muresk, WA, and of the CSIRO machine at Geelong, Victoria. Further funding would depend on the results.

Although I had maintained a casual interest in these developments, my own interests lay elsewhere. Apart from an active teaching programme, I had consulting contracts, and an interest in learning about the newly appearing microprocessors—computers in single integrated circuits which had occupied whole rooms just a decade before. However, the Corporation had requested an independent technical expert from the university for their review, and I was asked. I accepted eagerly.

In January, the PATS machine arrived at Muresk. Roy Leslie was diverted from his work to help Jim Brown set up the machine. It seems astonishing to me now that the first Australian Merino sheep was tried only days before the review date. Major surgery was required (to the machine) to allow the long neck of the Merino to be accommodated. Further adjustments were needed for the body too, with the result that the first shearing took place just the afternoon before the reviewers arrived. The next morning, it was apparent that there was a mismatch between the cutter and the comb—a slightly wider Australian comb had been fitted by mistake. Fortunately there were suitable alternative parts available to allow standard Australian combs to be fitted.

So after five years for the Lewises, the moment of truth had finally arrived.

Jim Brown started by placing his hand on a wooden board laid across the rollers of the machine. The shearing head of the machine started to move towards him, the rollers slowly climbing over his hand, followed by the sharp teeth of the shearing comb following the measured profile faithfully just millimetres above his skin. It was an impressive demonstration.

The board was removed and a sheep placed on the belt, with its legs stretched out at each end and secured in two pairs of clamps. The rollers turned, and with them the sheep. The leg clamps were mounted on rotating shafts so they could keep pace with the animal. Once on its side, a stiff rubber cover was laid over the rear legs to provide a 'landing strip' for the rollers to run down to the sheep skin. The machine was started and a dozen faces watched in suspense as the rollers, followed by the cutter, climbed up on to the side of the sheep, hidden from view by the vibrating wool. The sheep twitched slightly and the tension mounted. Finally, with the cutter nearing the end of its run, it lifted clear of the sheep and returned automatically to begin another blow where it stopped. The rollers hummed for a moment and the sheep was turned slightly. We all moved in to see what had happened. Anxious arms pulled away the shorn wool to reveal an evenly shorn strip with a slight jiggle where the sheep had moved. We moved back for another blow, and then another after which a reasonably clear patch of sheep had become visible. However, the next time the rollers moved to turn the sheep, it flopped over too far, and the next blow was way out of line with the rest, leaving a rather scruffy result. Some minor adjustments were made, and this wool too was shorn off.

We were then informed that this sheep was called a crossbred and was relatively easy for the machine to shear. So the next sheep put on was a Merino, with finer, denser wool and more wrinkles. The result was not nearly so impressive—the rollers seemed to run out over the top of the wool and the cutter seemed to be too close to the skin one moment and too far away the next. It strained to push through the wool, overloaded one of its motors, and stopped with the cutter pointing straight into the side of the sheep. Jim restarted the machine and tried again, but the shearing was ragged. A few more blows were tried, but the machine was obviously having great difficulty dealing with the sheep.

I was most impressed in all kinds of ways. First and foremost, I could see that the idea could be made to work, though not with this machine of course. I was most impressed with the care which had gone into the making of the equipment. Finally, I was impressed that the complex program of instructions for the machine had been set up with 'hard-wired' logic circuits.

I had become quite used to the idea of using a computer as the controller so it was a surprise for me that PATS had chosen this approach. It certainly was a major limitation to their design because changes which might have seemed sensible involved major reconstruction of the hardware, taking considerable time and effort.

Computer control

A computer can be used as a process controller which can also incorporate feedback control (p. 6).

A program in the computer can read input signals from one or more sensors, can calculate the value of an output signal which is needed to control the process and finally output the signal value through an electronic connection to the process. When this has been done, the program goes back to the beginning and repeats these three steps again and again.

The three steps take a certain amount of time for the computer to perform. As a very basic rule of thumb, this time should be less than one twentieth of the time the process takes to react to the change. Otherwise, the controller may not react in time to prevent overshoot.

One can imagine steering a car, only glancing at the road once every two seconds. After seeing that the steering wheel needs to be turned at one glance, and turning it, the car will probably have turned too far by the next glance.

One reason why engineers like to use computers as process controllers is that the control method is represented by a program rather than a special piece of electronic or mechanical hardware. Thus the same type of hardware can be adapted to control many different processes just by changing the software. Complex control techniques can be used with relatively simple and cheap equipment.

The principal problem with the machine was the profile measurement method—the wheels simply did not penetrate the dense fleece wool of Merino sheep. A further problem which occurred was the tendency of the wheels to sink further into soft parts of the sheep than the hard bony parts. Therefore, the cutter height had to be offset so the comb did not sink into the soft parts. However, the offset needed for this was too great for the bony parts over the rib cage—here the cutter was too far away from the skin. The difference in softness was much more pronounced in Merino sheep than the English breeds tested in Cambridge.

Jim also had recognized that the method of supporting the sheep was too simple, and a more elaborate method would be needed to shear more than just the back and sides of the sheep.

Sadly, neither I nor my fellow technical reviewers could recommend further development of the machine. Without conversion to computer control, a new profile measurement method, sensing of the skin at the cutter and better animal support, it would be very difficult to make significant improvements. Our reports were to be strictly confidential, but the disappointment was written across Norman's face.

In May, the Corporation announced that there would be no further funding for the SDSRC project. Norman immediately announced that the SDSRC would fund further development of the machine. The Lewises had to dig deep into their own pockets to pay for the machine to be shipped to the newly opened Melbourne office of PATS where work was resumed until the end of 1977. Jim Brown returned to England, so the work in Melbourne had to be undertaken by new engineers unfamiliar with sheep. By the end of the year with another $60,000 spent, Norman was frustrated and disappointed and resolved to conclude his personal interest in shearing robots.

Norman and Marjorie returned to their Kojonup farm and after a while focused once again on Marjorie's interests—fertility. On their own, they continued to persevere and were finally rewarded for their efforts. They continued to refine their flock, carefully culling the unwanted progeny in each generation. But still, the main stumbling block was the relatively high mortality of the lambs, compared to other breeds of sheep. In recent years, they have found that the trace element selenium in minute quantities has had a startling impact, and their lambing rate has climbed to 98% in a flock of 8000 ewes. A truly remarkable result and a clear 30% over the 'acceptable' or 'reasonable' figure. They still travel widely and command respect in the highest quarters of the wool industry. They have continued to lobby in support of the automated shearing research project and visit our laboratory when they feel able to. There is no doubt in my mind that their unique combination of determination, connections, influence, hard work and their obstinate refusal to be 'quiet' was the initial driving force which provided the momentum for robot shearing research to get under way. We have good reason to be grateful.

From fantasy to possibility

The CSIRO Division of Textile Industry in Geelong occupies a typical laboratory building in salmon brick and aluminium, a true product of the sixties. My fellow reviewers and I were met by David Henshaw with his neatly trimmed moustache, silvery hair and white coat, looking the quintessential CSIRO scientist. He took us downstairs to a basement room neatly panelled in varnished wood. In the centre of the room was a device that could only be described as a contraption. A shearing handpiece hung from the end of it, supported by a series of sloping metal tubes pivoted on a large frame about the height of a man. Wires of all colours draped across it, secured with sticky tape and leading to large rather old-fashioned looking electrical cabinets behind the machine. In amongst this was a coiled orange air hose, supplying compressed air to run the handpiece motor.

Laid out around the room were memorabilia of David's early experiments; photographs, charts, and dismembered pieces of equipment.

There was a temporary hush as David turned his machine on. As soon as he touched the comb, it suddenly came to life with a jittering and rattling noise, the rods whirling this way and that, the handpiece seeming to dance on his hand. Two sensing pads just behind and underneath the comb were reacting to

David's fingers; gentle pressure was sufficient to move the comb away. David moistened his fingers and demonstrated the same effect when he touched the points of the comb.

A sheep made a quiet entry to the room, legs splayed apart, its hooves slithering on the polished vinyl floor, its body tightly secured between the knees of a white-coated lab technician. Two or three people helped him heave it up so that it was sitting astride a beam carried by a small trolley on castors. The legs were secured into four clips and the head was strapped down over the front end of the beam.

The chatting groups of people around the room quietened expectantly as the trolley was pushed into position in front of the machine. An electric handpiece was used to shear a wide swathe by hand down the backbone from the head to the tail. I was struck by the contrast between the white fluffy fleece on this animal, and the hot greasy grey wool of the sheep at Muresk full of dust, grit and grass seeds.

The trolley was secured to the front of the machine with the shears pointing down just above the backbone. With a piercing growl the air motor came to life and the points of the comb descended into the shorn area along the backbone of the sheep. They landed gently on the skin and danced down the side of the sheep, lifting out a neat lump of shorn wool.

As at Muresk, anxious arms pulled back the wool to reveal the neatly shorn and intact skin beneath. The trolley was moved one space to the side and registered once again, against a notched piece of wood so that the next shearing blow would be just alongside the first. Soon a dozen blows had been shorn and the upper half of the side of the sheep was cleanly shorn from one end to the other. The sheep was tilted on its beam and another row of blows shorn to complete the shearing test. The entire side had been slowly but cleanly shorn with one noticeable ridge where the cutter had neatly climbed over a wart leaving deeper stubble on either side. An amazing demonstration!

In contrast to the quiet murmuring around the difficulties of the PATS machine at Muresk the chatter was jovial and optimistic but still restrained as another sheep was mounted on the trolley.

This was far more experimental than the PATS machine. Almost 'Heath Robinson'[8] with pieces of string tying it together, but it worked. Every now and again the cutter bucked and rocked forwards and backwards but somehow there was no more serious effect on shearing than slight ripples in the stubble. Every now and again the cutter refused to start. No doubt a weakness of the makeshift pneumatic drive.

The second sheep had a noticeable hollow in its side and the performance of the machine in climbing out on the ridge of the rib cage and then diving into the hollow of the stomach was quite astounding. It was a convincing demonstration

[8] Heath Robinson was a whimsical cartoonist who drew elaborate, but makeshift machines for achieving unlikely tasks such as picking up safety razor blades. His machines were often driven by long strings running over pulleys, turning levers or tripping over buckets.

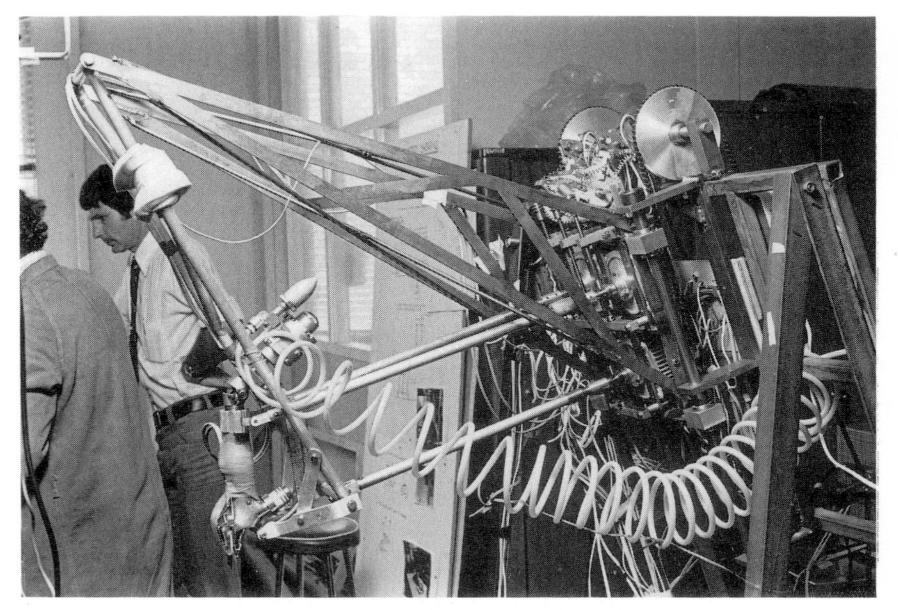

Figure 1.3 'A device that could only be described as a contraption' (p. 12).CSIRO shearing machine, January 1977.

of the effectiveness of feedback control. Here was demonstrable evidence of the possibility of sheep shearing robots.

Once again, as in the case of the PATS machine, David had chosen a hard-wired controller. As the external appearance of the machine suggested, the controller was even more primitive, consisting of half a dozen or so electronic amplifier circuits. Yet in spite of its crude construction and simple design it came out a winner on performance.

In reality, both the CSIRO and PATS machines had clear strengths and weaknesses and the talk at the moment was on how the expertise which had designed and constructed them could be brought together. David had avoided the problem of supporting the sheep for shearing any area apart from the side. Even though the PATS machine had its deficiencies at least it was possible to access most parts of the sheep.

David's control system could clearly do with a great deal of improvement to reduce the bucking and rocking motion of the cutter. There were unanswered questions about the effect on the resistance sensing of wool contamination and moisture—effects which would not worry the PATS machine.

I had been impressed with the performance of the PATS machine in Muresk but now I was excited. Of the two, David's was the one which had crossed my intuitive boundary between fantasy and possibility. There was a real shearing robot in my imagination; free to move in harmony with the sheep, able to follow complex curves just as the arm of a shearer does. David's machine, with its

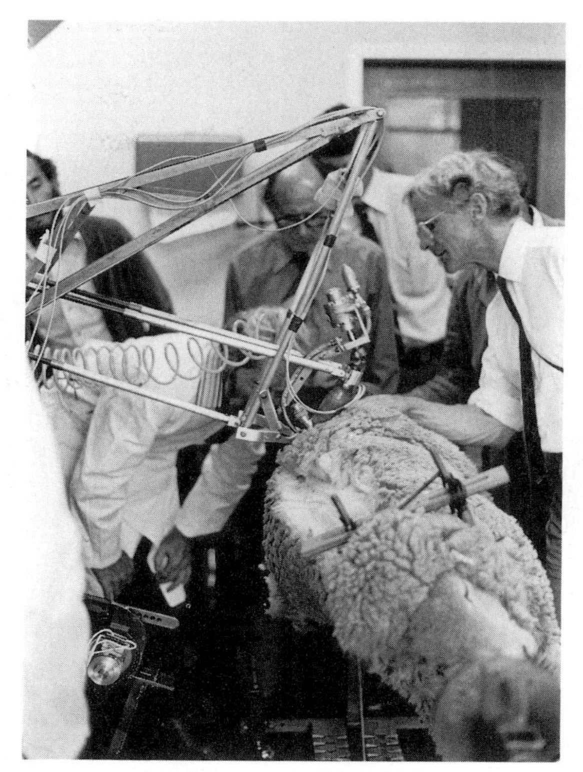

Figure 1.4 'Anxious arms pulled back the wool', first shearing January 1977 (p. 13). In background, (L to R), Paul Burrow (Melbourne University), David Henshaw (bending down), Lionel Stern, David Hindle (AWC), Alan Richardson (obscured), Alastair Mackenzie.

short, cramped movements, constrained to the straight line of its supporting frame, was just a starting point. It seemed natural to me to extend the idea into the six dimensions of free space.

I had turned the corner—first a sceptic, now excited and convinced that it could be done.

In retrospect, the report which I wrote in the following days was both understated and over-optimistic. I concluded that both the CSIRO and PATS machines were too primitive to develop further, but a composite design had a good chance of success. I suggested that a specification for an automatic shearing system should be drawn up so that intensive development could proceed with a realistic budget and time-scale. Collaborative research was needed; it wouldn't be practical to bring all the expertise and supporting facilities together. I advocated coordination between the different organizations involved. I warned that it might be hard to establish a satisfactory working relationship between the participants. At least I was right in that respect!

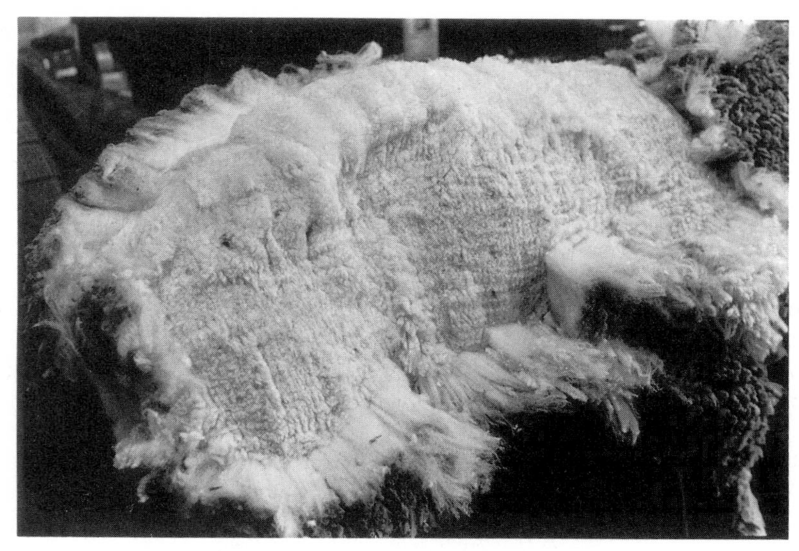

Figure 1.5 Back and hollow side of sheep shorn by CSIRO machine. The tail is at the left, shoulder at the right.

The wool industry

At the time, I was new to the wool industry and knew little about its history and policies. Few Australians appreciate its size, though most might have school memories of Australia riding on the sheep's back in the last century. Yet it is still one of Australia's foremost industries, and one of our most forward looking industries, despite its conservative image. A little background information will help set the rest of this book in perspective.

At the time of writing, the industry is enduring a devastating downturn caused partly by over-optimism in attempting to control the price, and an unfortunate coincidental withdrawal of our major customers from the wool market. Trade with China was suspended in response to the events in Beijing in May 1989, and economies in Eastern Europe are in chaos. Yet the industry has a long-term view and is maintaining its research and development programme, and increasing its spending on wool promotion around the world.

With a turnover which recently reached $5,000,000,000, it is the largest single industry apart from coal mining. With government support, the industry spends $40,000,000 per year on technical research and development. Much of the research is directed at long-term projects with payback periods measured in decades. Nearly $150,000,000 is spent each year on first class world wide promotion and advertising; this too is recognized to be a long-term investment. These policies are in stark contrast to most of Australian industry which spends little on research and development and uses advertising programmes based more on fantasy and desire rather than product quality.

The popular image of vast sheep flocks roaming the outback has been replaced by tens of thousands of smaller farms. The greatest numbers of sheep are concentrated around Kojonup in WA, Hamilton in Victoria, and Dubbo in New South Wales, typically on fertile rolling hills too steep for grain farming. The average flock size is about 3,000 sheep, and less than 30% of the 180,000,000 sheep in Australia are in flocks of 4,000 or more. After adding in lambs, about 220,000,000 sheep are shorn each year, mainly by about 12,000 professional shearers. The shearers work in teams of four or five, and travel from farm to farm with a classer, sometimes a cook, and work with four to five rouseabouts carrying the wool to the classer and pressing the sorted fleece lines. Farmers spend about $500,000,000 on shearing each year in direct costs, and an unknown amount on building and maintaining shearing sheds.

Patsy Adam Smith has written an excellent book *The Shearers* for readers who would like to know more about traditional shearing in Australia.

2

ORACLE - a shearing robot

ORACLE was our first robot though it was seldom called a robot. For years, it carried the dreary name of 'Hydraulic Test Rig' or just 'The Rig'. To many people, in the late seventies, the word 'robot' implied unemployment, de-skilling, labour replacement, and so on. We thought that our machine lacked the universality of a robot—it was just a shearer, nothing else. It was not until we started designing its successor that ORACLE finally earned its name—one to whom one turns for reference or advice.

This chapter presents the concepts from which ORACLE was designed and relates aspects of the construction phase and early trials. Later chapters describe the details of sensing, motion control and surface models.

The vision

In my mind, I could see a small cutter skimming over the sheep like a low-flying aircraft, hugging the hills and valleys, slicing the wool away in white waves of fleece. It was guided by a computer which precisely remembered where it had cut, and simply turned it round at the end of each blow, and made it skim along a new path exactly alongside the previous one. It was gentle, yet precisely efficient. I naively thought it could run at up to 50 or even 75 centimetres per second, and shear the whole sheep in about a minute of continuous movement. The sheep would have to be rolled over now and again of course, and that would take a little time. Still, it would be much faster than manual shearing and effortless.

The sheep would have to keep still. But then, all the sheep I had seen on the cradles at Muresk and Geelong had kept remarkably still. If the sheep struggled, the robot could backtrack a little and resume once the sheep was quiet.

The resistance sensing technique seemed to be full of potential, and as a sense of touch it would allow the computer to 'discover' the exact shape of the sheep progressively. The computer would start shearing with a reference path for the first blow. Then, as each blow was shorn, the computer would gather a more and more comprehensive 'map' of the sheep.

Tracks and path-following robots

Both the PATS and CSIRO machines had used straight tracks to carry their shearing equipment over the sheep. David Henshaw recognized the limitations of a straight track when he redesigned his machine in 1978—he chose a curved track to better match the side of a sheep (figure 2.1). Unfortunately, this intro-

Figure 2.1 Redesigned CSIRO shearing machine May 1978. The cutter was also a new development based on the same concepts as the cutter which was later fitted to ORACLE.

duced a further complication, and was little better than the original straight track.

Moving the mechanism at constant speed along a straight, or curved track of constant radius, produced constant speed at the cutter. But David's track had straight and curved sections so the speed varied—fast on the straight sections and slow on the curves.

Only some parts of some sheep matched the new track—sheep may look alike, but they differ to a surprising extent, particularly on the sides where David was attempting to shear.

Moving the track (or the sheep) sideways by a fixed amount also has limitations. On the thinner sheep, shearing blows overlap. On fatter sheep, there are gaps even though the sheep has been moved sideways by the same amount in each case.

I saw that we could escape the constraints imposed by a mechanical track by using a robot to carry the cutter, complete with the mechanism to allow height, pitch and roll corrections in response to sensors.

Playback or path-following robots

Playback robots, typically used for spray painting, appeared at about the same time as the better known Unimate robot. However, they are programmed in a slightly different way.

Light and manoeuvrable, they can be led through a sequence of spray painting movements by a skilled painter. The positions of the arm joints are recorded continuously, by a tape recorder or a computer. When switched to playback mode, the arm joints are driven by powerful motors to follow the exact movements recorded during the teaching phase. The robot replays the painter's motions, whether there is something to paint or not.

The early point-to-point robots, like the Unimate, were jerky and looked clumsy. The movements of the different arm joints were not well coordinated. This image of robots has been remarkably persistent—just ask a child to mimic a robot and watch the arms stiffen and jerk about. Playback robots move with the grace of humans—the humans whose motions they replay.

We could have mimicked the movements of a shearer using a playback robot—but that would only have worked for shearing if all sheep were identical. There was another difficulty. The only commercially available playback robots were used for spray painting and they were too light and flimsy for shearing. They could carry lightweight spray paint guns but not a shearing cutter, and certainly could not push the cutter firmly through the wool (a shearer has to push hard for much of the time).

Even if shearing movements could have been stored for one particular sheep, and then modified for each different sheep by measuring its length and width, we foresaw other problems to overcome. Sheep are alive, and they move. Their shape changes as they breathe, and occasionally wriggle. The shape changes just by pressing the cutter on to the surface. No amount of prediction beforehand could precisely anticipate the profile of a living moving animal.

A new approach was needed, and a new robot. Fixed tracks were no use; whether mechanical, or preprogrammed in a playback robot's memory. We needed a robot which could plan its movements 'on the run' and learn about the sheep being shorn as it was working. To realize this, I needed a robot which could work in the six dimensions of free space motion. And this meant a certain degree of mechanical complication. I was prepared to accept that, knowing that the rewards would be worthwhile and that working in six dimensions was no more inherently difficult than working in two. Others were more keenly aware of the complications, and had some understandable difficulty in seeing the images I conjured up in my imagination.

Since that time, I have refined my explanations. The following explanation presents some fundamental aspects of robot kinematic calculations in a way which illustrates their elegance and simplicity. This material has been useful for training team members to program robotic shearing, as well as for teaching undergraduate students. This explanation is based on resolved motion rate control (Whitney 1969) though there are many other schemes for computing the inverse kinematics of a robot. The advantage of this approach is that it introduces the concept of the Jacobian matrix which was fundamental to high-speed adaptive control for sheep shearing. The kinematic computations for ORACLE were implemented with a direct closed form solution initially, and later a pseudo-inverse solution (Trevelyan *et al* 1983). Later the SM robot was controlled using variants of resolved motion rate control. Several more elegant schemes have been devised in recent years; Siciliano (1990) presents a wide ranging review of these.

The boxed sections which follow provide a simple introduction to robot arm kinematics calculations which I have used to explain the basics of robot motion control. Readers familiar with kinematics computations can safely skip them. I have used this material for introducing robot programming concepts without relying on mathematical derivations, though occasional references to mathematical notation have been inserted.

The six-dimensional computation problem

We normally think of space as having three dimensions, but for a robot free space has six dimensions. What do we mean by dimensions?

For us, each space has length, breadth and height. Since none of these depends in any way on the others, we have three dimensions for space. Expressed in another way, a robot which you might build to freely move a ball anywhere in the space around it needs at least three movements or joints. Depending on the design, the robot may need more joints, but three is always the minimum number.

Robots used for positioning tools need six degrees of freedom or joints because a tool must not only be positioned correctly, but it must also be at the correct *angle* as well. You can deduce this for yourself by the experiments which follow.

The practice of navigation is one of movement by numbers. We use numbers to represent position, heading and our destination. On a ship, these numbers are:

for position: longitude
latitude
for direction: compass angle.

These numbers are sufficient for navigating on a two-dimensional surface. To work in six dimensions a robot needs more.

For position, we now need three numbers. We usually measure position by coordinate values along three axis directions all at right angles to each other—X, Y and Z. We call the position of the point at the end of the robot **p**.

Orientation in space is more complicated, as we shall see. Although you can represent orientation by three numbers, it is often more practical to use six or nine. We usually represent orientation by the directions of three lines extending in three directions from the end of the robot.

The cutter is the end of a shearing robot, and the three directions we use are forwards, left, and up relative to the comb. The three directions are called **i**, **j** and **k** respectively. Three numbers represent each direction, being the proportionate distances along the X, Y and Z axes.

Thus we have:

for position: X, Y and Z coordinates for **p**
for orientation: X, Y and Z coordinates for **i**
X, Y and Z coordinates for **j**
X, Y and Z coordinates for **k**.

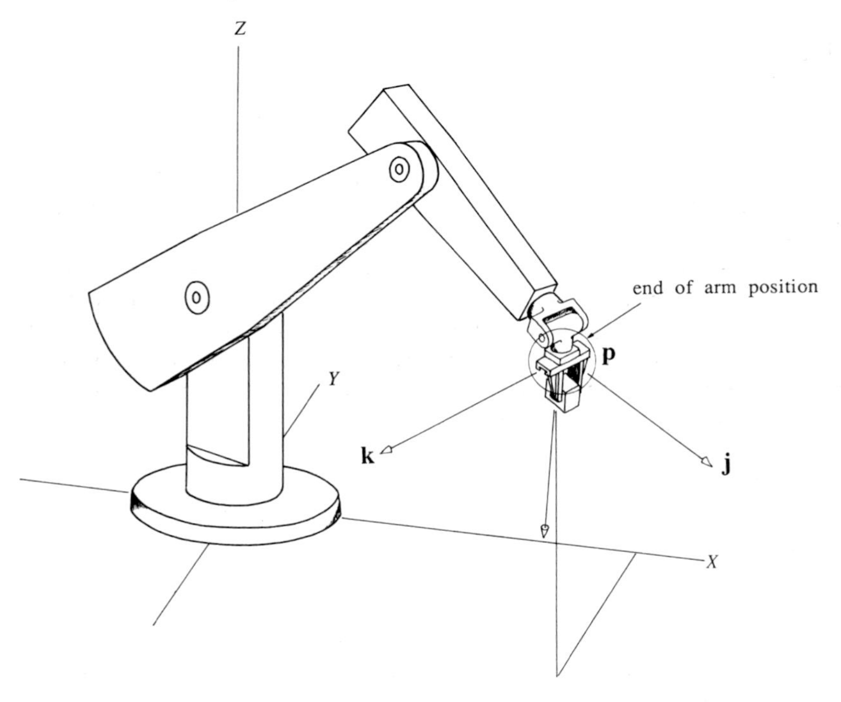

Figure 2.2 Robot arm—coordinates of end effector position and orientation.

Figures 2.2 and 2.3 illustrate **p** and **i**, **j**, and **k** in the context of a conventional and then a shearing robot. Robot engineers have chosen the simple term ***pose*** to combine the terms 'position' and 'orientation'.

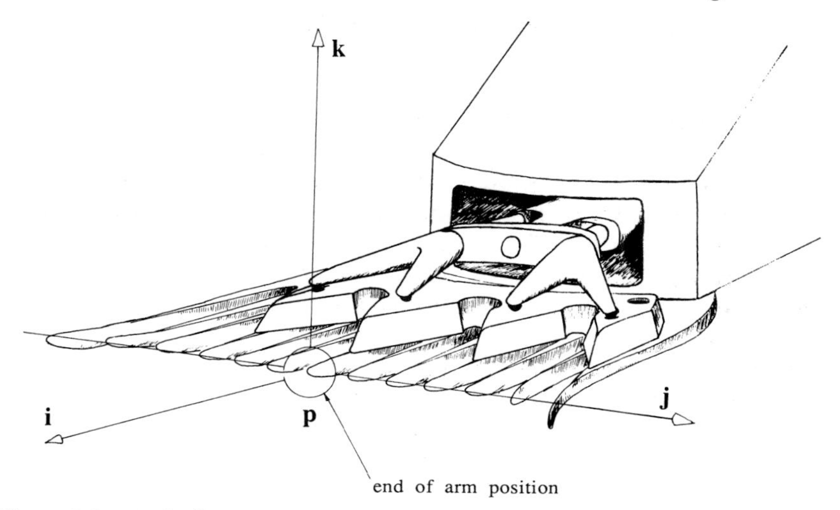

Figure 2.3 End effector of a shearing robot arm showing coordinates of position and orientation.

Experiment 1
Use an empty matchbox and a long thin screwdriver. Pierce a hole at the centre of each face of the box using the screwdriver. Now pass the screwdriver shaft through the holes in the ends of the box. The box can slide along the shaft, or rotate around the shaft. Two different motions—two degrees of freedom. Holding the box still with one hand, withdraw the shaft and poke it through the holes in the sides of the box. Now the box can slide or rotate in two more ways—two more degrees of freedom. Now poke the shaft through the two remaining holes and you will find two more degrees of freedom—six in all.

The problem gets harder.

Rotations are more tricky than they seem to be! Try this.

Experiment 2
Place a matchbox on the table, label side up. Then push the box 10 centimetres away from you, and then push it 10 centimetres to the right. Note its position on the table.

Now repeat the experiment, starting with the matchbox in the same position, but reversing the order of the moves, i.e. 10 centimetres to the right, then 10 centimetres away. The box should be in the same position as it ended up before.

This experiment shows that the final position of the box does not depend on the order of the movements. In fact, if both moves were applied simultaneously, the end position would still be the same. You may think that this could be deduced by common sense, but even if you did, try the next experiment anyway.

Experiment 3
Write letters A to F on each face of the box. Then place the box on the table, label side up. Tip the box over 90 degrees away from you. Then turn the box over 90 degrees to the right. Make a note of which face is uppermost now. Return the box to the starting position, label side up again, and this time reverse the order of the rotations, i.e. 90 degrees to the right, then 90 degrees away from you. You will find that a different face is uppermost!

We can conclude from this that rotations depend on the order in which they were applied. If the rotations are applied simultaneously, then the result is different again!

Robot motion

Our aim is to find a way of calculating the movements of the joints in a robot arm so that the attachment at the end of the arm can be moved to any pose we wish. Figure 2.4 illustrates the joint angles for a typical robot arm.

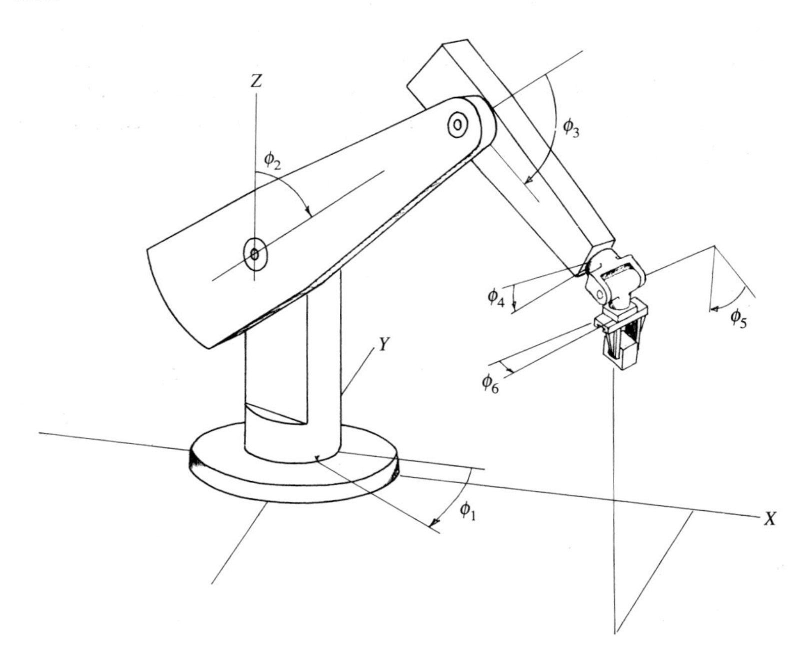

Figure 2.4 Robot arm—joint movements. This arm has 6 rotary joints thus each joint movement is defined as an angle.

Experiment 4

Repeat experiment 2, but just move the box to the right in as straight a line as far as you can. Do this on a large table. Watch how your arm moves—as the box gets further away from you, your elbow has to straighten out until eventually you cannot reach any further. All the joints in your arm have moved through different angles, just to move the box in a straight line. If you were to move the box in a more complex manner, then the movements of your arm joints would be more complex again.

The problem is simplified, at last!

Fortunately there are simple methods which make it easy to work in six-dimensional space. Unfortunately, they are mostly written in mathematical terms which are difficult to understand. Note that it is the language which is difficult—not the method!

The secret is to forget about large movements, and just focus on very small movements of the box, and of your arm. Try this experiment.

Experiment 5

This one needs a little care and attention. Hold the box in the air over the table in any pose you like. Now flex your elbow just a little bit each way as shown in figure 2.5. Try to keep all the other joints rigid when you do this—it's not easy. Notice how the box moves a little, this way and that, and rotates a little too. Then try each of the other joints in your arm, one at a time, in turn. The box moves in a different direction each time you flex a different joint.

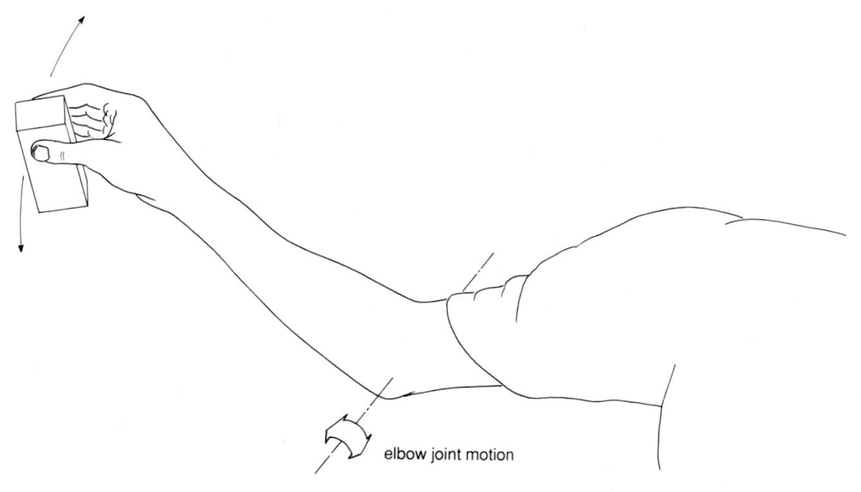

Figure 2.5 Human arm—effect of elbow joint movement, keeping other joints fixed—refer to text.

You could do the same for the different joints in a robot arm, and the end movement directions for each joint can easily be calculated. In effect, we find the change in the end effector position for a unit change in each joint angle. The result of this is a Jacobian matrix, referred to as **J**.

To get the robot arm to move the box a short distance in any chosen direction, it is then only necessary to find the combination of joint movements for which the resultant end movements add up to the chosen direction. This is the same as solving six equations in six unknowns. Each of the joints is then moved by the calculated (small) amounts.
Mathematicians represent this by:

$$\delta \mathbf{f} := \mathbf{J}^{-1}\, \delta \mathbf{pose}$$

where δ**pose** is the incremental change in pose and $\delta\phi$ is the incremental change in each joint. **J** needs to be inverted but there are several useful techniques for avoiding the need to invert a full 6 x 6 matrix.

Then the process is repeated for each motion step. Because the joints all move to slightly different positions each step, the Jacobian will be slightly different, so the proportionate movements in each joint will be different. By repeating the process over and over again, each time for just a short movement, the computer controlling the robot arm can cause it to move in any desired direction, almost exactly, for as far as the robot arm can reach.

Again, by restricting ourselves to small movements, the problem that we discovered with rotations disappears, almost. When the rotations are small enough, the order in which they are applied has only a very slight effect.

Dealing with the two ‘almosts’ above requires some slight elaborations to the method to make sure that the slight errors do not accumulate.

The essence of the solution

Using this type of method, we can calculate how to move any given robot arm to move our box in any chosen direction or rotation, or combination of the two. There is no greater degree of difficulty in working in six dimensions than two. We simplify the problem by dividing the desired movement into many small steps.

In practice, we obtain sufficient accuracy for shearing with movement time steps of about 0.075 sec. At this rate, the small errors which arise from the assumption that the Jacobian does not change significantly in that time are insignificant for shearing. We calculate new actuator positions more frequently by linear interpolation—a simpler calculation.

And sensors too. . .

In the process of solving the kinematics we calculate the inverse of the Jacobian matrix for correcting the planned movements of the arm to take sensed position errors into account. Given that the sensors will be used to calculate a small correction to the arm pose, the same equation can be used (though an additional coordinate transformation is usually needed).

$$\delta \mathbf{f} := \mathbf{J}^{-1}\ \delta \mathbf{pose}$$

This is a simple calculation which can be repeated hundreds of times per second if necessary. Rarely does the arm move far enough that we need to do the more lengthy calculation of the Jacobian more than ten times per second. In any case, small errors which result from this simplification will be corrected by sensor feedback

Thus, an apparently complex task, in six dimensions, can be reduced to simple computations which a modest computer can handle.

A more detailed account of these calculations appears in chapter 4.

The bandwidth problem

There was one more major hurdle to cross before we could really start. Typical robot arms then (and now) were much too sluggish to lift the cutter over the skin creases and wrinkles at a reasonable shearing speed.

Frequency response and bandwidth

The frequency response of a device indicates how it will respond to inputs which vary at a given frequency.

A typical control system, such as a robot actuator or whole arm, will respond well at low frequencies, with the output following the variations of the input signal closely. However, as the frequency of variations in the input signal increases, a stage is reached when the elements of the control system react too slowly, and the output no longer follows the input.

Often the performance of a control system is summarized by giving the highest frequency at which the output still follows the input reasonably well. This is called the *bandwidth*. If we were to say that a given robot actuator has a bandwidth of 10 Hz, then we would imply that it will follow an input signal varying up and down at up to ten times per second, but not so well at higher frequencies. At 20 Hz, its output response may be less than one tenth of the response at low speed.

To shear fast, the cutter would have to track the hollows, ridges, bumps and wrinkles of the skin. Some error would be allowed—a couple of millimetres. The skin is soft in many parts; a shearing comb can be pressed down a centimetre or so, quite firmly, without cutting the skin. But on the hard and boney parts of the animal, the error must be no more than two millimetres.

A human arm has a bandwidth of 3 to 6 Hz, but this can be as low as 1 Hz when carrying a reasonable shearing tool at maximum reach. A good hydraulic manipulator arm has a 6 to 10 Hz bandwidth. Much larger and heavier actuators, and a more rigid structure are needed for a faster response. But the increased weight requires still more strength for equivalent stiffness, leading to compounding increases in weight, size and ultimately cost. We decided that 6 Hz would be a reasonable target with a reduction to 3 Hz when the arm was fully outstretched.

Joris de Schutter, a Belgian robot researcher, suggested a simple and useful rule of thumb at a conference in Santa Cruz in 1987. It matches our experience well, and is useful to illustrate the issues which we resolved by educated guesswork in 1978. Rules of thumb like this are very valuable, and can often be just as precise as complex computer simulations which take enormous effort to prepare.

Joris' rule, adapted for surface following error, states that:

$$\upsilon \approx \omega \delta$$

where υ is the approach speed,
ω is the controllable bandwidth (rad/sec), and,
δ is the allowable tracking error.

The approach speed is related to the traverse speed across the surface by the relative slopes of the bumps. Assuming slopes up to 45°, the approach speed will be about the same as the traverse speed.

If your arm is similar to mine, you can verify this rule for yourself. Try to move your finger across an irregular surface such as a rumpled rug or bedclothes, with your arm outstretched, keeping a constant height above the surface (figure 2.6). Concentrating just on the gap between the finger tip and the surface, go as fast as you can without touching, keeping the gap at about 10 mm, but no more than 20 mm. The maximum following error must be less than 10 mm. You should be able to keep up a traverse speed of between 100 and 300 millimetres per second without bumping into the surface. It may be less if you're tired at night or more if you're perky in the morning. With an arm bandwidth of about 3 Hz, the rule predicts a speed of 180 mm/sec.

With our shearing error limit of 2 millimetres and a bandwidth of 6 Hz, the

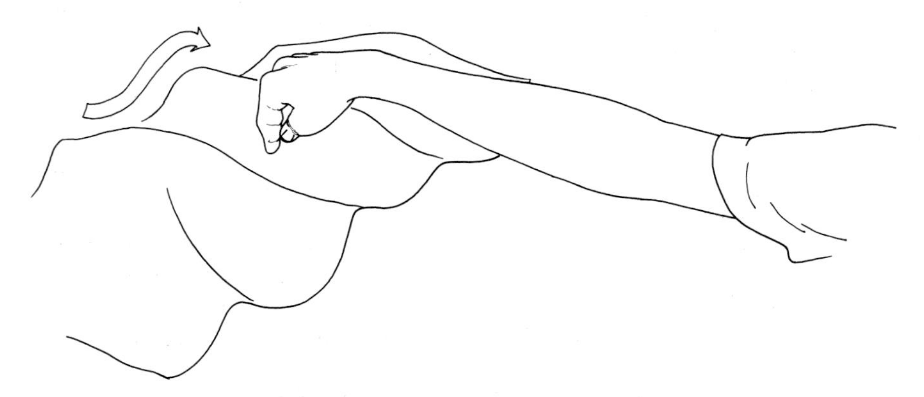

Figure 2.6 Following a surface profile by hand—refer to text.

rule suggests a shearing speed of about 70 mm/sec. This is about 50% faster than the CSIRO and PATS machines—a useful correspondence. The rule demonstrates that we would need a bandwidth of about 50 Hz to shear at our target speed of 600 mm/sec.

One might question the validity of this rule because electric robots move at speeds of up to 5 m/sec, with a positioning repeatability of 0.1 mm. This apparent contradiction is a useful one because it serves to emphasize just how different our task is. A factory robot repeats the same movements over and over again. Each position is precisely known in advance; the robot can slow down at exactly the right instant to stop at exactly the right position. But we have to follow an unknown, uneven, moving surface which can never be exactly located or predicted.

We still appear to have a major contradiction. A good shearer moves the cutter at up to 800 mm/sec, yet the bandwidth of his arm is about 25 times less than Joris' rule suggests is necessary. The next experiment will help to resolve this and leads us to a useful solution.

This time, hold a pencil in your fingers so that they act like springs to keep the end of the pencil lightly pressing on the surface you are tracking (figure 2.7). Now you no longer have to track the unknown surface so accurately with your arm. Your hand can be close to or far from the surface, but your finger springs will keep the end of the pencil touching the surface. Now you can sweep your arm much faster because the end of the pencil tracks the bumps much faster than your arm could.

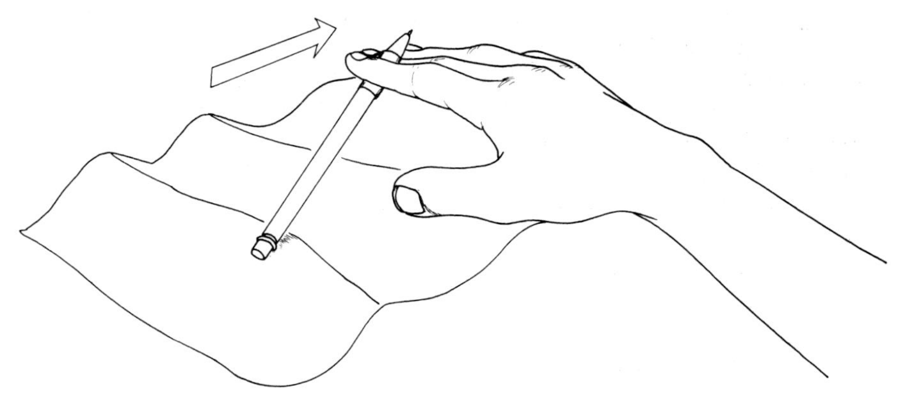

Figure 2.7 Following a surface profile with a pencil follower—refer to text.

A good shearer firmly presses the handpiece against the skin of the sheep just like you were holding the pencil. A firmer force is usually needed, but the principle is the same. Of course, the shearer also uses accumulated experience, vision, touch, proprioception[1] and hearing to judge the right degree of firmness,

[1] The sense of proprioception provides feedback on the positions of joints in the arm; in the human sense it is provided by nerves on muscles and ligaments.

and yet avoid skin cuts—most of the time. Because the pencil or handpiece can move in the fingers or hand respectively, the allowable tracking error for the arm is much greater, and the higher speeds are therefore possible.

One could use a spring to hold the cutter against the sheep; we considered the idea but soon discounted it. The cutter was likely to weigh about 1.5 kilograms, and the spring would have to be quite soft to give sufficient latitude for arm position errors. I had visions of the cutter bouncing up and down as the robot tried to 'land' it on the surface. And if a cut started, the spring would keep the cutter pressing down—there would be no way of lifting the cutter away quickly. It wouldn't be practical at high speeds.

A hydraulic actuator could simulate a spring, with appropriate control system and sensing, and could rapidly move the cutter away from the skin if necessary. A properly designed actuator would be capable of the required bandwidth—50 Hz. The actuator would respond to the sensors, keeping the cutter against the skin. The robot would carry the actuator, but need not position it accurately; the sensors would compensate for arm positioning errors. Unlike a real spring, the actuator would never allow the cutter to bounce or jump uncontrollably.

Hydraulic actuator (figure 2.8)

The actuators of a robot roughly equate to our muscles. They respond to computer signals and allow the computer to control the position of each of the joints in the robot's arm.

The central element of a hydraulic actuator is the cylinder with a moving piston—this can be as small as your finger or as large as a car, depending on the force needed. Hydraulic systems work with high fluid pressures, typically 100 to 300 times atmospheric pressure, so even a small actuator can develop enough force to lift a car.

Oil flow to and from the cylinder is controlled by a servo valve which responds to a small electric current. A good servo valve used for a robot actuator can handle high-pressure oil at up to 100 litres per minute and can adjust the oil flow in less than 5 milliseconds. An electronic transducer is used to measure the position of the piston in the cylinder, and the signal from this is used by a feedback controller to adjust the oil flow from the servo valve so that the piston moves towards the required position. The position signal is also used for monitoring the performance of the actuator.

Depending on the characteristics of the components, hydraulic actuators can be used to move heavy loads very quickly and precisely. Thus, the act of changing a number in a computer program can cause a large and heavy piece of machinery to move to a new position in a few hundredths of a second.

When properly designed, hydraulic actuators are safe and reliable. Nearly all commercial aircraft rely on them for moving the elevators, flaps and ailerons in response to the pilot's control column movements.

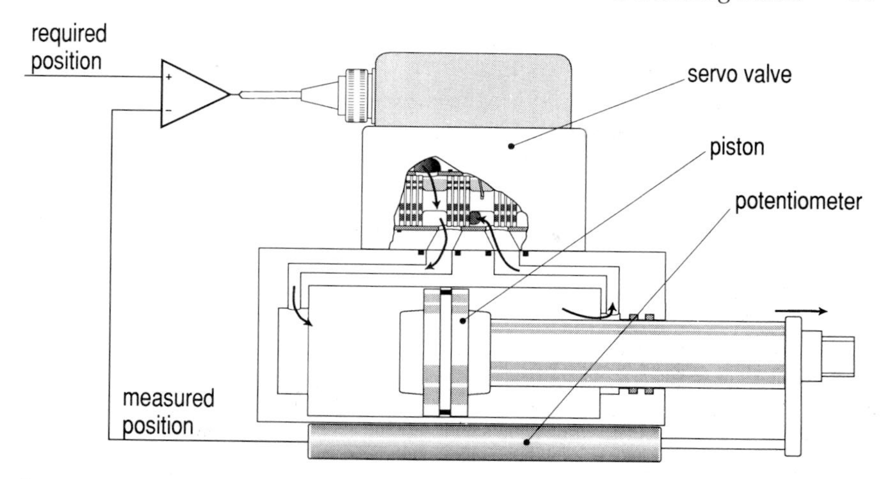

Figure 2.8 Schematic diagram of servo hydraulic actuator. The position of the cylinder is measured electronically and compared with the position demand from the computer. The calculated error causes the servo valve to admit high-pressure oil to one end of the cylinder to push the piston towards the correct position. There are many different possible configurations—the actuator can be linear (as shown) or rotary; the feedback transducer can be a potentiometer (as shown and used on ORACLE), a linear variable displacement transformer (LVDT), an optical encoder, magneto-strictive rod, etc. The electronic comparison can be performed by the computer. However, the principles remain the same.

The stroke of the follower actuator needed to be just long enough to follow the bumps and valleys which the arm could not keep up with. We can use Joris' rule again to calculate the stroke. We want to move the cutter at 600 mm/sec. Given the worst case arm bandwidth of 3 Hz, the stroke needs to be 33 mm up or down—a total of 66 mm. I allowed a little more than twice that, just to be safe.

Ten years passed before I realized that we missed a tantalizing opportunity which could have saved years of work. Not only did I miss the turning, but the next step kept the gate firmly locked until our second robot was built. But please read on as there are always many paths to good solutions, and you can only see the best with hindsight.

My follower attachment would keep the cutter on the skin, but not necessarily at the correct angle. While the orientation bandwidth would not need to be as high as 50 Hz, it was likely to be more than 6 Hz—the arm bandwidth. Orientation control seemed to be computationally expensive so I chose a virtual centre wrist mechanism to carry the cutter on the follower actuator. This way, the fast orientation changes would not need equally fast arm position adjustments to keep the comb at the right position.

Virtual centre mechanism (figure 2.9)

A virtual centre mechanism provides a means for rotating a part about a point, without having to have a pivot at that point. A model train running on a circular track provides a good example: the train rotates about a point at the centre of the track, but no pivot is needed at the centre.

Apart from the ORACLE wrist, we used virtual centre mechanisms for pivoting legs on our sheep manipulators (figure 7.19 p. 190), figure 9.20 p. 274).

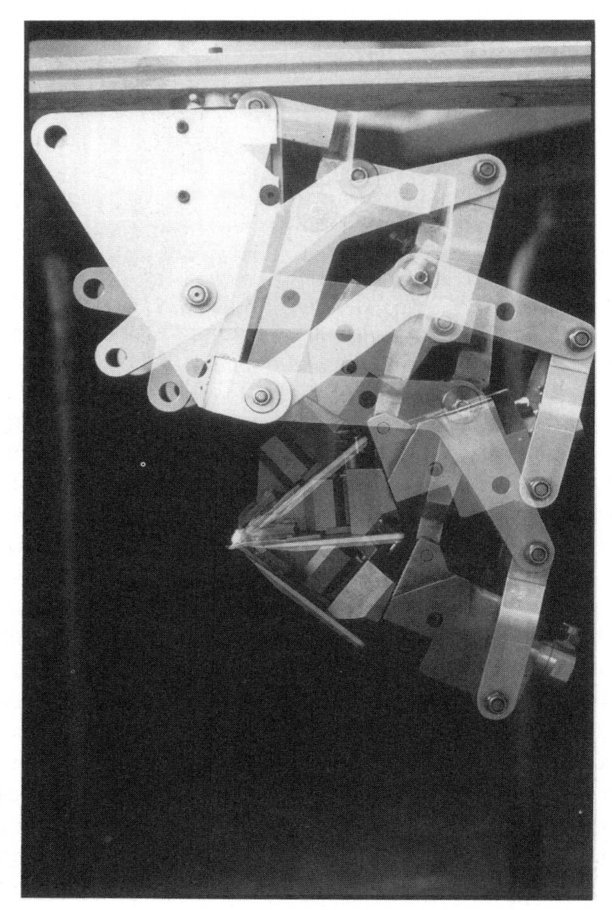

Figure 2.9 Virtual centre wrist mechanism used for ORACLE robot. The multiple exposures illustrate rotation about the comb centre point. Rotation is also possible about the cutter roll axis and the yaw (here vertical) axis; both of these also pass through the cutter centre point (see also figure 2.10).

My vision was now complete. Some of the details were a little misty but, in essence, I knew I had something which would shear a sheep. At the same time I was not without apprehension. The mass of the virtual centre wrist and cutter assembly, coupled to the follower piston, was likely to be about 25 kg and the follower actuator would need to develop a peak force of about 1 tonne to produce the required acceleration (40g). The 'taming' of such awesome power to provide a surface following force of 1 N (about 100 grams weight) was a remarkable demonstration of the effectiveness of software feedback control. Nevertheless, what if plus 1 tonne had accidentally become minus for an instant? Our software turned out to be very reliable, but I had the feeling that there had to be a better solution.

Genesis of ORACLE

Once the review of the PATS and CSIRO machines was completed, Roy Leslie returned to a study of sheep restraint devices while I developed the control system ideas. Jim Blair and Prof. A-W compiled an application for $75,000 to start building a machine to test these ideas. Months of delicate negotiations followed. CSIRO were reluctant to include us, but the Corporation were strongly in favour of our contribution. By October 1977, CSIRO had emerged as project leaders, under David Henshaw's direction. Roy Leslie and myself (part-time) were to begin work on a computer-controlled machine for experimental shearing. Yet CSIRO regarded us as a sub-contractor for the design of a computer control system for their own machine. By now, they appreciated the ease of changing software in a computer control system. On the other hand, we thought that the mechanical design concept which CSIRO had adopted was a major constraint; unless this was removed, the full potential of computer control could never be realized. We also knew that we needed our own experimental equipment—no one knew enough to design a computer control system without having to experiment with it first. We had to persuade CSIRO to let us build a test rig for our own use.

Even if David had acquiesced, it would have been clear to his CSIRO colleagues that our machine would soon eclipse anything which could be usefully achieved at Geelong if we were allowed to proceed. Even though the Corporation had nominated CSIRO as project leaders, it might be a title in name only.

David Henshaw visited us for the first time in December and we were able at last to give him a close look at our ideas. We showed him a trial hydraulic actuator Roy Leslie had built as soon as we had received word to proceed. It was awesomely powerful and fast as it hammered from one end of its stroke to the other in a fraction of a second. We took him to the darkened computer room to watch a simulated shearing robot shearing simulated sheep, tracing delicate patterns across the glowing green screens. We looked at a real sheep, particularly some of the harder areas for shearing like the belly and crutch. We laid it upside down, pulling legs this way and that; then on its side, curling its neck around, for Roy had already realized that manipulating the sheep's body was the only avenue towards complete shearing. By the end of the day David seemed to be catching some of our enthusiasm. It was as if he realized then which way the project had to go.

David revealed his new design for a low-power cutter. This was good news for us. A conventional cutter needs up to 500 W of power and the only available motors were heavy and bulky. We could lighten the cutter mechanism for our robot and the smaller size would make it easier to design. So we commissioned David to build one for us to use on our rig.

By the end of February 1978, CSIRO had conceded that we needed to build a robot. Our concession was to focus initial work on the back and side of the sheep—the same area as CSIRO were shearing. This turned out to be a near fatal mistake, one which was to cause us many technical problems later. In retrospect, we ought to have designed a robot capable of shearing other areas too, and then simply used it initially 'as directed'.

Exit CSIRO

David Henshaw had redesigned his machine to run longer shearing blows from the backbone to the bottom of the side but for this he needed a much longer track. He also needed a curved track which would allow the cutter to start almost level at the backbone, and then curve over the edge of the rib cage and down the side (figure 2.1 p. 19). The control system was improved too, with more robust electronic sensors under the comb and a more compact arrangement of the electric motors.

Sadly, the new arrangement never met David's expectations. The curvature of the track introduced speed control problems as the motor carriage rounded the corner and then moved off down the straight section. The sensors did not seem to work as well as expected. A special capacitance proximity sensor (p. 40) was designed and built under contract by AWA Pty Ltd, a large electronics firm, but its sensitivity was less than David had hoped. It just never seemed as good as the first machine 18 months before. In CSIRO's final technical report, it was written off with little more than a paragraph and a few comments of which only one had a positive note to it.

As soon as CSIRO had announced their intention to review their own work in October 1978, we were conscious of increasing pressure on us. There was little prospect of shearing with our rig before the review started. By September we were aware of the disappointments at CSIRO. We also knew that CSIRO were evaluating the economic prospects for shearing robots, but we were not at all happy with their analysis. Even though Roy Leslie and I were deep in the technical details of constructing our own machine we were apprehensive about the prospect of CSIRO deciding to withdraw from the project. What would be the reaction of the Corporation if they decided to pull out? We felt we were in a weak position with our own rig still under construction.

CSIRO reviewed all their results—technical and economic—by the middle of October, and their report finally arrived on my desk late in October with a request from the Corporation for my comments. I was immediately dismayed by the tone of the summary and conclusions:

> 'it is difficult to see a basis for regarding automated shearing as a worthwhile area for investment in the light of present knowledge or expectations'.

With great hope and optimism I finished my reply by predicting that our experimental results would soon be available and it would be premature to assign any great significance to the conclusions of CSIRO's report!

Two weeks later, the Corporation reluctantly accepted CSIRO's withdrawal but asked us to proceed as quickly as possible. It was all too easy for us to have felt undermined by the CSIRO decision to withdraw from the project. In retrospect, it was the right decision, and it left us in charge of our own destiny—an enviable position. We had felt very vulnerable without experimental results. But we were newcomers to wool research, and did not understand the determination of the Corporation to press on.

David Henshaw left CSIRO not long after, and commenced an active political career, representing the dockland areas of Melbourne in the Victorian state parliament.

Rereading the full CSIRO report 11 years on I found some interesting comments made by David Henshaw in his own conclusions which I missed. I found justification for what I had always supposed were the real reasons for their withdrawal. In David's own words:[2]

> 'The work described in this report, taken together with the more refined development work undertaken by the team at the University of Western Australia, suggests that automated shearing is technically feasible. However, it is not clear that any practical system would be commercially viable in terms of yielding a return to investors. Indeed, economic analysis, using assumptions which must be considered optimistic rather than pessimistic, indicate that it might well be more expensive than manual shearing. It may be that different approaches for a more refined analysis would lead to a different conclusion. In either case, *this division of CSIRO does not have the expertise or resources available in areas such as electronics, digital processing, or hydraulic control*[3] to justify participation in the exacting work which would be required in further developments. In this respect it is suggested that the teams at the University of Western Australia and Musheep[4] are better situated.'

We inherited three jewels from the CSIRO crown. The first was the *existence proof* that automatic shearing was possible. Though intangible, it was, and still is, the most important of the three.

[2] CSIRO Review of Automated Shearing—Technical Report—Conclusions. CSIRO Division of Textile Industry, October 1978.

[3] Author's emphasis

[4] Melbourne University Sheep Handling Experimental Establishment located at the University and at Werribee near Geelong. This group is part of the Department of Civil and Agricultural Engineering of Melbourne University.

The second was *resistance sensing* for controlling the comb height. It was David's idea, and served us well for ten years of shearing experiments. Yet David himself had already realized that it was not the ultimate answer. It took ten years to find a better one.

The third was David's design for the cutter, but just a few months later we were to discover a major flaw. Yet the subsequent investigation of its failure was to lead to the first real understanding of the physics of wool cutting.

Construction and testing

Neither Roy Leslie nor I knew much about robotics or hydraulics when we started so we designed ORACLE from basic principles of mechanical and control engineering. Fortunately the Department of Mechanical Engineering was well equipped for robotics; a workshop with highly skilled technicians and superbly maintained precise machine tools. Through on-going research in the dynamics and control of very long ore carrying trains, the department was also well equipped with minicomputers; both Roy and I had accumulated an understanding of operating systems programming, computer interfacing and data communication techniques.

We decided to tackle the design in stages, gaining experience along the way. Roy started with the follower actuator, and I tackled the basic software for following an unknown surface and following previously recorded motion paths. Both of these were ready to demonstrate to David Henshaw by the time of his visit in December 1977.

With new confidence, Roy started to wrestle with the compounding design complexities of the virtual centre wrist mechanism. I use the term 'wrestle with' because the process of selecting and designing a special mechanism like this involves endless iterations of guess work, drawing, cardboard models, analysis, redrawing and calculation, only to find some detail which has to be changed. It may be a retaining screw which would knock against a bearing at one end of the movement, or a part which would snag wool or a bearing too small to stand the loads. Figure 2.9 (p. 32) shows a multiple exposure photograph of the mechanism which was constructed to Roy's design; the cutter rotates about the centre point of the comb tips.

Computer and software

We chose a Hewlett Packard 21MX-E minicomputer with 64k bytes of memory, similar to others already in the department. With a peak throughput of just under a million instructions per second, and about 50,000 arithmetic operations per second, it was slower and had less memory than the original Apple Macintosh home computer. At $25,000 it was expensive but we expected that computers would be much cheaper by the time our research was finished.

I decided to structure the software as multiple independent real-time tasks, each operating at a different frequency. The highest frequency level would read sensor signals and adjust the follower actuator set point at an update rate of about 250 Hz. The next level would calculate set points for the other actuators at 125 Hz. A slower level still would calculate the inverse Jacobian matrix, solve the arm geometry, and calculate motion control in Cartesian space, to follow the recorded path. This principal real-time level would run at about 14 Hz. Finally, motion planning programs and a program to interpret operator commands would run in background. The choice of update rates was an educated guess. But like many other aspects of the first software, they remained almost unchanged through the life of the project.

These update rates were too fast for a commercial real-time operating system, though the RTE real time multi-tasking system supplied by Hewlett Packard was efficiently written. A special purpose operating system was needed. Fortunately, I had already developed a suitable operating system for another project; my system simulated the RTE environment with minimal efficiently written code for just a few tasks (Trevelyan 1981).

The structure which emerged in the first two weeks has served us well. It was changed in 1988 to implement recursive motion control—see chapter 11.

Arm mechanism

With four joints in the wrist and follower actuator, we needed at least two more joints—six being the theoretical minimum. But the theory ignores the effects of mechanical restrictions on joint movements. Our best attempt at the virtual pivot mechanism provided ±45° of rotation to compensate for slope changes. Even if we had found a way to extend this, we would have found that our fast response actuator would have become much less useful at greater angles. Our four-joint 'surface follower' mechanism still needed another five degrees of freedom for ideal performance, a total of nine joints in all. The only compensation was that the motions needed to provide the extra freedom could have a lower bandwidth—3 to 6 Hz would be enough.

With more than six joints, the arm mechanism would be kinematically redundant. Part of the redundancy arises from the need for fast dynamic response—the bandwidth problem described above. The follower can be treated as a specialized shearing end-effector for a five-joint robot, and in this way the apparent redundancy does not pose additional control problems. Unfortunately the weight of the structure, controls and services needed for nine actuators would be a problem, so I looked for ways of using less.

The follower assembly was longer than I expected; the distance from the cutter to the mounting point would be about 500 mm. The workspace for the arm carrying the follower assembly had to be large enough to include the envelope of the follower as it moved around the sheep.

Even a large sheep without its wool looks quite small. With twelve months wool, up to 130 mm long, a small sheep looks big. Now, to imagine the space needed for moving the follower around the sheep, think of a sheep with 650 mm

Kinematic redundancy

In the context of robots, the term redundancy usually refers to the number of joints in the arm. A redundant robot arm has more joints than the minimum number needed.

Recollecting that a robot arm which is to move a tool to a given pose will need at least six joints, we can conclude that a seven-joint arm has some redundancy. This is not necessarily the case, but explaining why now would confuse rather than clarify.

With the tool in a particular pose on a six-joint arm, it is impossible to move any joints in the arm without moving the tool. But if there is an extra joint, then it is possible to move some of the joints in such a way that the tool position remains unchanged.

Your arm has redundancy; place your hand firmly on the table top, palm down. You will find that you can move your elbow without having to move your shoulder or hand. Your arm has, in effect, seven joints between the shoulder and the wrist.

To my knowledge, there are no commercial robots which have redundancy—the extra joints are more costly and increase the weight of the arm. The arm is more complex to control too. There are a few research robots which have redundancy—the sheep shearing robots ORACLE, and later SM, are both redundant.

of wool. It would be huge. That is the working space of the arm which we needed to carry the follower actuator.

One theoretically attractive solution was a vast horseshoe-shaped arch over the sheep. This arrangement is used by medical CAT scanners. Only two movements would be needed—one around the arch and the other to move the arch along the sheep. While this would give access all over the upper surface of the sheep it presented overwhelming mechanical difficulties. Apart from the sheer size and weight, the sheep centre would always have to be close to the arch centreline—and sheep simply aren't cylindrical. Further, a large structure would be needed around the sheep—it would be difficult to gain access for loading and removing the sheep from its cradle.

In the end I chose a mechanism which would allow us to shear the back and one side of the sheep (figure 2.10). Responding to the politically imposed focus on the back and side of the sheep, and also to increasing urgency, I chose a design which would be simple, cheap, and quick to assemble. After all, this was to be a test rig—not at all a prototype shearer. We had always assumed that the next machine might look quite different. I built a model using my children's Fischer Technik[5] parts. Only four extra actuators were needed making eight in

[5] A construction kit used for mechanical prototyping, similar to Technical Lego or Meccano, but more robust.

all. It was ungainly, but my linkage used components familiar to farmers—welded steel and bearing blocks for the pivots. I naively assumed that farmers would relate to it better. It would be a stark contrast to Roy's precision machined surface follower mechanism. Only later did we learn how important appearance can be.

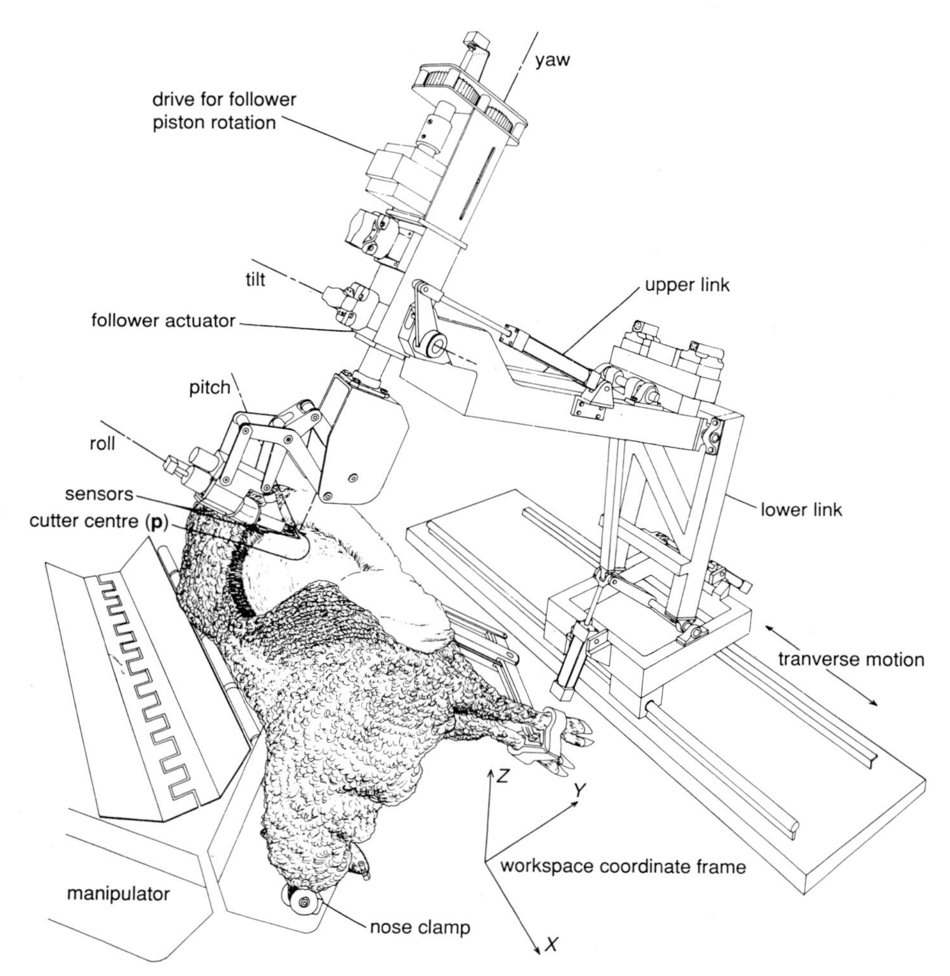

Figure 2.10 ORACLE robot mechanism—hydraulic and electrical services have been omitted for clarity. This drawing shows the second cutter/follower assembly which was not fitted to the robot until December 1979.

It was March 1978 and Jim Blair raised a concern that was to echo through our monthly project meetings. He proposed building a simpler test rig so we could have some experimental results before the Corporation reviewed the project at the end of the year. He was also worried about spending $200,000 before a single piece of wool was shorn. I argued strongly to continue with the ORACLE design; I thought that we might otherwise become trapped with a second-class machine, like CSIRO. I cannot remember Roy Leslie's view, but he probably supported me. Jim and Prof. A-W deferred to us as our salaries depended on the review so we had the most vested interest. We pressed on, but the apprehensions remained.

Electronics and wiring

There was a secondary benefit flowing from our choice of hydraulic actuation; the servo valves worked with simple amplifiers which even mechanical engineers could design. My incidental experience with computer interfacing left only the sensors as an area where electronics expertise was needed. Duncan Steven, a lecturer in the Department of Electrical Engineering, provided the help needed and invented a robust, compact and sensitive capacitance sensor with a range of about 15 mm (figure 6.2, p. 151).

Capacitance sensors

A capacitor is an electronic device which stores a small amount of electrical energy by placing an electrostatic charge on two parallel metal plates with an insulator between them. The greater the capacity, the more energy can be stored. The capacity is proportional to the area of the plates and inversely proportional to the distance between them.

A capacitance sensor exploits the same principle but the aim is to measure the capacitance and hence deduce the distance between the metal plates. The plates do not have to be metallic—they can be any electrical conductors. To measure distance to the sheep skin we assume that one of the plates is the skin of the sheep and the other is mounted under the back of the shearing comb.

The practical problem with capacitance sensing is the extremely small capacitance which has to be measured—this requires circuits of great sensitivity. Chapter 6 presents details of electronic circuits for capacitance sensors.

Our grant allowed us to take on an electronics technician, Rob Greenhalgh. Rob took on the construction of the computer interfaces, sensors, and servo control circuits. Although the electronic circuits on ORACLE were quite simple, the wiring was not. Eight hydraulic actuators needed three wires for each of two tracks on the potentiometers, three wires for each of two pressure transducers, and two wires for the servo valve. Then there were the sensor circuits too, power supply wires, monitoring points, the remote control button box—the list seemed endless. All the wires came together in a junction box at the end of the trolley Rob had built to carry the computer and electronics circuits. It was made in a hurry, as it ought not to have been, and was too small. But we only discovered that when two-thirds of the wires were in place, and undoing them all seemed worse than pressing on.

When we decommissioned ORACLE years later, we found that the trolley was too wide to pass back into Rob's workshop without first being dismantled. When we asked Rob how he moved it out, he simply smiled.

Assembly and test

Two of the mechanical technicians in our departmental workshop were assigned to Roy to make the mechanical parts and assemble them (the large components were subcontracted to a local workshop for welding and machining). They were soon busy covering my 'agricultural' linkage with hydraulic hoses and tubing. As the weeks went by, my neat linkage disappeared under a growing mass of valves, valve blocks and tubing. The plumbing for a single actuator looked simple enough, but with eight actuators to be joined to their respective valves, check valves, pressure relief valves, filters and so on, pipes were running everywhere.

Most of the 'robot' had been assembled by September 1978. Just as the tension around the imminent withdrawal of CSIRO from the project mounted, so did our own problems. First among these was a series of protracted delays in the delivery of key components. Perth is at the end of a long supply line from the 'technology' centres of the USA and Europe. Parts which would be available from a local warehouse in 'Silicon Valley' could (and sometimes did) take months to reach us through a series of agents and 'middle men'.

By the beginning of October, enough components had arrived to begin the first tests with hydraulic power. The first pressure test on a new machine is always a moment filled with apprehension. Countless pipes have been bent and fitted to manifolds and hose fittings—the hoses have all been carefully cut to length and had their special fittings attached. Stacks of valves, relief valves, check valves and adaptor plates have been bolted together, with all the holes and oil passageways correctly aligned. Then every part has been completely stripped and scrubbed clean, dunked in an ultrasonic cleaning bath and rinsed again to remove microscopic metal filings which would clog and scratch the delicate servo valves. Then oil has been flushed through the pipes for hours to clear out the last of the dirt and contamination. Finally, Roy carefully removed the blanking plates, bolted the delicate servo valves firmly in place, and turned on the main pump.

As Roy gently increased the oil pressure, we anxiously looked for the faintest signs of a leak—an oil smear growing on the side of a valve, a drip appearing at a connection.

At 25% of full pressure, we stopped and checked for leaks. Roy tried the manual control levers on the two main arm actuators, but the pressure was still too low to move the heavy linkage. Still no leaks.

Roy closed the pump bypass valve, and the pressure climbed. Full pressure, and still no leaks. The two main actuators were still fully retracted. Cautiously, he nudged one of the manual control levers—no response. A little further, and then a touch more brought a squeal of protest from the relief valves as the main arm mechanism leapt up. Roy, startled as we all were, let go the valve handles, and the arm silently sank back to its rest position. The manual control valves were just a little touchy, but that could easily be fixed.

Roy gingerly adjusted each of the servo valves. Each valve had an adjusting screw which could be used to slightly open the valve one way or the other. Roy gently turned each of the adjusting screws; each actuator in turn moved gently but firmly to the end of its stroke, raising the arm up under its own power for the first time.

The noise of the pump was something we would all have to get used to. Driven by a 5 kilowatt electric motor, the air was filled with a beating whining hum. Talking was an effort, and as the pump reached the maximum pressure, there was a sudden loud clunk and change in the beat as the automatic bypass valve opened. When the pressure in the high-pressure accumulator fell, the bypass valve closed with another clunk and the hum strengthened as the pressure rose again. The clunk-clunk every few seconds irritated all of us, and probably ORACLE too, so Roy soon installed a more expensive proportional control valve which continuously adjusted the pump for the right flow.

With the pressure testing over, peace returned to the laboratory for the long job of connecting the computer and the electronics. ORACLE used three heavy cables with 40 wires each. Ordinary computer cables or telephone cables contain many more wires, and are far lighter and more compact. But a robot moves, so the cable has to be designed for repeated bending, rolling and twisting, being trodden on, and carried alongside oil hoses which crack like a whip when pressure waves radiate away from the control valves. Once again, our supply problems were critical. The 'ideal' cables were available, but on a six to nine month delivery from the USA and a minimum length of 1,000 metres. We needed just 30 metres. So we had to make do with an oversize industrial cable designed for mining equipment. It was robust but we had to bend it more than it was designed for, and that meant that the wires inside would break, sooner or later.

Once again, moments of apprehension for all of us. The first test under computer control had to be taken slowly, the servo valve cables being plugged in one by one. In spite of all precautions, there was every chance that a valve connection had been accidentally reversed. If this happens, the valve passes oil to the wrong end of the cylinder and the actuator moves faster and faster away from its intended position until it thuds against the end stop! Sure enough, two connections had to be changed over.

Finally, with all the valves connected and checked, the first moves could be programmed. At long last, after months of preparations, the arm lifted smoothly up, across, down and back again. Days of tuning followed to readjust the actuators, correct minor software errors and to check the arm for correct positioning. The first of many parties of visitors arrived to take a look. One of the first groups included the Vice-Chancellor of the university, Bob Street, a professor of physics and a keen supporter of the project. ORACLE seemed to sense his affinity, and at the start of the demonstration sequence, leapt up towards a very startled Vice-Chancellor, stopping centimetres away before gracefully completing the programmed sequence. We were just as surprised, but ORACLE obstinately refused to repeat its sudden leap when we later tried to find out why it had put on such a performance.

It was March 1979 before the last of the long-awaited hydraulics connectors arrived. When Roy announced that he was ready to power the surface follower mechanism for the first time, we all gathered around the robot to watch. After so many months of waiting, the robot was finally complete, and we all expected to be shearing our first sheep within a few days (figure 2.11). But we were to be disappointed.

Following our success with the surface follower actuator, Roy had designed special miniature rotary actuators for the wrist joints. Although they had been completed months before, we needed special miniature hydraulic hose fittings which had taken so long to arrive before they could be tested. One actuator steadfastly refused to move, no matter what pressure was used and the other was very 'sticky'. Roy soon realized that we had a major problem—the friction inside the actuators was mainly due to the pressure acting sideways on the shafts, and the force on the vane would always be less than the hydraulic pressure.

Roy discussed the problem with our technicians who took the parts away and cut minute oil passageways around the curved edges, from one side to the other. Two vanes could be used, and the oil pressures on each side could be balanced. The yaw actuator needed to turn more than 180° so an extra vane could not be used. But Roy was able to smooth the parts enough to remove most of the sticking. Three weeks later, we tried again, but soon realized that the oil passageways were so narrow that the bandwidth would be much less than we had hoped for.

Meanwhile, another cloud had appeared on the horizon. Reports from CSIRO had raised doubts over the cutter mechanism on which we had all pinned our hopes. Shearers had found it very hard to push through the wool, even at low speeds, and it would not work at all well at the speeds we wanted the robot to run at. Yet the surface follower mechanism had been entirely designed around the cutter, with its small motor. To fit a conventional cutter to the robot, we would need to completely remake the surface follower mechanism, a minimum of six months work. But then we also had misgivings about our own rotary actuators so Roy decided to start work on a new design.

March passed into April, and we desperately needed results for the first Australian Wool Harvesting Conference in May. I fitted a dummy sheep in front of the robot: a steel drum covered in foam rubber, brass mesh and conductive plastic sheeting to simulate sheep skin. My first attempts to shear the drum were far from encouraging. The approach was smooth enough, but as soon as the cutter touched the plastic, it jumped up and down, bucking wildly, and before long I had gaping tears in my simulated sheep, a broken comb, and injured pride. Even the steel drum had been punctured! Repairs took two weeks, and by then I had made several changes to the software. Days before the conference, I was able to make a videotape showing ORACLE shearing its dummy sheep.

In all of this we were joined by Stewart Key, a recent UWA graduate, whom we had persuaded to leave the corporate ladder of the Ford Motor Company and return to WA. We also took on our own mechanical technician, Ian Hamilton. Ian had learned his trade as an aircraft toolmaker in Britain and we soon appreciated our good fortune to have such a high degree of technical skill in the team.

We mounted our first sheep in a rush, and made scant preparations for what was to have been trial number 1. I followed my routine procedure for the dummy sheep tests—all we wanted was one blow. But it was too rushed. I mistyped the command to start shearing and ORACLE lunged forwards towards the sheep. Roy lunged almost as quickly for the emergency control levers, but not quite quickly enough. The cutter pressed heavily into the side of the sheep before Stewart rescued the unfortunate animal. Happily, the sheep made a quick recovery in Jim Blair's back yard, but we went to the conference without shearing results.

We put on brave faces, and Prof. A-W skilfully parried the disparaging comments from the conference floor by reciting the following poem.[6]

T.T.T.

Put up in a place
where it's easy to see
the cryptic admonishment
T.T.T.
When you feel how depressingly
slowly you climb
it's well to remember that
Things Take Time.

We need not have been so worried—CSIRO had withdrawn and the Corporation had placed their hopes for robot shearing with us. The conference turned out to be a great encouragement, particularly as it came just as our difficulties seemed to be overwhelming (Trevelyan and Leslie 1979; Leslie and Trevelyan 1979).

Yet worse was to come.

I needed more time to rework the software to avoid another frightening plunge towards the sheep. I rearranged the way robot movements were programmed so that there was no possibility of mistyping a command in the tense atmosphere of a trial (Trevelyan, 1981*a*).

Meanwhile, Roy took the cutter mechanism off the wrist and tested it on a sheep to make sure it would work. He soon discovered that it would not work at all! Stewart and Roy soon realized that the sand and grit in the sheep's wool

[6] *Grooks* by Piet Hein, Hodder Paperbacks, 1969, p. 5. (by permission)

was to blame. The cutter had been tested in Geelong on nice clean white sheep, washed by frequent showers of rain. Our sheep had come from semi-desert pastoral country and the wool was full of dust and sand.

Roy finally made the cutter work by bending the pivot, forcing the comb and cutter together more tightly. It seemed an unthinkable mutilation of such a precisely and carefully made machine, yet it was effective. For the first few months of shearing trials at least, we would have to make do. At least the experience confirmed our earlier decision to redesign the surface follower and fit a more conventional cutter as soon as possible.

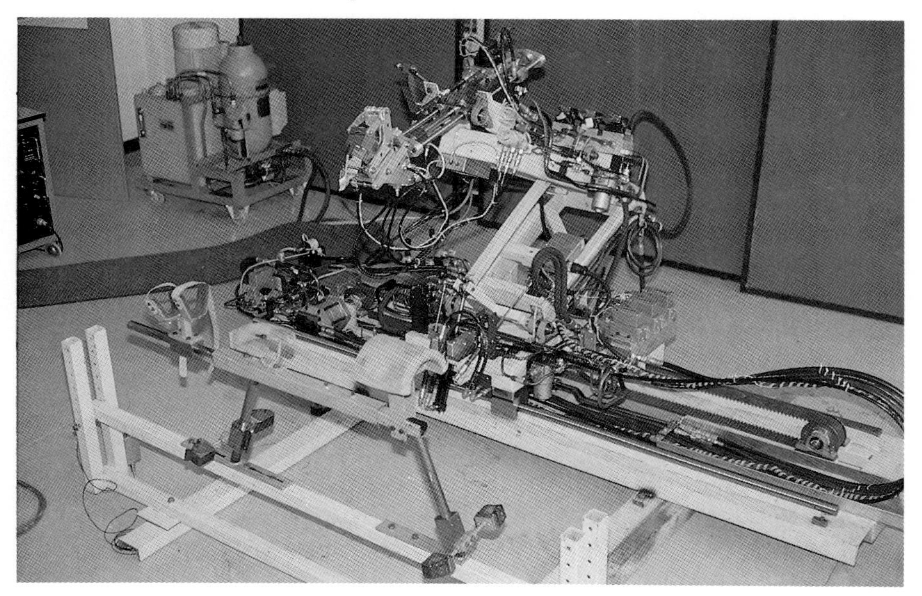

Figure 2.11 ORACLE robot ready for first shearing trials. Note the sheep restraint in the foreground with the head rest at the left-hand end.

Shearing tests

It was the last of the delays. On 25th July 1979, we were once again prepared for trial number 1. A sheep arrived, trussed up with three legs tied together to prevent it from jumping out of the back of Roy's aging utility.[7] It was carried in and untied, firmly held between Roy's legs, and then lifted up by willing helpers, high on to the restraint bars. It rolled alarmingly to one side, was righted, turned its head curiously, this way and that. The legs and neck were too long, and the clamps had to be readjusted. Finally the animal was immobile, panting gently, patiently.

Roy had shorn most of the wool from the rear end and far side of the sheep so that we would have a better view, and the robot could land for the first time on shorn skin.

[7] a pickup truck

The robot was slowly raised and moved across to the sheep under manual control. When the cutters were just above the shorn patch at the rear end, the position was recorded by the computer. The robot backed away automatically, and we all took a deep breath. This was it.

We exchanged glances and a silent 'let's go' passed round the ring of expectant faces. I typed the command to start the automatic approach to the sheep, and looked around for one last check. My finger hovered over the RETURN key for a second or two while I glanced at Roy who was holding the manual control box. Rob Greenhalgh held a second emergency stop button in his hand, standing well back, but ready to act. Stewart had his head down behind the video camera. My finger tapped the key, and the cutters moved almost imperceptibly towards the sheep. The cutter motor came to life, a shrill whine above the loud humming of the pump, and then there was a gentle nibbling, and rocking as the sensors reacted to the skin for the first time. A second later and the robot was moving away again. A black smear of dirty oil marked the cutter's path across the skin—the sheep seemed quite unaware of what had happened.

We all broke out in smiles—but the real test was still to come. The robot was moved manually once more to mark a new landing point which would take the cutter into unshorn wool for the first time. I typed the start command, and once again the robot was slowly closing on the sheep. Again the gentle nibbling, this time for real as the cutter seemed to nibble and chew the wool off. As it pitched back and forth, the wool worked its way up into the wrist linkage where it was squeezed before falling away on the underside (figure 2.12).

A cheer broke out as we glimpsed a perfectly shorn strip left by the robot for the first time. Another blow was requested, alongside, with the same result.

We moved in and carefully collected the wool, dividing it among us—our first robot shorn wool. I still have some on my desk, alongside one of the still photos taken that day.

The sheep was taken down and trussed up once more to be returned to its paddock outside the animal house.

We were on our way, just a few months late, but elated to have achieved such a perfect first result.

Exploring limits

With ORACLE shearing, exploration could begin. Well, almost. The first successful shearing test was an important milestone and the next was to repeat the results obtained by CSIRO. Only then would we be exploring unknown domains. It was one thing to shear a couple of blows, quite another to shear the back and side of a sheep reliably, at any speed.

High-speed shearing tests would have to await the new cutter on the surface follower mechanism being assembled in the workshop, but there was plenty to get on with. There were countless small frustrations to overcome. The trial records abound with the phrase 'trial abandoned due to.....'.

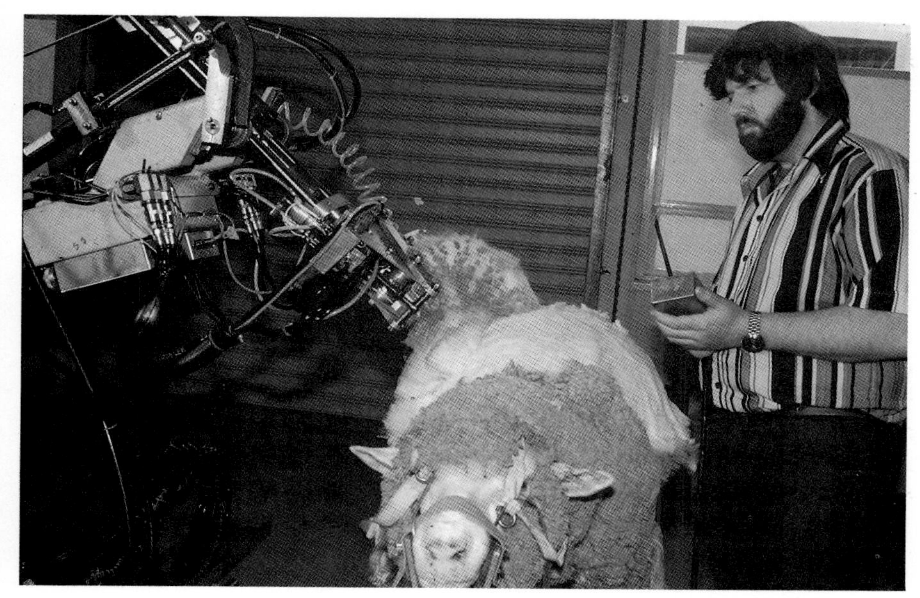

Figure 2.12 First shearing trial, 25th July 1979. Roy Leslie was holding a small teach pendant; only the 'pause', 'resume' and 'emergency stop' buttons operate during shearing.

A loose screw fell out and the capacitance sensors fell off. One of the 20 or so wires to the manual control box broke, and Rob Greenhalgh needed a day's work to locate the break and repair the cable. Wool became snagged in the wrist mechanism which had to be partially dismantled to remove it. A simple modification was needed to prevent a recurrence, but this meant another two days wait.

The cutter mechanism was particularly unreliable. Cutter problems would be with us for years to come, and had we known that, we might have fixed the problem there and then. But we wanted to avoid cutter problems. We had no wish to become involved in research on cutters as it seemed that other experts knew far more than we did. So we just dealt with the problems on an *ad hoc* basis. Disassemble, regrind, reset and reassemble. A day's work.

The rotary yaw actuator stiction[8] problems which held us up for so long (p. 43) were just beneath the surface. From time to time they reappeared, just to disappear frustratingly. Again, the new follower mechanism would provide a more permanent solution so there was little point in trying to fix the current one—we just persevered.

Another seemingly trivial problem—the potentiometer on the drive belt which measured the traverse position of the carriage seized. This puzzled us—

[8] combination of friction and sticking, often causing jerky movement.

we looked at a new replacement and the shaft was not a tight fit. There was a flexible coupling so that slight misalignments could be accommodated. So with a shrug, we just replaced the potentiometer and presumed that it wouldn't happen again. It did, several times.

All these problems were transient, fleeting, slight blemishes which we could deal with, or just put down to a hurried fix here, a patch there. Frustrating—yes but not a real concern. Soon we began to see the more fundamental difficulties which we had expected in one way or another, and it wasn't long before there were some unexpected ones too.

Our first shearing blows ran along the firm muscular side of the back, just a little way down from the backbone. The object was to discover how well we could control overlap without any explicit knowledge of the surface beforehand. Each blow was defined by:

(1) a starting position obtained by moving the robot under manual control until the cutter was just above the desired landing point on the sheep,
(2) a surface normal vector obtained by using the robot to measure another point further off the surface at the landing point,
(3) a direction vector parallel to the sheep, and,
(4) the desired shearing distance along this vector.

Once the first blow had been shorn, the next blow was planned automatically alongside the first blow with a given degree of overlap. The surface normal at the landing position was estimated from cutter attitude information remembered from the first blow. In effect, this was a strategy for a blind shearer with perfect motion recall. The results were surprising. A blow repeated by the robot would follow the same path on the sheep skin. With just a nominal amount of overlap programmed in the computer, wool would be neatly shorn with just wisps left between successive blows. More amazing though was the sight of the wisps remaining after the shearing pattern was run a second time over the shorn skin, even though the sheep had wriggled a bit. Clearly, blow positioning on the sheep was repeatable, at least under these conditions.

The sheep kept remarkably still, most of the time, but when they did struggle, there were some amusing and unexpected results. First, the sensitivity of the follower was surprising, even to us. As a sheep writhed and wriggled, more to relieve discomfort after lying still for a long time waiting for us to get moving than from any pain, the cutter rode up and down as if glued to the skin. Some sheep were genuinely ticklish. But then, when the cutter was not working properly, it caused some discomfort as it pulled the wool rather than cutting it. At other times, the cutter would rhythmically rise and fall, following the breathing of the sheep.

We replayed the robot movements on a screen—fourteen images per second (figure 2.13). Sheep movement due to breathing showed as a series of small periodic ridges; struggling showed as jumbled chaos, a jagged disruption of an otherwise orderly record. Remembering that we had programmed each blow to follow the trace left behind by its predecessor, you can now visualize the conse-

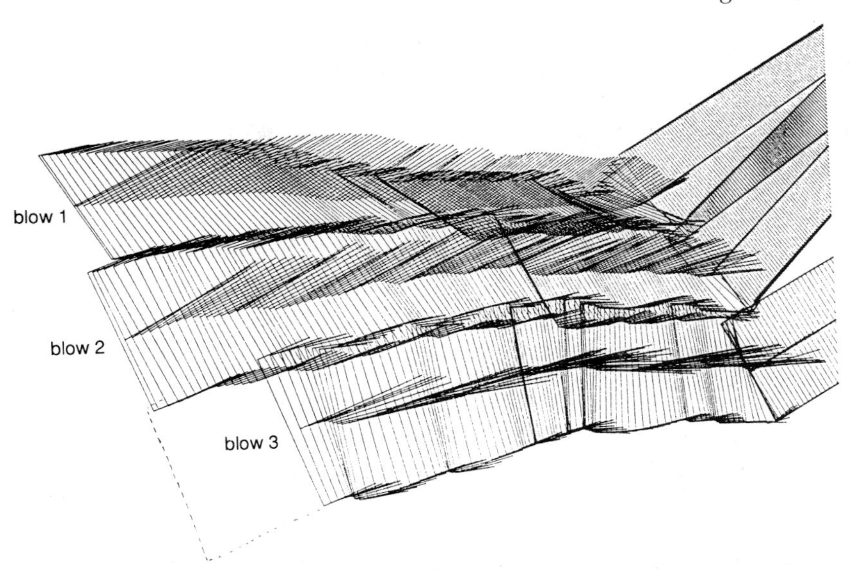

Figure 2.13 Robot movements replayed after shearing trial 6, 4th September 1979. Each stick diagram represents one motion step (0.072 sec). The first blow is the top one; the robot landed at the right-hand end. A minor sheep twitch in the middle of the second blow causes follow-on effects in the third blow at the bottom.

quences. The first blow would appear smooth enough; apart from a brief heaving and wriggling, the sheep would remain calm. But some distance along the next blow, with the sheep still, the robot heaved and wriggled, following the trace precisely inscribed in its memory! After we realized that our software was performing so faultlessly, it was amusing to watch.

Once we had demonstrated the basic mechanism for planning blows automatically, we attempted to implement our continuous shearing pattern. Instead of stopping at the end of one blow, and repositioning to start the next blow next to the first landing point, we turned the cutter on the sheep and followed the trace of the first blow backwards. This worked well for two or three blows, but as we extended the pattern down the side of the sheep we discovered some problems.

The consistency of the sheep flesh varies enormously from one part to another. We built up our confidence on the firm muscular rump quarters only to lose it on the soft wall of the abdomen. Here, when the sheep relaxed, the skin transformed into a loose membrane containing fluid, losing all its resilience. As the comb nosed in, seeking electrical contact, the skin retreated. As the capacitance sensors reacted to the warm closeness of the bulging membrane behind, the computer correctly pitched the cutter forward to dive the points deeper down. Then, when the sharp prongs finally made contact through the dense side wool, the comb would start to rise out again. But now the wool clung to the teeth, holding the skin in place, the comb rising higher and higher. Finally the skin fell away, and the comb stopped, and started to plunge in again. With hindsight, it is easy to understand what was happening; at the time, we could only be thankful that the sheep was completely unharmed. The only effect on the sheep was the rough and ragged swath of half shorn wool.

Part of the problem lay with the sheep support. The animal was comfortably supported at the brisket between the front legs and the crutch at the rear. The 'tummy' hung from the ribs and backbone. When the animal relaxed, this could sway some distance either side—it clearly had to be restrained. Roy added more support with webbing between two bars shaped to hold the bulging abdomen. Part of the problem lay with sensing.

We attempted our original unidirectional shearing pattern, with slightly better results. The continuous shearing pattern was retained for just the first two or three blows.

As our trials progressed, time passed and the wool on our flock of sheep grew longer and a new problem appeared. The cutter assembly was about 220 mm long, and with 120 mm high walls of wool on each side, its turning ability was severely impeded.

We decided to abandon our forwards and backwards shearing pattern. Each minor improvement served only to make the real difficulties all the more apparent. There was no room between the walls of unshorn fleece to turn the cutter, and the bluff shape of our cutter mechanism snagged the shorn fleece and stopped it from smoothly peeling away. After two or three blows, we had to stop shearing as the cutter head became lost in a growing bundle of tangled locks of fleece.

It was time to think again.

At least we had proved that memorizing the blow paths would allow successive blows to be shorn with minimal overlap. So we returned to David Henshaw's approach, shearing down the side of the sheep. We had the advantage, of course, that ORACLE could shear the opening blows along the top of the side, opening up a clear swath to start the downward blows.

And we were handsomely rewarded. Time after time, almost the entire side of the sheep was cleanly shorn, albeit slowly.

And so, by the end of the year, we had reached the stage of shearing nearly all of the back and one side of the sheep, using only a notional starting point, an initial shearing direction and a set of 'fencing' planes (figure 4.24 p. 110) and of course the sensor inputs to guide the comb. In December and January 1980 the new follower mechanism was installed on the robot and we could begin to shear faster. Soon we were to discover the advantages of encoding more sheep knowledge (chapter 3), refine our motion control techniques (chapter 4), and improve our sensing (chapter 6). Over the following years, we often looked back at the photographs and videotapes recording those five months of shearing and wondered whether all the work in between had achieved anything. It had of course, but sometimes it seemed that we couldn't shear nearly as well.

Epilogue

ORACLE served us well for six years of shearing trials. Its reputation as the new breed of Australian shearer spread, as figure 2.14 shows. Apart from its one lunge at the Vice-Chancellor it was remarkably well behaved in all that time. The oil hardly ever leaked. It ran off its track once—a runaway software error. The results of our five hundred shearing trials were coalesced into its successor in 1984 and 1985 (see chapter 8). Our decision to mount the hydraulic connections on the outside was a mistake—we never made significant changes. But our decision to place the wiring on the outside was a good one—we had many wiring failures and had much to learn about designing wiring for flexure and vibration.

Figure 2.14 Cartoon which appeared in New Scientist, March 1984. (By permission)

Introduction to chapters 3, 4, 5 and 6

Developments in surface modelling, motion control, sensing, and wool severance span the years since ORACLE was commissioned to the present day, though most of the foundations were laid between 1980 and 1982. Therefore I have departed from the narrative style to present a coherent account of these important techniques and I return to the chronological account in chapter 7. These chapters represent the results of countless small steps, trials and experiments, frustrations, despair, and occasional breathtaking inspirations, in the same pattern as the other chapters relate.

The reader will find references to both ORACLE and SM though the latter is not described in detail till chapter 8.

3
Software sheep

Impatience and curiosity quickly led to surface models.

Our early trials with ORACLE were often marred by unstable cutter motions caused by difficulties in sensing the skin and controlling the cutter attitude. When shearing dense wool over the soft flesh on the side, the cutter would rock forward and back, alternately trying to push the comb points on to the skin and then lifting them away.

From the start we had always conceived of some kind of forward looking sensor which would allow the robot to 'see' the skin ahead of the cutter, instead of having to watch from behind with the proximity sensors. With forward sensing, ORACLE would be less reliant on resistance sensing which provided a very erratic signal. But we knew we would have to wait for one to try. The University of Adelaide was only just beginning to work with ultrasonic devices, and we had little time to develop such sensors ourselves.

Curiosity and some impatience led to an experiment to see how effective forward sensing could be. We programmed the computer with an approximate model of part of a sheep, so it could calculate a simulated signal from a forward sensor and we used that data during a shearing trial.

The first model represented the back and part of the side of the sheep with an ellipsoid like a football. It was crude, but the simulated sensor readings were easy to calculate. The results from the first shearing tests were outstanding. By calculating the cutter orientation from the football model and using the capacitance sensors for height control, the robot smoothly glided across the sheep, with no sign of the instabilities which we normally encountered. The resistance sensor signal was only used if it suggested that the comb was pressing too hard. We altered the software so that we could use the sensor signals to calculate minor corrections to the cutter attitude calculated from the model. The robot glided just as smoothly, but this time the tips of the comb followed the skin much better.

The shearing behaviour was so much better that we decided to develop the idea further. The model had to be more accurate—an alternative surface representation was needed. I looked up the classic texts, but found that my mathematical reading age would need a major boost. I also suspected, without being sure, that the computation time needed for the textbook methods would be too great. Most of the texts discussed surfaces for computer-aided design and manufacture of high-precision mechanical parts like turbine blades or aircraft wings. We needed a rather different approach—less precision and fewer calculations.

Surface representation

The need for rapid computation of surface data suggested a simple approach. The conventional method of representing a surface is to use a vector function of two parameters t and u: Faux and Pratt (1979) provide an excellent survey of conventional approaches. A simple method which suited the typical shape of a sheep was to use section curves in parallel, equally spaced section planes. The first method which we tested defined the section curves with single-span cubic splines. Ruled surface elements filled in the gaps between the section curves.

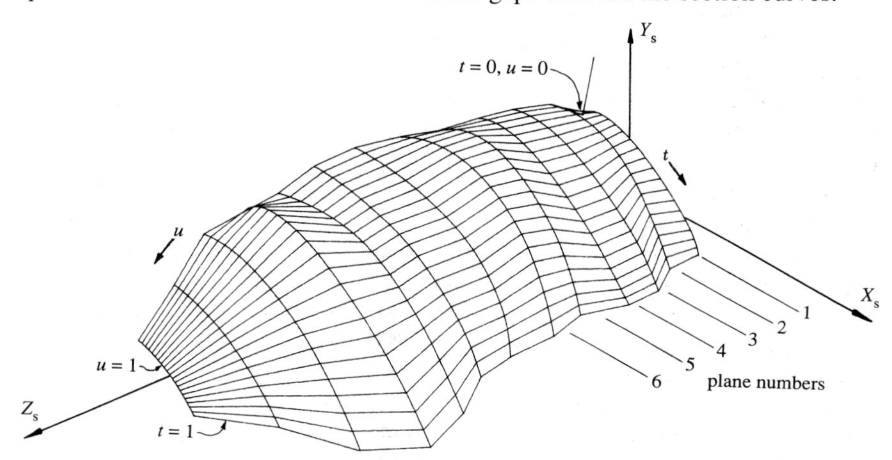

Figure 3.1 Surface model used for early shearing experiments, c.1980. Each curve is a single-span cubic defined in the X_s - Y_s plane. Variations in parameter distribution result in wavy lines of constant t values. The surface coordinate frame is shown. The line projecting from the surface at the upper right corner shows the point $t = 0, u = 0$.

Each point on the section curve in plane i was defined by:

$$\mathbf{q} = \mathbf{q}_i(t) = \begin{vmatrix} c_0 & c_1 & c_2 & c_3 \\ c_4 & c_5 & c_6 & c_7 \\ u_i l & 0 & 0 & 0 \end{vmatrix} \begin{vmatrix} 1 \\ t \\ t^2 \\ t^3 \end{vmatrix} \tag{3.1}$$

where u_i is the u coordinate value for plane i and l is the length of the patch normal to the section planes, $0 \leq t \leq 1$, $0 \leq u_i \leq 1$.

Points on the ruled surface element between planes i and i+1 were given by:

$$\mathbf{q}(t,u) = (1 - \lambda)\mathbf{q}_i(t) + \lambda\mathbf{q}_{i+1}(t) \tag{3.2}$$

$$\lambda = \frac{u - u_i}{u_{i+1} - u_i} \tag{3.3}$$

λ varied from 0 to 1 in proportion to u between each pair of section planes.

The coefficient array $c_0 \ldots c_7$ was defined by a table of values for each section curve.

For the first model, we measured cross section profiles of the back and side of a sheep on the cradle. Interactive curve fitting techniques were used to determine the coefficients for each curve. The model looked fine, but to use it we needed to solve a tricky computation problem.

The crux of the problem was this: the cutter would never be exactly on the surface, nor would it follow a predetermined path across the surface. Each blow was to run along the side of the previous blow. After allowing for steering errors which arose when trying to follow the path of the previous blow, we realized that we needed a method of calculating the projection of the cutter on to the model surface along the local surface normal vector. The cutter could be presumed to be close to the surface (within 50 mm). The computation has to be iterative as there is no direct solution, and the surface normal can only be calculated directly if the projection is already known.

For early experiments we simplified the problem by using a two-dimensional iterative technique, working just in the closest section plane to the cutter position. Initially the results were disappointing. There were jerks and shudders as the cutter crossed from one plane to the next—some form of smoothing was going to be essential. Four weeks later, I was able to repeat the successful results of the first trials of the ellipsoid model, this time over the whole side of the sheep. Prof. A-W had meanwhile coined the term 'software sheep'.

The simple surface model had some limitations. The section curves lay in parallel planes, equally spaced. It needed to be more or less lined up with the long axis of the sheep. Otherwise, the section curves became too complex to represent with single-span cubics. It was never intended to represent the whole surface of the sheep—just the part that the robot could access in one shearing position.

When we started to test shearing on different parts of the sheep, we set up different models for each part, and called the models 'surface patches'. Thus the whole sheep surface was represented by a collection of 'patches' (figure 3.2).

It is not obvious how 'simple' patches could represent the whole sheep; they did not! All that they represented was the shape of each part of the sheep when it appeared in front of the robot. So, when the sheep was laid on its back to shear the belly, we were only interested in the shape of the belly—the back and sides were not represented. Once we turned the sheep over we were no longer interested in the shape of the belly—just the back and sides.

At one time, it was suggested that we should somehow represent the internal tissues of the sheep—the bones, fat, muscles, etc. That way, we might be able to predict the surface shape more accurately. We have not attempted that, yet.

Improved shape representation

Single cubic splines represented section curves sufficiently well while we used ORACLE for shearing trials. However, when we attempted to enlarge our surface

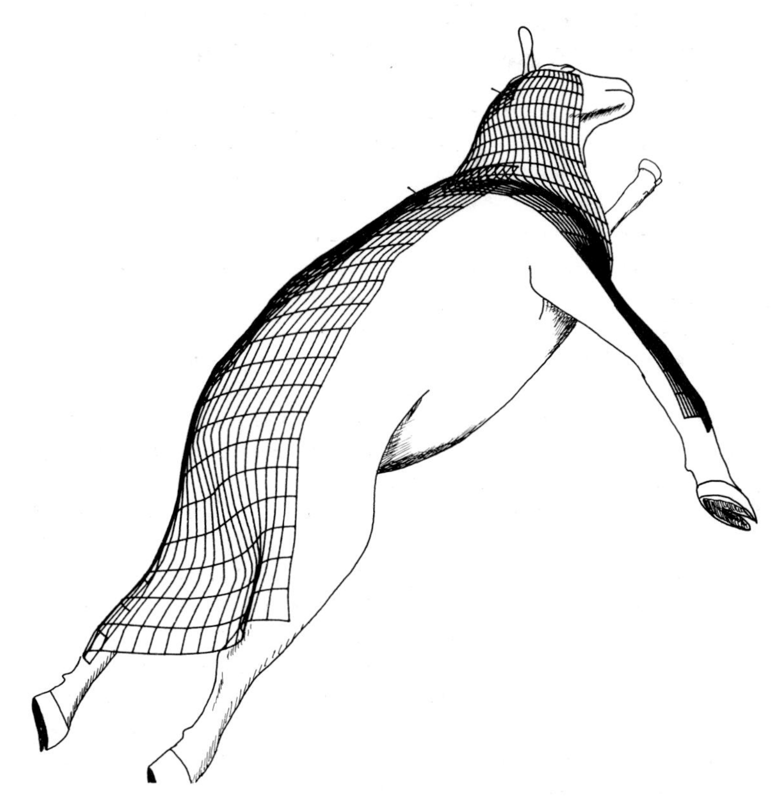

Figure 3.2 Surface models for two overlapping patches, *c.*1990. Eight control points (not shown) define each section curve. Each patch has a different frame and section plane direction. Note that the model exists between the rear legs and under the chin where the sheep does not. The model has become more a 'guide surface' for robot movements than a strict representation of the sheep surface.

patches to make use of the greater workspace of the SM robot (chapter 8) we found that we needed a more detailed representation. In 1985 we adopted a compound curve—six quadratic splines connected end to end. A marginal increase in computation time was offset by the greater accuracy of representation. The splines are formed from second-order Bezier curve segments joined at the mid-points of lines joining each pair of control points. (figure 3.3)

Points on each Bezier segment are given by:

$$\mathbf{q}(t') = 0.5\,(1-t')^2\,\mathbf{r}_{j,i} + (0.5+t'(1-t'))\,\mathbf{r}_{j+1,i} + 0.5t'^2\,\mathbf{r}_{j+2,i} \qquad (3.4)$$

where $\mathbf{r}_{j,i}$, $\mathbf{r}_{j+1,i}$, $\mathbf{r}_{j+2,i}$ are adjacent control points for section plane i, and t' is a segment parameter which varies from 0 to 1 on each segment.

Note that the curve parameter t is arranged to vary from 0 at the start of the first segment to 1 at the end of the last segment. The two end segments are extended roughly as far as the two end control points, by adding an extra 0.5 range in t'.

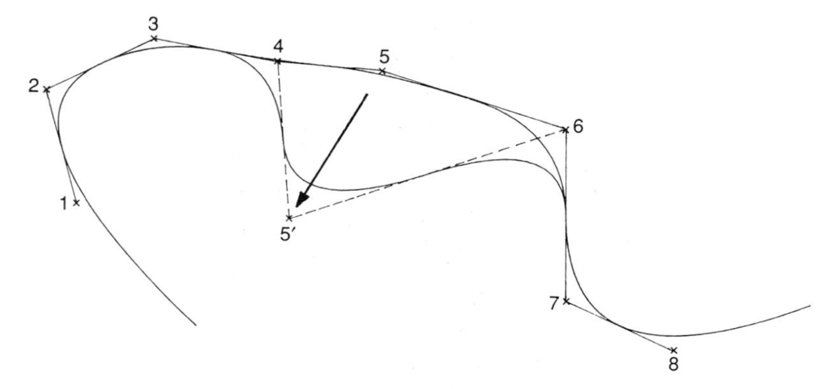

Figure 3.3 Six-segment quadratic spline curve defined on eight control points. Moving a single point (from 5 to 5′) modifies part of the curve, leaving the rest unchanged.

This arrangement still provides a compact and easily computable curve, yet with a useful degree of flexibility. The curve shape is governed by the control point positions, and can be intuitively modified by moving the control points as demonstrated in figure 3.3.

The surface model for a given patch is defined by up to 25 such curves—each one defined by an array of 8 control points. Since the control points are located on equally spaced planes, only two coordinates are needed for each. Although the representation of each section curve was different, the method of generating ruled surface elements between the section planes remained unchanged (eqn (3.2), figure 3.4).

The remaining parameters which define the model are:

(1) surface origin position;
(2) axis frame;
(3) plane spacing;
(4) number of planes.

Figure 3.1 depicts a surface model and shows the local Cartesian frame X_s, Y_s, Z_s which defines the section plane orientation.

Surface models provide simplifications

Any position on the surface can be defined by just two coordinate values. A path on the surface can be defined by two or more points, each defined by t and u values. All we need to define a shearing blow is this path information. From the surface model, we can calculate how to move the cutter to the start of the blow, along the blow, and away from the sheep at the other end (figure 3.5). Thus the two dimensions of the surface model convey all the information needed to control the robot movements in six dimensions—a great simplification.

The use of surface coordinates is not new of course. Navigators on the earth's surface have worked in terms of two surface coordinates for hundreds of years—latitude and longitude.

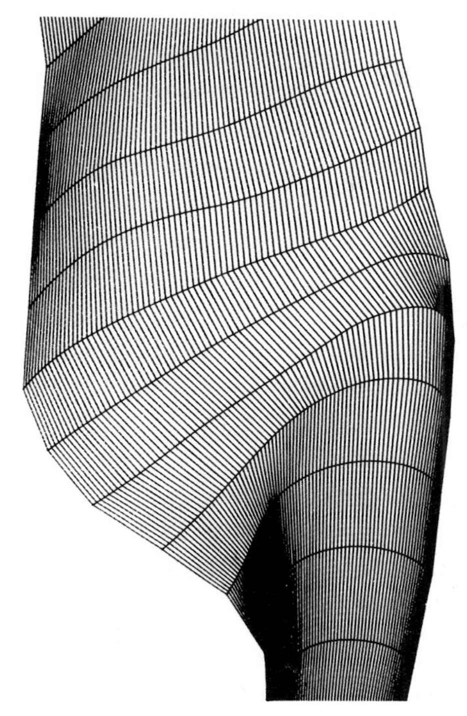

Figure 3.4 Surface model for front leg and armpit in belly shearing position (sheep is lying on its back). Closely spaced lines of constant t reveal the slope discontinuities between the ruled surface segments.

We soon realized that the software sheep could provide another major simplification. By careful choice of patches, we could align the models to each new sheep such that the surface coordinate values for each part of the sheep's body had the same values on every sheep. Thus a shearing blow programmed for one sheep would automatically be placed on the same part of the next sheep, even though the sheep was a different size, and possibly in a different position. The blow lengths would also be adjusted automatically. There would be no need to adjust the shearing instructions to lengthen the blows for a long sheep.

In effect, the surface model allows us to account for a major element of the sheep to sheep variability.

And so, from a makeshift experiment to test the effect of forward sensing came our first major breakthrough. Not only to stabilise cutter control, but also a map to know the sheep by, and one which provides the robot with a sense of proportion.

Coordinate transforms and kinematics

In order to use the surface model, we need two basic transformations. We need to calculate the cartesian position of a point on the surface so we can command

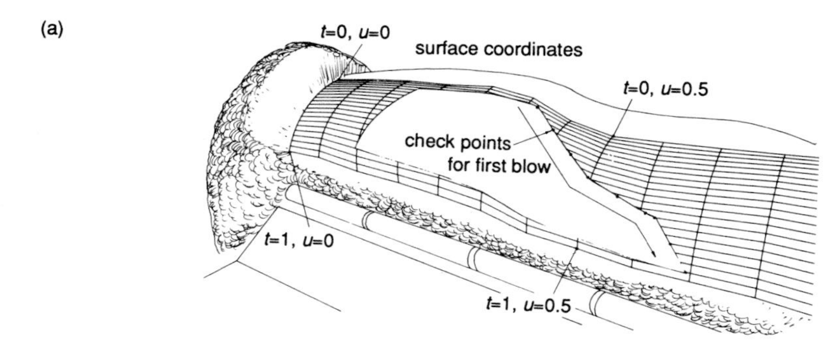

Figure 3.5a Surface model with shearing area defined by surface points. One side of the area is defined by the line of the first shearing blow.

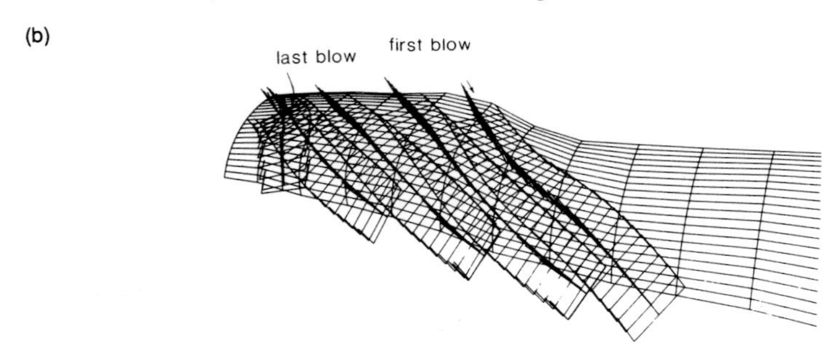

Figure 3.5b Robot movements for shearing area defined in figure 3.5a. All the movements are generated automatically, mostly from the surface model data.

the robot to move there. We also need the reverse calculation—given a robot position near the surface, we need to calculate the projection, or the closest point on the model.

A simple visualization has led to a useful insight into the algorithms which perform these transformations.

Imagine the pole from a baby's set of rings—it has a flat base and can be moved across a surface as shown in figure 3.6. One of the rings can slide up and down the pole.

We can think of this as a robot arm—a rather strange mechanism, but the sliding ring has three motions (discounting rotation about the pole). The pole base can be moved in t and u on the surface, and the ring can slide along the pole perpendicular to the surface.

First, the forward kinematics of this robot: we can calculate the position of the ring, given the values of t, u and h. Next the inverse kinematics; given the position of the ring in space, we can calculate t, u and h using the motion rate control method (Whitney 1969) just as we would for a more conventional robot arm. The algorithm which we have used was not based on motion rate control. However, now that the close similarities are apparent, it can be explained more easily in similar terms.

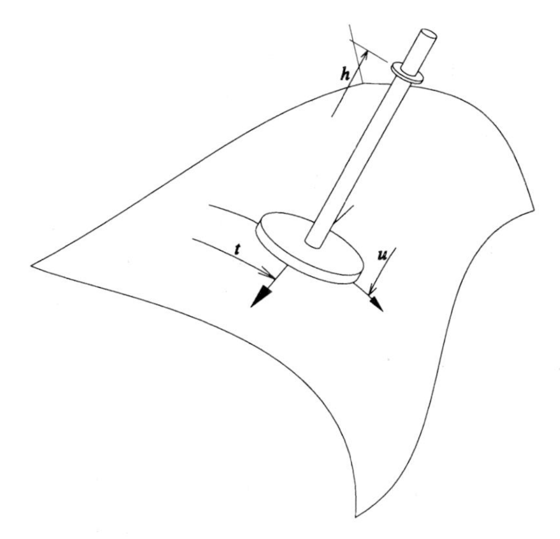

Figure 3.6 Imaginary surface robot—refer to text.

Forward surface kinematics

First, the function **surfp**(t,u) which calculates the Cartesian position of a point on the surface. The coordinates of the point are t,u. The vector which results from this function has three coordinates—x,y and z. In its basic form, this function encodes eqns (3.1) to (3.4), depending on the representation being used.

A variant of the function calculates the Cartesian position of a point at height h above the surface point at t,u. If h is negative it is below the surface. In using the terms 'above' and 'below', I mean along a line perpendicular to the surface at the point. To use this variant, we add the value of h as a third coordinate **surfp**(t,u,h). The computation of the surface normal is needed to do this and a description of the method follows.

The function **snorm**(t,u) calculates the direction of the 'surface normal vector' which is perpendicular to the surface at the point in question.

We need to look a little more closely at the structure of the model to appreciate some subtleties at this stage. Remember that the surface model is defined by cross-section curves in parallel section planes. Ruled surface elements fill in the gaps between the curves. They are called 'ruled' because they are generated by an infinity of straight lines between the curves (figure 3.4). The u coordinate represents distance from the origin, so the function **surfp** linearly interpolates between adjacent section planes using eqn (3.2).

The **snorm** function works by calculating the positions of a cluster of four points around the point t,u. The points could be calculated by **surfp** but considerable computation time savings can be achieved by optimizing this calculation. The surface normal is calculated by forming the cross product of lines joining opposite pairs of points.

The simple ruled surface has breaks in slope at each section curve. It was these breaks (slope discontinuities) which resulted in erratic robot behaviour when we first tried using spline curves for the software sheep. The effects of the breaks can be conveniently overcome by spreading out the cluster of points in the *u* direction (figure 3.7). This way, the calculated surface normal varies smoothly across the section planes. By this simple step, we have effectively increased the order of the surface in the *u* direction, just for the normal vector.

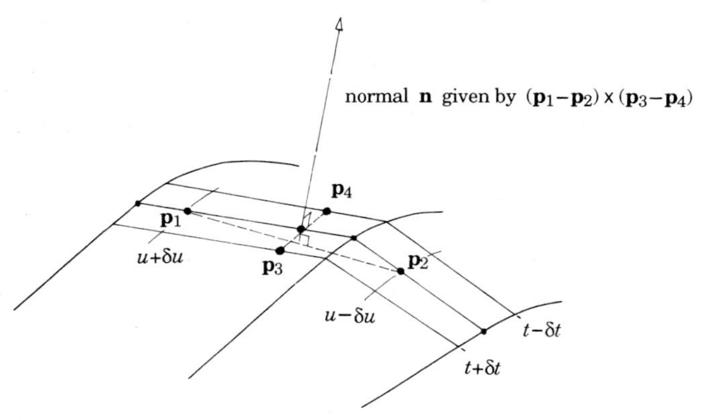

Figure 3.7 Surface normal computation method which results in continuous variation of surface normal across section planes—refer to text.

Now we have the forward kinematics we need to understand how to calculate a pose for the cutter at a surface point. Remember that the pose will have two parts—a position (*x,y,z* coordinates) and an orientation frame (three vectors **i**, **j**, **k** each with *x*, *y* and *z* coordinates). The position at the start of the blow can be calculated with a single **surfp** call:

$$\mathbf{surfp}(t,u,0) \tag{3.5}$$

assuming that the cutter will land on the model surface, and that the sensors will pick up the skin by the time it reaches that point.

The orientation is slightly more complex, depending on how it is specified. Suppose we want the cutter to be lined up for shearing along a line to another point t_1,u_1. Then the **i** vector can be calculated by

$$\mathbf{norm}(\ \mathbf{surfp}(t_1,u_1) - \mathbf{surfp}(t,u)\ ) \tag{3.6}$$

The function **norm** normalizes a vector to unit magnitude.
To line the cutter up with the surface, we calculate the **k** vector with **snorm**(*t,u*). Then, the **j** vector is given by

$$\mathbf{j} = \mathbf{k} \times \mathbf{i} \tag{3.7}$$

where the **x** represents a vector cross product.

We could imagine many other ways of specifying a pose relative to the surface model. One way which we have often used is to specify orientation by angles relative to the *t* direction (the direction of increasing *t*).

Inverse surface kinematics

Given **p** we need to find *t*,*u* and *h* such that **surfp**(*t*,*u*,*h*) gives **p**. For early experiments we adopted an iterative method for calculating the value of *t* closest to the projection of **p** in the nearest section plane. The value of *u* was calculated directly from the *z* coordinate of **p**. This method was unsatisfactory for more complex surface shapes when we attempted to represent more of the sheep surface.

An investigation of bi-quadratic patches to represent the area of the surface near the cutter (Sharrock 1983) led to the idea of a 'surface tracking algorithm'. This is a recursive predictor-corrector method, and relies on accurate initial values of *t* and *u*. This is not a problem since each shearing blow is commenced at a known *t* and *u* position on the surface.

Return to the child's toy visualization of figure 3.6. The pole can move in *t* and *u* on the model surface, and the ring can be raised to height *h*. We can calculate the (Cartesian) directions in which the ring will move for small changes in *t*,*u* and *h*. This gives us a Jacobian matrix for the device. Then we can use the motion rate control method to calculate the approximate values of *t*,*u* and *h* for a given ring position.

In effect, we have a non-linear mapping between two coordinate spaces—Cartesian space on the one hand and surface coordinates on the other. This is the same as the robot arm mechanism kinematics—the mechanism is a mapping from cartesian space to joint space and is also non-linear. The same mathematical techniques can be applied.

This is a useful connection because it reveals some difficulties which can arise (figure 3.8). If **p** (the cutter position) is inside the model surface (assumed to be convex), close to the centre of curvature, then the pole position is hard to determine. If **p** moves to the centre of curvature, then the pole can be positioned

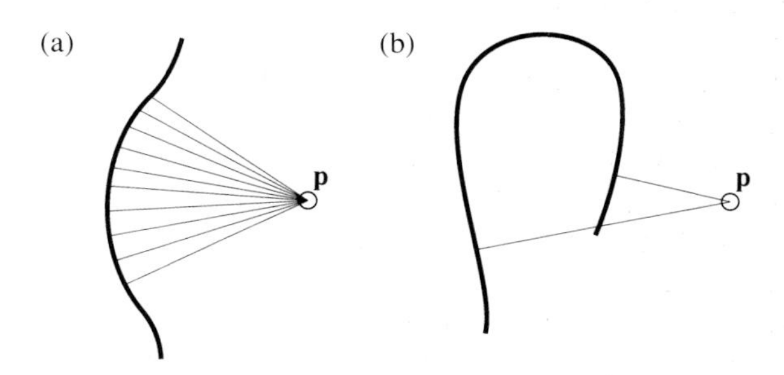

Figure 3.8a Multiple solution case for inverse surface kinematics. **p** is at a local centre of curvature—an infinite number of solutions exists for *t*,*u*. This corresponds to a singular configuration of a robot arm.

Figure 3.8b Multiple solution case for inverse surface kinematics. A finite number of solutions exists for *t*,*u*. This corresponds to the case of multiple solutions for a robot arm.

at an infinity of different points, and yet the ring remains at **p**! This is the equivalent of a kinematic singularity—singularities are discussed in more depth in chapter 8. The Jacobian matrix will be singular under these conditions, and the motion rate control technique will fail.

Another problem can arise with convoluted surfaces where multiple projections are possible. The pole could be at two or more distinct positions yet the ring would still be at **p**. This is equivalent to a mechanism which can provide the same end-effector position with different arm configurations.

Surface tracking algorithm

While it is similar to inverse kinematics, and can be simply explained that way, the algorithm is constructed in the form of a predictor-corrector technique. Computation of the surface normal leaves the following by-products which will be used:

$\frac{\partial t}{\partial u}$ $\frac{\partial s}{\partial s}$ are the ratios of t and u variation to surface displacement respectively,

t u are unit vectors in the directions of increasing t and u,

n is the unit surface normal vector, and

r_t r_u are the curvature radii in the t and u directions.

First, the known change in **p** is used to predict the change in t and u. This is the prediction step.

Next, the surface normal at the predicted position is calculated. Then the predicted values for t and u are corrected by similar computations.

The prediction step commences with the following computation:

$$\mathbf{e} := \mathbf{p}_{\text{new}} - \mathbf{p}_{\text{old}} \quad \text{(change in } \mathbf{p}\text{)}. \tag{3.8p}$$

The correction step commences with:

$$\mathbf{e} := \mathbf{p}_{\text{new}} - \mathbf{surfp}(t,u) \quad \text{(correction vector)} \tag{3.8c}$$

From this point on, both the prediction and correction steps are identical.

The height off the surface is given by

$$h := \mathbf{e} \cdot \mathbf{n}. \tag{3.9}$$

Factors f and g allow for the fact that **u** and **t** are not perpendicular in general:

$$f := \mathbf{u} \cdot \mathbf{t} \tag{3.10}$$

$$g := \mathbf{u} \cdot (\mathbf{t} \times \mathbf{n}). \tag{3.11}$$

The change in u is given by

$$\delta \mathrm{u} := \frac{r_u}{g(h+r_u)} \{\mathbf{e} \cdot \mathbf{u}\} \frac{\partial u}{\partial s}. \tag{3.12}$$

The change in t is given by

$$\delta \mathrm{t} := \frac{r_t}{g(h+r_t)} \{\mathbf{e} \cdot (-f(\mathbf{t} \times \mathbf{n}) + \mathbf{t})\} \frac{\partial u}{\partial s}. \tag{3.13}$$

Note that if h is $-r_t$ or $-r_u$, at a singularity, du or dt will be infinite. To prevent this occurring, a modified value of height h' is used in eqns (3.12) or (3.13) such that h' is always greater than $-0.5r_t$ or $-0.5r_u$ respectively (this applies to a concave surface—opposite signs apply for a convex surface).

When a parallel pattern of shearing blows is required, with one blow following the edge of another, these computations need to be extended to calculate the projection on the surface model of two or three points on the cutter, at different positions relative to the centre point **p**. The surface projections are needed to mark the path traced by the edge of the cutter on one blow; the path will be followed by the other edge of the cutter on the next blow. These calculations can be simplified by exploiting the assumption that the normal vector and curvature parameters vary linearly in a small region close to the projection of **p** on the model.

The corrections related to curvature do not account for local surface twist. However, the degree of twist in the models which we have used is small compared to curvature. The method could be extended to compensate for twist.

It would be satisfying to be able to provide a detailed comparison of the surface tracking algorithm with the resolved motion rate control algorithm used for inverse arm kinematics or other more recent algorithms. However, this remains to be done; the often conflicting demands of an applied research project have prevented this so far. Both algorithms have been demonstrated to work to an acceptable degree of accuracy and robustness in their respective applications. And that is sufficient to achieve our main objectives—shearing sheep.

Computational complications

Although it sounds simple, some 1,200 or more lines of code are needed to implement the surface kinematics calculations. Why so many? It is instructive to list some of the complications which need to be accounted for.

First, the solution process is approximate, and may not always converge. Under some conditions, the calculation of the error at the correction step may result in a corrected position which is worse instead of better. The usual causes are singularities and multiple solutions as discussed above.

Although the surface model is defined for values of t and u within a fixed range, the model surface exists for values beyond this range. Therefore, the practical implementation of the algorithm needs to allow for this. A preprogrammed robot following fixed paths could ignore this. But the variations due to sheep will inevitably cause calculation results outside the normal ranges from time to time and much of the extra coding has been included to ensure that the algorithm is robust under these circumstances. By 'robust' I mean that the calculation results are 'sensible' even though the inputs may be outside accepted limits.

Finally, the algorithm is written as a number of distinctly separate stages, each of which produces an intermediate result. The intermediate results can be recorded if any unexpected error conditions arise. If it were possible to write a completely correct, robust algorithm without any errors, more compact coding could be used. For the time being, we have to allow for the probability of errors in design or coding, and thus the actual program has more lines of coding than are strictly necessary.

The surface modelling system

The technical description of the surface representation is now complete. However, this represents the end product of a system which we use to create models, manipulate and view them. The surface modelling system comprises about 100,000 lines of code.

The major elements of the system are as follows.

1. Surface model computations, transformations, as described above.
2. Patch store—maintains a number of patches in memory for rapid access when required. Provides facilities for grouping patches in data files.
3. Graphics—provides a visual display of surface models (and robot movements). Provides for viewing controls (direction, distance, position, appearance) and display of blow patterns.
4. Measurement—measurement of sheep surface shapes and generation of a surface model from measured data.
5. Alignment—global and local distortion of surface models to match the size and shape of a particular sheep. Includes facilities for subdividing and joining parts of patches.
6. Prediction—analysis of population of measured patches, determination of significant predictors of shape, generation of regression models, and prediction of surface models for particular sheep. An optional module uses data acquired during shearing to refine the prediction model.

Measuring surface models

It's not easy to measure a sheep! This may be self-evident, but beyond the obvious difficulties of keeping it still enough, there are important points to keep in mind. I need to discuss them now because they affect most of what follows.

In a later chapter, I shall show how we keep sheep still enough for shearing and measurement. Suffice to say that sheep can be remarkably cooperative and forbearing in keeping still for long periods of time during our experimental sessions. Furthermore, they don't change their shape unduly over a period of an hour or so, though they can change dramatically in two days.

We measure sheep to be able to predict the shapes of other sheep.

The first step is to manipulate the sheep into the correct shearing position for the patch we intend to measure. We normally follow the standard shearing sequence because the shape of the sheep will be affected by wool which the sheep might by lying on. The patch must now be shorn before measuring the surface. Even if the robot is used to shear the wool, it is still finished by hand to make sure we start with a cleanly shorn surface.

The next step is crucial. The outline of the patch is drawn on to the sheep with a felt pen, by hand. This is important because the operator defines the boundaries of the patch, relative to key features of the sheep. The position of the patch relative to the key features is also recorded, because it will be used to locate the patch on another sheep. One difficulty is that the positions of the features can depend on human interpretation—for example, which point defines the knee joint? There may be insufficient clearly defined features to relate to a surface model.

We use a hand digitizer for measuring sheep surface shapes (figure 3.9). There are several optical measurement devices which could be used to collect surface coordinate data automatically. However, there are two problems which must first be considered. Coded patterns of light stripes have been used for biological shape measurement (Ng and Alexander 1986) but small surface movements due to breathing can disrupt the decoding algorithms. Laser scanners could also be used though they are more expensive and we need to assess the

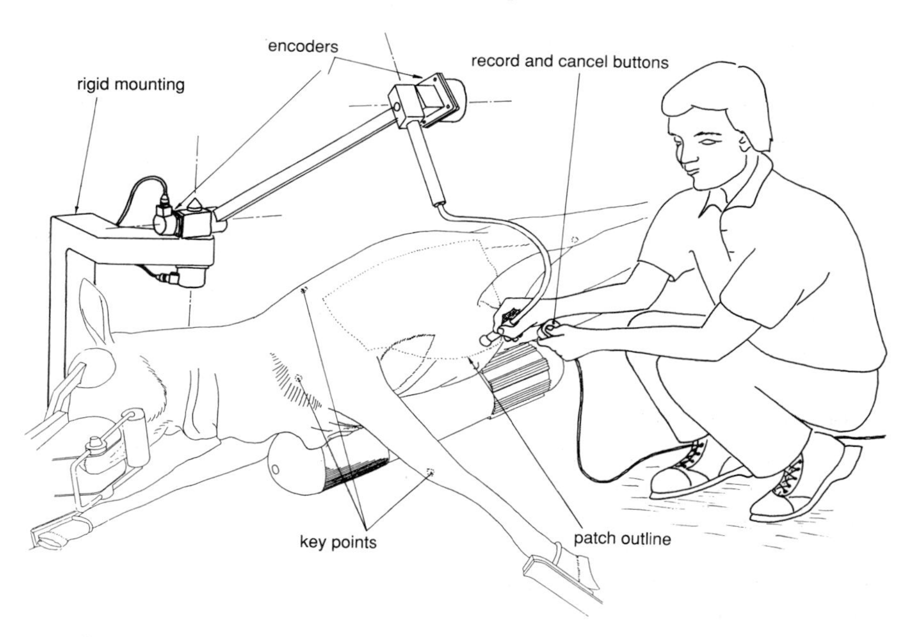

Figure 3.9 Measurement of sheep surface shape using a digitizer. The digitizer consists of an *RRRR* kinematic linkage. Encoders measure the joint angles for the first three joints. The last joint allows the hooked end link to swivel for access to otherwise difficult locations—it does not affect the measured position. The ball end is optional and can be removed to reveal a sharp point. The errors arising from the radius of the ball at the end of the hooked link are removed by software. The ball is easier to slide over the sheep's skin than the sharp point (and less provocative!).

effects of woolly surface texture. To use these devices effectively, we need to locate the key surface features of the sheep and the patch boundaries automatically—by no means an easy task and one of our outstanding research goals.

Consistent definition of the surface patches is important because we rely on finding parts of the sheep at fixed coordinate points.

The choice of surface model Z_s axis is also important (figure 3.1, p. 53) because it defines the section plane orientation. Computational stability is poor if the angle between the surface and the planes is less than 30°.

Once the Z_s axis and the patch boundaries have been defined, the operator moves the digitizer pointer over the surface, more or less following section curves. The surface measurement software collects the data and automatically arranges a large number of position measurements into data points for the equally spaced section planes. A degree of automatic smoothing can be selected for the *t* and *u* directions independently (figure 3.10, Kovesi 1981).

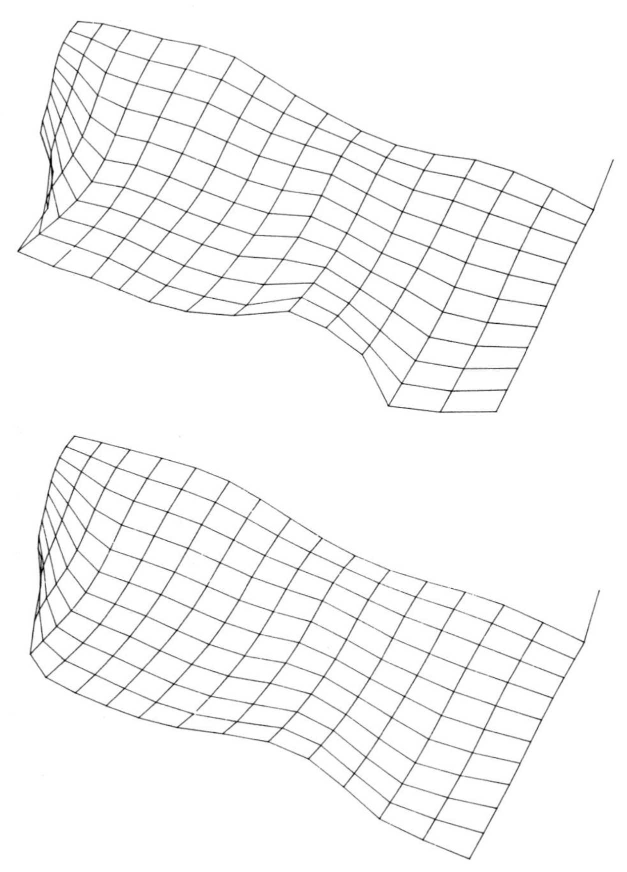

Figure 3.10 Effect of smoothing algorithm—the lower surface has been smoothed. The degree of smoothing can be varied.

One of the technical problems which required some careful consideration was the distribution of the parameter *t* along each section plane. There is a need to represent the shape of each section curve, but this can conflict with the need for a similar distribution of *t* values on each section curve. Figure 3.1 shows a mild case of what happens when this problem is ignored—lines of constant *t* appear to zig-zag along the sheep. The smoothing algorithm reduces parametrization variations, and improves the internal consistency of the model.

We devised an automatic parametrization algorithm which allocates a larger range of parameter values to the more sharply curved parts of each section curve, resulting in better representation of the curves.

Another consideration which has become increasingly important is the use of the model for programming shearing movements. When shearing sharply curved parts of the sheep such as legs, the cutter attitude is calculated almost entirely from surface model data. The model is thus a 'guide surface' for the robot rather than a representation of the sheep surface. We have found that it is convenient to add 'flaps' and other modifications to the edges of the surface model to help with certain robot movements. In effect, this is a subtle change of the definition of the model. Originally a map of the sheep surface, it has become a 'guide to desired robot movements close to the sheep'. Some parts of the model may not represent the sheep, but do provide guidance information for the robot.

Because of this, the model is usually modified by the operator after it is measured. Parts of the model may still represent the sheep surface, but other parts may not (figure 3.2 p. 55).

Editing and checking

We have developed a simple tool for modifying surface models after measurement and smoothing. An operator can use a graphics workstation to modify section curves by moving control points with a mouse or digitizer pad. Figure 3.3 illustrates the effect of moving a control point. The editing program presents a similar display on the workstation screen. The operator works on one curve at a time, with the two adjacent curves also displayed for reference.

Once completed, the model can be checked by using the robot or the digitizer to designate a point near the sheep. The surface coordinates of the point (t,u,h) are displayed on a small display screen near the sheep. As the digitizer is moved (or the robot is moved by its teach pendant) the displayed coordinate values are updated.

Our surface measurement sessions are usually time-consuming. At some stage, we will need to automate surface model measurement, particularly for statistical prediction. This will not be easy.

Predicting surface models

Once we can measure the shape of a shorn sheep, the next step is to predict the shape of an unknown, unshorn animal.

This problem can be subdivided. Roughly speaking, there are two classes of patches—those with more or less constant shape, and the others. The head, neck and legs of the sheep tend to have almost the same shape on every animal. Even the size differences between breeds and differently aged sheep are remarkably small. These patches can be modelled quite satisfactorily by using a standard model shape, and distorting it to match the animal.

Other patches, mainly on the body of the sheep, vary in shape and size. A single animal will change shape too—particularly with changes in feeding. A nice round sheep with smooth skin can appear next year as a lean hollow animal with wrinkled skin, if it is hungry, thirsty or sick. Farmers and shearers use the term 'condition' to describe this—the former condition being 'good' and the latter 'bad'. Shearers (and robots) much prefer sheep in good condition.

Shape prediction

In the early stages of the programme, we were interested in shearing the body of the sheep. The legs and head were thought to be less important. Once we had established that simple sizing of standard sheep shapes was not going to be appropriate, we investigated ways of predicting the shapes.

Sheep in a flock may look similar, but a full year's growth of wool can hide big differences between individual animals (a full fleece is between 90 and 120 mm deep). Robyn Owens, a mathematician, devised a statistical method for predicting surface models in 1981 (Trevelyan *et al* 1982, Owens 1984). She measured the weight and seven other overall size parameters and used these to predict the coordinates of each of the two hundred control points needed for each patch using non-linear multiple regression models for each coordinate of each control point.

She started by measuring thirty or so sheep. Each surface model for each part of every sheep was measured manually. This task took weeks to complete. The data was transferred to a mainframe computer where she analysed it using statistical packages, and finally, the regression coefficients for predicting surface models were transferred back to the robot computer. Six to eight weeks later, after false starts, measuring errors, computer problems and other frustrations, we were able to predict the shapes of unshorn sheep. The measurement effort was the major disadvantage of this approach. Even if we only wanted to change part of a model, the whole process had to be repeated.

Robyn chose a non-linear regression model for each coordinate value:

$$y_k = c_0 + c_1 \ln m_1 + c_2 \ln m_2 + c_3 \ln m_3 + \ldots + \varepsilon \qquad (3.14)$$

where

y_k	is an x or y coordinate of a control point to be predicted,
$m_1, m_2, m_3, \ldots$	are sheep measurements such as weight, width, etc, and
$c_0, c_1, c_2, c_3, \ldots$	are regression coefficients, and
ε	is the prediction error.

Before we changed from the surface representation of eqn (3.1) to the compound curve eqn (3.4), we discovered that predicting control point coordinates provided more consistent results than predicting the section curve coefficients of eqn (3.1). The choice of logarithmic conversion is suggested in Bookstein (1978).

For this method to work, sheep had to be repeatably positioned on the cradle.

This was one of the major considerations governing the design of the ARAMP manipulator (chapter 7). Later, this requirement was relaxed when we were able to use a machine vision system (chapter 10) to measure the sheep position.

Within these limitations, we were able to predict models to a position accuracy of about 10–25 mm and a surface slope of about 6°. We found that the weight of the sheep was always the single most effective measurement, contributing about 80% of the prediction. When the vision system was introduced in 1987, Graham Walker found that the area of the image of the sheep, looking from above with the sheep lying on its back, was almost as effective as the weight measurement (Walker, 1987).

Learning

The statistical method has an important extension, which could be valuable. The regression model can be refined by measuring the errors in predicting surface models on new sheep. The robot finds the errors when it shears the sheep, so it is possible, in theory, for the process to learn about sheep shapes and improve itself with practice.

The method is based on recursive estimation (Owens 1982, 1983, 1985). We evaluated its use, and found that the mathematical arrangement of the estimation algorithm requires some careful attention. Otherwise, some of the elements of the covariance matrix become so small that the calculations become numerically unstable. Owens (1985) cites Anderson and Moore (1979) as a useful reference for estimation algorithms. (The covariance matrix stores estimates of the variability of different measurements and coordinate values. It is a by-product of the regression calculations, but is needed for the recursive estimation procedure.)

Prediction of shape-invariant patches

We developed many ways of modifying models to match parts of the sheep which have more or less the same shape. We have used these methods far more than the statistical approaches because they are much easier to set up and modify. We have even used them for predicting body patches which do vary in shape, though we know that the statistical methods work better.

It would be tedious to describe all the methods we use. Instead I will describe how some of the methods can be applied to predict a particular patch of the sheep.

The patch is used for the flank of the sheep. The sheep is more or less on its back, and the patch covers a rear leg and the side of the body down from the belly. Figure 3.11 shows a sequence of manipulations which lead to the final patch.

Figure 3.11a

The patch has been repositioned by matching it to two key points. The key points on the original patch and the measured points on the new sheep are used

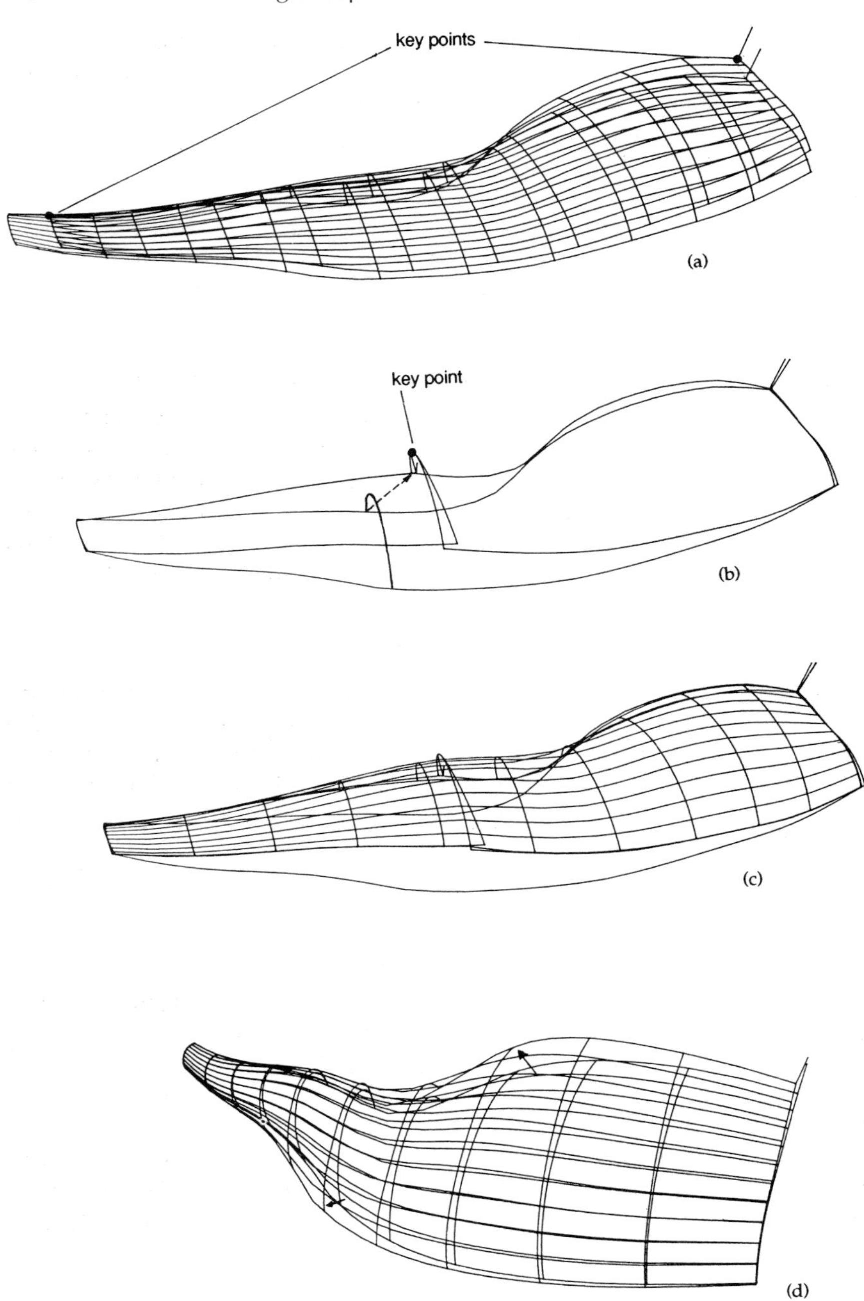

Figure 3.11 Steps in the prediction of a shape-invariant surface patch—refer to text.

to calculate a coordinate frame shift and rotation and a stretch (or in this case a squeeze) so that the patch is realigned and adjusted for length. The section curve data remains unchanged—only the patch frame and length change. This operation is referred to as 'hanging'.

Figure 3.11b
The patch is divided into two sub-patches which are hung between the ends of the patch and a third key point in between. The two sub-patches roughly represent the rear leg part (to the left) and the body part (to the right).

Figure 3.11c
The changed sub-patches are then joined and blended. Blending alters the section plane data in the region of the join in a manner similar to smoothing (figure 3.10). New section curve control points are calculated by linear interpolation because the original plane spacing is changed by the hanging prior to the join.

Note that these two steps are not the same as a simple bending of the patch which could be achieved more simply. The reason for subdividing and rejoining is that the length of the rear leg part does not vary in proportion with the length of the body part.

Figure 3.11d
The patch is fattened (or thinned) locally in two places called fattening points. At each point, a fattening displacement is specified normal to the surface. The closest section curve is then rotated about the end furthest from the fattening point and stretched to match the displacement. Then all other section planes are given the same rotation and length change. The fattening is then given a local effect by recalculating all control points by:

$$\mathbf{r}_{\mathrm{j,i}} := \lambda \mathbf{r}^{*}_{\mathrm{j,i}} + (1-\lambda)\mathbf{r}_{\mathrm{j,i}} \tag{3.15}$$

$$\lambda = 1\sqrt{1 + s_t(t_{\mathrm{j,i}} - t_{\mathrm{f}})^2 + s_u(u_{\mathrm{j,i}} - u_{\mathrm{f}})^2} \tag{3.16}$$

where $\mathbf{r}_{\mathrm{j,i}}$ is control point j on section plane i before rotation and stretching, $\mathbf{r}^{*}_{\mathrm{j,i}}$ is control point j on section plane i after the rotation and stretching, and s_t and s_u are suitable scaling factors. (3.16) defines a bell-shaped weighting function. Manipulation of the control points in this way generates comparable manipulations of the section curves.

This example illustrates the two basic types of distortion operations which can be applied to surface models—changes to the coordinate frame and length (affecting the whole model) and changes to one or more control points (affecting just a chosen part of the model).

On the useful shelf. . .

Two significant developments have been discarded; they have been removed from the surface modelling system. While they are presently kept in reserve we, or others, may yet find a use for them and therefore they deserve a mention.

Dictionary of patches

Statistical analysis of the surface shape data which was collected for a large number of sheep suggested that the regression model (eqn 3.14) would work well for restricted size ranges of sheep. Different regression models would be needed for different breeds and size ranges of sheep. However, this would require an extensive, well-organized measurement programme to collect sufficient data.

The dictionary was to contain a number of patches measured from a range of sheep sizes and breeds. An unknown sheep would be measured as it would be for statistical prediction. However, instead of using a regression model to calculate a new surface model, we use the measurements to select patches from the dictionary belonging to the most nearly corresponding sheep. These patches would then be finally distorted to match the exact size of the unknown animal.

The technique was implemented for one shape-invariant patch, but never tested.

It has since been discarded because we have restricted most of our shearing trials to medium-size Merino sheep, but could well be taken up in the future.

One extension to the idea is the possibility of forming a patch from a weighted mean of patches from a number of 'nearby' sheep in the population in the dictionary. This would help to smooth over gaps in the population distribution.

One potential problem exists, and it is similar to the automated sheep measurement problem. Since surface models were first used, they have become 'guide surfaces' rather than surface models. Therefore, it is very important for the models to retain certain shape and parametrization characteristics, and to represent surfaces which are not the same as the sheep surface. The dictionary approach is attractive because it can easily be extended; as new sheep are measured, they can be added to the dictionary without having to recalculate large regression models. But to do this effectively, we need a way of deriving the 'guide surface model' from data representing the actual sheep surface.

On-line adaptation

The statement at the beginning of this chapter that 'single cubic splines represented section curves sufficiently well while we used ORACLE for shearing trials' was not strictly correct. We were looking for improved shape representation methods which could make use of data collected by the robot sensors during shearing.

We experimented with two methods for adapting the shape of surface models during shearing. Both methods revealed the same fundamental problems. First, because the surface model changes between one shearing blow and the next, the relative alignment of blows is changed. Depending on the circumstances, this can outweigh the advantages of a more accurate surface model. It also introduces significant software complications; what happens if the sheep struggles? Is the adapted model representative of the sheep before or after struggling, or neither? Second, the robot cannot easily tell which part of the sheep it is touching—if there is a model error, is it a position error or shape error or both?

We chose to consign on-line adaptation methods to our scrap heap because of these complexities. We pursued alternate methods since they were simpler. We would only revive these methods if all other approaches failed.

Deviation function

The shape of the surface of the earth is defined by a compound surface. The basic shape is defined as a slightly squashed spheroid. This defines the shape of the ocean surface at mean sea-level. Then the solid surface of the earth is defined by the height (or depth) deviation from mean sea-level.

We applied this approach to sheep surface models by storing a two-dimensional table of height deviations from the base surface which was represented by eqns (3.1) and (3.2). This method provided a greatly enhanced ability to represent complex surface shapes (figure 3.12).

When shearing commenced, the deviation table was blank. As each blow was shorn, a strip of the table was filled in with deviations measured by observing the position of the cutter relative to the surface model. Unknown sections of the table were estimated after each blow, and then smoothed before starting the next blow (Trevelyan 1982).

By a strange coincidence Charles Weatherburn, the foundation professor of mathematics at the University of Western Australia, developed much of the theory for this type of surface representation in the early years of this century (Weatherburn 1927, p. 250).

Sliding adaptation

This method was used for shearing legs. Figure 3.13 illustrates the scheme. As the robot shears along the leg, the position of the cutter indicates the position of the leg. The section curves of the surface model can then be moved by sliding them within their section planes to realign the model to correspond with the measured robot position. This method was much simpler than the deviation function method.

Geodesics and cyclides

Towards the end of 1980, Tony Nutbourne visited us for a month thanks to the efforts of Jim Blair and money from the Corporation. Tony was well known for his work on spline curves for computer-aided design at Cambridge (UK). It was a most stimulating experience and Tony left us with a delightful report written in a style to which I can only aspire. I have quoted from the relevant part:

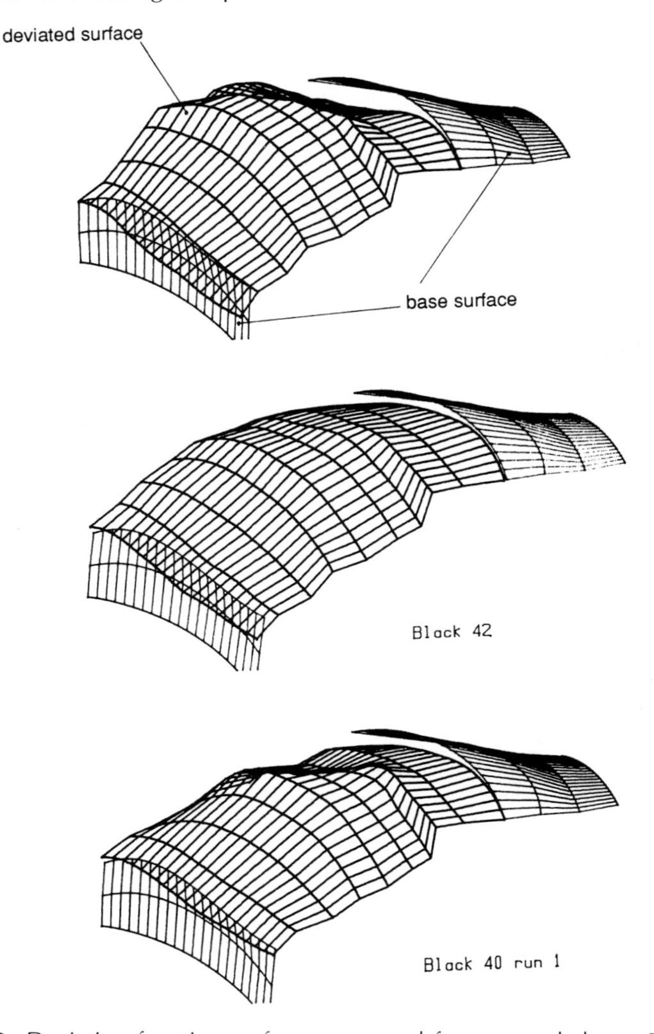

Figure 3.12 Deviation function surfaces measured from several sheep. The base surface (defined by spline curves) is seen at each end—the deviated surface is in the middle sections. The height deviation is stored as a rectangular array of numerical values at equally spaced *t* and *u* increments.

I have taught the team to think of an alternative approach, not arguing that this is better than the Cartesian approach, but inviting them to experiment with it more fully. It depends on recognizing that there is almost a one to one correspondence between the geometry of the rig steering motions and the differential geometry of a surface blow.

To understand this it is necessary to learn the rudiments of surface differential geometry. This has never been taught in undergraduate courses in engineering, and is therefore unfamiliar to most engineers. Most of the terminology used was invented by Frenchmen, notably Monge and Bonnet in the middle of the last century, and is singularly baffling. Who, untutored, would believe that geodesics alone among curves have no geodesic cur-

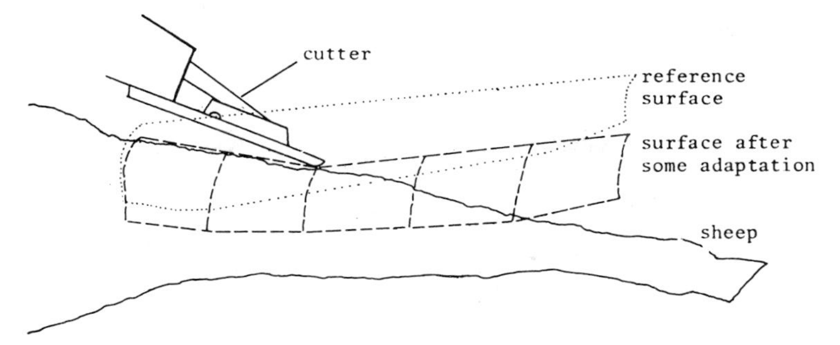

Figure 3.13 Sliding adaptation of surface model—refer to text.

vature; or that geodesics have geodesic torsion equal to their torsion whereas for most (but not all) other surface curves these quantities are different? A considerable part of my month's work was devoted to gentle persuasion about such matters.

Shortly after my arrival in Perth, I painted lines of curvature on a shorn sheep to examine the surface orientation in detail. I had it in mind to describe the shape by means of principal surface patches. Such patches have right angles at the corners, and at any point on the surface the maximum and minimum normal curvatures of the surface are in the direction of the patch sides. I thought that I might be able to use cyclidal patches taken from cyclide surfaces. These were invented by a 17-year-old Frenchman called Dupin in 1801. They were researched in some detail by Clerk-Maxwell and Cayley who both wrote papers 'On the Cyclide'. For a century they have been forgotten. I have resurrected them because they are surfaces whose lines of curvature are circles lying in different planes. They embrace the cylinder, cone, torus and many surfaces of revolution so that they have considerable potential use in engineering. They are cursed, however, by singular points where the surface is locally spherical or plane and for which all directions are principal. I found several of these points on the surface of the sheep, and I decided that my own research had not reached the stage where I could tackle the problem this way and still leave the team with a useful technique at the end of a month.

Open issues

Once we had built ORACLE, the 'software sheep' was one of the first outcomes of our research. As the project moves out of its research phase, the final threads of our research still involve surface models. The ultimate performance of the robot is limited to a major extent by the surface model accuracy. We are still working towards improving the surface model accuracy, but our main focus is on improving the measurements we need to predict models. This is most likely to come from machine vision—see chapter 10.

4
Movement

Movement is the essence of robot software and adaptation is the essence of sheep shearing. Software provides the means. Yet there is more to it than just moving the cutter. People need to be able to program the robot in terms of simple concepts with reliable results. Further, a shearing robot needs to move gracefully; sudden starts or jerks disturb the sheep and accelerate wear to joints and actuators. A graceful robot is aesthetically more pleasing and conveys a reassuring sense of finesse and competence.

I introduced our vision of a shearing robot and the elements of robot motion control in chapter 2. The next chapter described surface models which encode implicit knowledge of sheep. Now we can explore robot motion more fully. Adaptation is an integral part of this; so too are sensors but I have left the details of sensing devices to another chapter. For the moment it will be sufficient to regard the sensors as information sources for adapting robot motion.

This chapter focuses on the calculation of motion. The structure of the software which mechanizes the calculations is also important but we shall explore this in more detail in a later chapter as it is closely connected with shearing skills and error recovery.

I will take a user's point of view first and explain how we program robot movements. I believe we have devised reasonably simple programming concepts which can readily be grasped. The rest of this rather long chapter can be skimmed.

Programming motion

One day it will be possible to program shearing movements graphically to bypass tedious coding to specify movements. We would like this ourselves but informed readers will appreciate that it is a development item; the effort required to write this software is hard to justify in research. Yet it is essential to understand coding. The concepts need to be understood, even if one has the luxury of a graphics screen to develop the shearing program.

A simple shearing program follows. It consists of a set of commands stored in a text file; any line beginning with ! is a comment and is ignored.

```
! a simple shearing demonstration
lp belly;
blow [0.25,0.75    0.40,0.25];
tr park2.c;
```

This program loads a patch called 'belly', shears a single blow along part of the middle of the patch, and returns the robot to its park position by means of a procedure 'park2.c'. This procedure consists of further movement commands which return the robot safely to its park position from any likely shearing position. We do not usually use it—normally we program a reposition movement to the start of another blow or some other position.

The blow is defined by the start and end points in surface coordinates [t_0, u_0, t_1, u_1], in this instance [0.25,0.75 0.40,0.25]. The mass of geometric information needed to plan and execute the shearing blows, and their associated reposition movements, is obtained from the surface model—later we shall see just how that happens. Furthermore, since the surface model will have been adapted for the particular sheep, the blow will be placed correctly on each different animal presented for shearing.

Note that a curved reposition movement will be planned automatically to move the cutter to the start of the blow. Reposition movements are inserted between successive blows too. Further, the blows can be curved; we add more control point pairs and the cutter follows a smooth curved path from one segment to another. Figure 4.1 shows the pattern of movements which the following sequence generates and figure 4.2 explains the cutter symbol. The second shearing program (p. 78) generates the pattern of movements shown below.

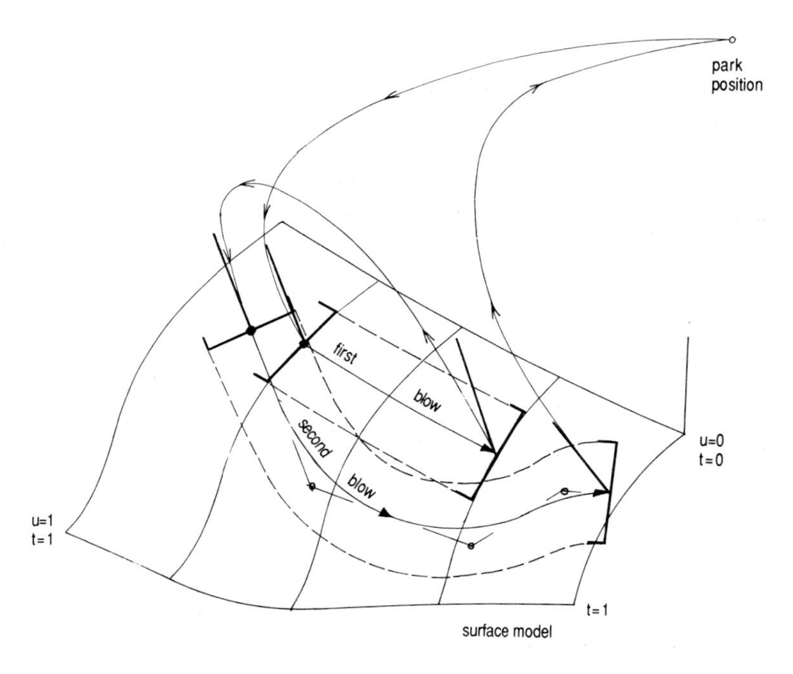

Figure 4.1 A simple shearing sequence with two blows. Figure 4.2 explains the cutter symbols. Refer to text for details.

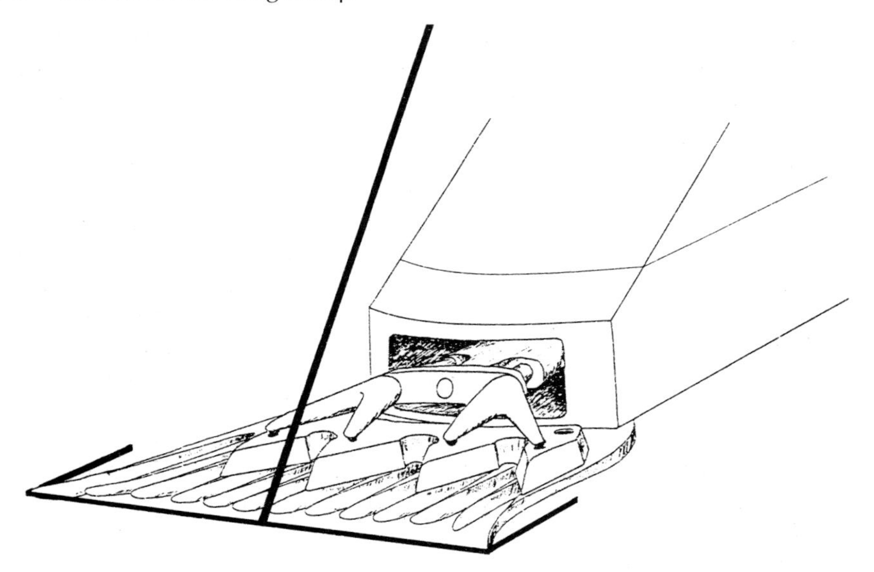

Figure 4.2 Symbol used to represent the cutter position in diagrams.

```
! a simple shearing demonstration, extended;
lp belly;
blow [0.25,0.75 0.40,0.25];
blow [0.30,0.85, 0.65,0.57, 0.85,0.22, 0.50,0.10, 0.50,0.03];
tr park2.c;
```

Many assumptions have been made; a standardized landing on the sheep skin, a default shearing speed and so on. Up to 20 parameter values are set for a typical shearing blow in a real shearing pattern. Typical parameters which need to be set for each shearing blow are:

shvel	shearing speed
pofs	pitch offset

The effects of changing some of these parameters will be explored later. Each parameter is defined by a **set** command:

set shvel:0.25; set shearing speed to 0.25 m/sec.

Shearing programs are rarely as simple as the example I have shown. This is not surprising. The command language has been evolved and designed to represent *shearing style*—it needs to accommodate the procedural knowledge used to shear sheep. Yet the example is sufficient to illustrate how we have simplified the programming concepts. We believe we have reached the stage where an interested person can readily learn to program the robot, or alter existing programs, with predictable results. We believe this is an important achievement.

Interpreting commands

The commands are processed by an interpreter which processes them sequentially; we call the interpreter program 'CLARE'. Most of our major software modules carry the names of children born to team members.

CLARE handles simple variables and vectors and can evaluate arithmetic expressions and combinations of vectors. A number of special functions can be accessed, such as surfp and snorm mentioned in the last chapter. Each movement command results in a call to a planner—a software module which plans a movement of the robot. There are four planners—three for repositions and one for shearing blows.

In the absence of repositioning commands, the shearing planner calls the automatic reposition planner to move the cutter to the start of the blow. Trevelyan *et al* (1984) contains further details of the automatic reposition planner.

The shearing planner collects the current parameter values from the interpreter store, and inserts the *t,u* coordinate pairs specified in the blow command. After a validity check, the planner passes the data to the motion software and sends a request to commence a shearing blow.

The motion software, called TOM, runs in real time—it runs repeatedly, each motion step spanning a time interval of 0.08 sec. TOM calculates how each robot actuator is to move over the coming motion step. The movement is first computed in Cartesian coordinates and we then use resolved motion rate control to calculate joint movements.

As soon as the sheep skin is sensed, the cutter follows the skin surface; the path of the cutter will be different from the planned path. Path-following errors are reduced at each motion step by steering the cutter so that the projection of the cutter on the surface moves towards the planned path. This gradual process of removing errors is also used in reposition movements.

HEATHER interpolates the actuator movements and updates the servo set points every 0.01 sec. The interpolator also modifies the position demands to correct sensed arm position and attitude errors, some of the dynamic errors in the actuators, and drift errors in the hydraulic controls

TOM takes the effect of these corrections into account by using the forward kinematics to calculate the actual pose of the cutter at the start of each motion step. This data can be accessed by interpreter commands and used to modify the planned sequence. However this is rarely required; normally TOM and HEATHER can automatically adapt to the differences between the real sheep and the surface model.

Programming reposition movements

Reposition movement planning is the closest we come to 'conventional' robotics; the basic concepts are common to most robots with off-line programming capabilities. There are refinements specifically for sheep, of course. The redundancy[1] of the robot can be ignored—it is treated as a six-joint robot for programming purposes.

While simple straight repositions are described initially, they are rarely used. Fast, efficient and graceful movements will be curved and our programming techniques are aimed at smooth curved motions.

[1] ORACLE has 8 actuators, and SM has seven—both are kinematically redundant arms.

We use 12 numbers to define a pose—3 for position, and 9 for orientation. These are handled by CLARE as vectors. The vector **p** defines the present end-of-arm position, and the frame **As** defines the orientation.

In our programming language, vectors are defined by element lists enclosed by square brackets. A vector can represent several elements of another vector:

> **[0.5,0.7,0.03]** is a three-element vector. If used for position, the numbers represent distances in metres.
>
> **[p, As]** is a 12-element pose vector formed by combining the position (3 elements) with the current orientation frame (9 elements).
>
> **[i,j,k]** is a 9-element vector corresponding to **As**, the current frame.

Combining vectors

Vectors can be added or subtracted, provided they have the same number of elements. Thus:

> **p + [0,0,0.05]** defines a position 0.05 units above the current end-of-arm position.

Vector operations include addition, subtraction, scalar multiplication, transformation, cross and dot products.

Straight reposition movements

The SM robot has two types of reposition movement—straight and curved. First I will show how to define a straight line movement.

> **move pose;**

where **pose** is the new pose (12 elements), comprising a position vector and orientation frame. A position vector may be used for a pose (omitting the frame); the orientation remains unchanged.

Examples:

> **move p**; Move to current end-of-arm position—do nothing in effect since robot is already there.
>
> **move (p + [0,0,0.05])**; move to a new position 0.05 metres above the present position. Note that round brackets are needed since there is an addition operation specified.
>
> **move [1.2,0.0,0.2]**; Move to the position 1.2 in X, 0.0 in Y, 0.2 in Z.

Calculating a frame

Before we can include a frame, we need to know some easy ways of specifying orientation. A frame is a special set of numbers, and we have built in several ways of calculating a frame.

Euler angles

The first method for specifying a frame is by three angles. Imagine that the cutter is horizontal and pointing in the X direction. **i**, **j**, and **k** are lined up with X, Y and Z. Now rotate the cutter 30° about **i**; **j** and **k** have also rotated 30°. Next rotate 45° about the new **j**. **k** is changed again, but now rotate 50° about the latest **k**. We call the three angles Euler angles, after the great mathematician. The built-in function **eframe** calculates a frame given three Euler angles like this:

eframe ([30,45,50])

The parameter is a three-element vector specifying the rotation about the **i**, **j**, **k** axes respectively.

Euler angles are awkward to work with, so we have implemented other ways of specifying frames.

Frame rotation

A new frame can be calculated by rotating an existing frame about an axis through a given angle:

rot(A, axis, *angle*);

A is an existing frame. We could use **As**—the current cutter orientation as the existing frame.

axis is a three-element vector defining the axis direction, e.g. **[1,1,5]**, or [(**i**+**j**)].

angle is the angle of rotation in radians.

Examples:

rot(As, i, (30/57.2958)); Cutter rotated about **i** by 30°.

rot(As, [0,0,1], (180/57.2958)); Cutter rotated 180° about a vertical axis.

rot(eframe([30,45,50]), [0,1,0], (–90/57.2958)); Can you work this out? The frame to be rotated has been calculated with the **eframe** function.

Now we can add rotations to our movement commands:

move [[1.5,0.2,0.2], eframe([30,0,0])];
move [(p + [0,0,0.05]), rot(As, [0,0,1], (90/57.2958))];

Surface frame

A frame can be set up relative to the surface model.

sframe([*t, u, angle, pitch*])

sets up a frame with **j** tangential to the surface at *t*,*u*. and **i** pointing in a direction ***angle*** to the local **t** direction, but angled down by angle ***pitch*** and **k** is normal to **i** and **j**.

Curved repositions

After that, the reader could be excused for thinking that curved repositions must be even more complicated than orientations. Thankfully no. By borrowing the concept of B-splines from computer-aided design, we devised an elegantly simple way to write graceful curved movements. The appendix to this chapter provides some of the basic theory of B-splines as used for path planning.

A curved reposition path is written by defining points along the curve—these are called control points. The curve does not necessarily pass through the control points, rather it heads towards them before veering off to the next control point. Here is a simple example:

startmove [dep, 0.3, 3];	**dep** is departure direction, speed 0.3 m/sec, order 3 B-spline.
movetowards [1.2, 0.2, 0.4, 0.45];	intermediate control point, optional new velocity 0.45 m/sec.
movetowards [1.0, –0.2, 0.2];	
endmove [1.0, –0.2, 0.0, arr];	destination point; **arr** is arrival direction. The endmove command is omitted if movement is followed by **blow** command.

The **startmove** command defines the departure direction from the start of the curve; the curve starts at **p**. Each **movetowards** command specifies an additional control point. The **endmove** command specifies the end of the curve and the arrival direction.

The integer 3 in the **startmove** command is the order of the curve. It can be 4 instead—the shape of the curve will be slightly different—the path will veer more gradually towards control points.

If two successive control points have the same position (for order 3, three successive points for order 4) the path will pass through that position with a sharp change in direction.

Orientation frames can be added to the end of each **startmove**, **movetowards** or **endmove** command. The cutter orientation varies from one frame to the next along the curve—for details refer to later sections. A different speed value can be added after each position vector—the speed along the curve varies smoothly from point to point. Figure 4.3 illustrates a typical curved reposition movement.

Blow programming

Like repositions, shearing blows are programmed as B-spline curves. Thus a blow path defined by a number of t,u coordinate pairs will be treated as a third-order B-spline, similar to the section curves of surface models (figure 3.3, p. 56). Several parameters normally need to be defined for each shearing blow to define a given style of shearing.

The landing on the skin is an important part of the blow, but it is the final part of a reposition movement. Thus, the **endmove** command is omitted; the last control points are defined by the landing[2] parameters.

Landing parameters

Three angles define the approach; *glide* defines the slope of the 'glidepath' to the landing point, *approach* defines the glidepath direction relative to the shear-

[2] The cutter can be visualised as a vehicle flying over the sheep surface—hence the references to aerospace terminology!

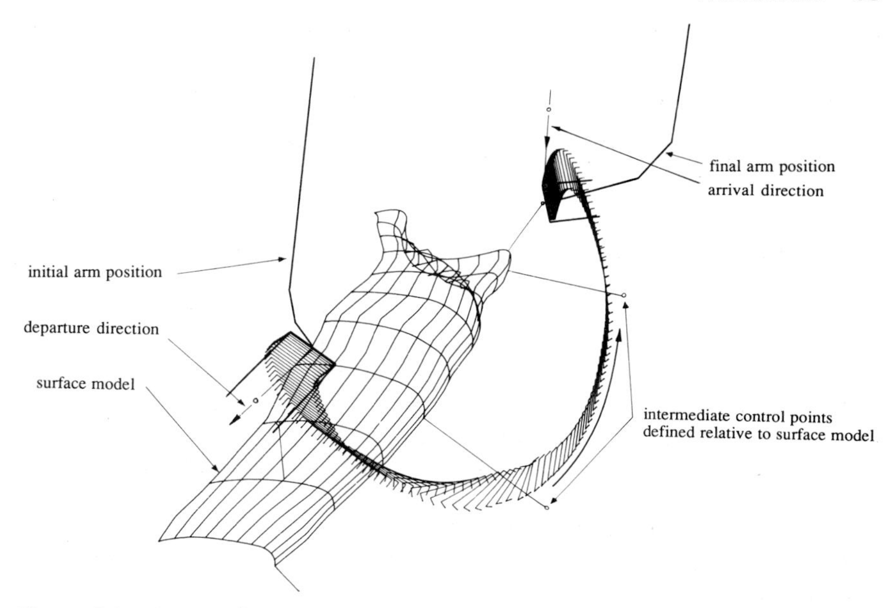

Figure 4.3 A typical curved reposition movement. The reposition path has been defined by control points relative to the surface model. The code used to specify the movement follows:

```
startmove [ (-0.1*i + 0.07*k), 0.1, 3, As ]
movetowards [ surfp([0.5, 0.1, 0.05]), As ];
movetowards [ surfp([0.0, 0.3, 0.2]), eframe([90,0,-90]) ];
movetowards [ surfp([0.0, 1.0, 0.2]), eframe([90,0,-90]) ];
endmove [ surfp([0.13, 1.0, 0.1]), 0,0,-0.1, sframe([0.07,1.0,60,10]) ];
```

ing blow, and *heading* defines the direction of the cutter relative to the shearing blow as it descends to the landing point (figure 4.4).

```
set glide:35 & approach:20 & heading –25;
```

is a multiple assignment command which defines an approach path pointing 20° to the left of the blow direction with a slope of 35°. The cutter will point 25° to the right of the blow direction during the landing approach.

pofsland is a pitch offset angle for landing—sometimes the comb needs to point slightly up or down.

overshoot specifies the depth of the aiming point for the landing below the surface model. Uncertainty in the sheep position means that the aiming point is normally some distance below—typically about 15 to 25 mm. Naturally, the landing descent stops when the cutter touches the sheep skin. Normally, the presence of the skin is detected by the capacitance sensors before contact, and the landing path 'flares out' reducing the landing impact.

runin specifies the length of the straight final approach along the landing direction. This can be very short, or long if the cutter has to approach past an obstacle such as a heap of shorn wool. *landdist* specifies a distance over which

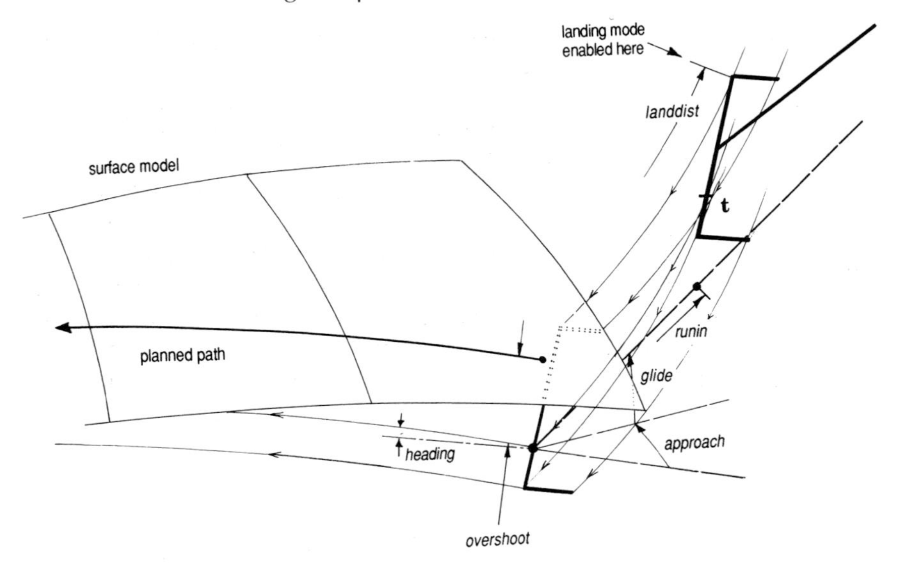

Figure 4.4 Landing parameters which can be set by the programmer. Note that **t** is the path tracking point. Allowance is made for shearing offsets, such as this, in planning the landing path.

the robot moves in 'landing' mode at the end of the reposition movement. This is different from *runin*; in landing mode, the velocity is constant *pnvel*, and the motion control software is ready to make the transition from landing to shearing. *landdist* is typically about 0.07 m.

Shearing parameters

During shearing the cutter attitude is controlled partly from sensed attitude errors and partly from the surface model. The degree to which this happens is set by gain factors; a messy approach. Ultimately, I would like to see a selection of 'modes' with preset gain factors.

gppitch, *gproll* are the proportional gain factors for the surface model. If the value is high the surface model information dominates, otherwise the sensor information dominates. *gdpitch*, *gdroll* are the derivative gain factors for the surface model. These are usually set so that the rate of change of the surface normal direction is transformed into an angular velocity for the cutter. This is a feed forward element in attitude control.

gpitch, *groll* are the proportional gain factors for the sensed attitude errors. When they are high, the cutter responds quickly to sensed errors.

Different shearing conditions require different combinations of gains. For example when shearing along a leg we set *gproll* high and *groll* low because the sharp curvature distorts sensed roll errors; we need to use the model. We set *gppitch* to a medium value, and *gpitch* high as long as at least one capacitance sensor will 'see' the leg passing beneath it. If we can be confident that the surface model defines the leg well we can set both *gpitch* and *groll* low.

depmax defines the maximum distance which the cutter can go *below* the model surface. This is essential when shearing poorly supported parts of the body which can be pushed down by the cutter. The cutter pushes the sheep down until *depmax* is reached—from then on, no more deflection is allowed.

We set a tracking offset *toff* to define the point on the edge of the comb which follows the path of the blow. The windrow[3] offset *woff* defines a point on the comb which will define the edge of the shorn area, or windrow line. When shearing parallel blows, this line becomes the path for the next blow. A third offset *soff* defines a 'surface point' on the comb—the point at which the straight comb edge will be 'ideally' tangential to the curved surface model (figure 4.5).

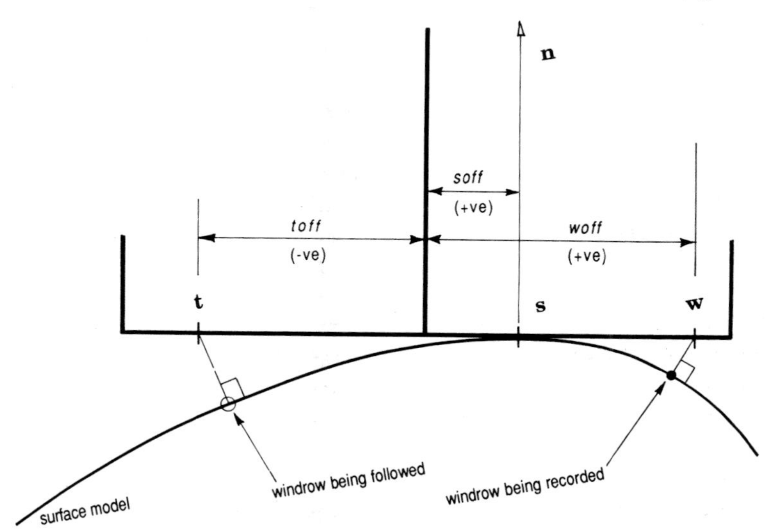

Figure 4.5 Shearing offsets—idealized tracking. **t** is the path tracking point, **w** is the windrow recording point and **s** is the surface tangent point.

Apart from shearing with parallel blows, these offsets are essential for shearing sharply curved parts of the sheep. Wool can only be combed close to the skin; the fibres of wool grow parallel there but as they grow away from the skin they tangle. Therefore, the cutting section of the comb must be kept close to the skin. When using only one side of the comb, we place *toff* and *soff* on that side; *woff* is usually irrelevant (figure 4.6). The landing approach is offset to take *toff* into account.

We often need shearing attitude offsets *shyof*, *shpof* and *shrof* for yaw, pitch and roll respectively. *shyof* in particular is important for combing. We usually run the comb with a yaw offset to help pull the skin to one side as it moves

[3] Windrow—the edge of the unshorn wool. Derived from the haymaking term—a windrow—a line of heaped, cut hay or stubble left to dry in the wind.

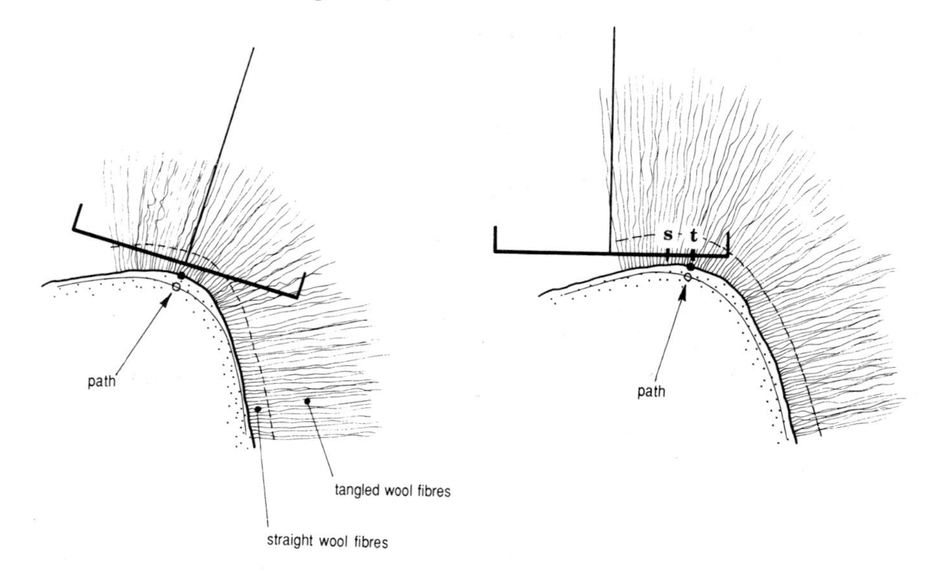

Figure 4.6 Shearing a sharply curved surface—tracking and surface offsets are needed.

forwards. This helps to stretch the skin and smooth out folds, lessening the chances of skin cuts. *shrof* is used whenever extra pressure is needed on one side of the comb, and *shpof* is used occasionally to shear with the comb points slightly down or up. For example, when shearing down the leg, we angle the points down; if the points run flat on the skin, the comb rides out over wool fibres which tend to lean down the leg.

Sample sequence
The following commands are part of the shearing sequence for the neck of the sheep. They have been reproduced from a command file used for shearing trials during 1990. Comments are preceded by ! Note that the variable *Tswref* is a vector which has three components: *toff*, *soff*, and *woff*.

```
! Shear down under the chin

startmove [(i*.13 + k*.03), 0.2, 3, rot(As,z,1.1)];
set shvel:0.12;
set force:6 & groll:2.5 & gpitch:5.0 & depmax:30 & xcurs:10;
set gproll:0.15 & gppitch:0.2 & gdroll:0.8 & gdpitch:1.0;
set glide:15 & approach:35 & heading: 0 & runin:0.05 & overshoot:0.02;
set pofsland:0.1;
set shyof:0 & shpof:0 & shrof:0 & Tswref:[ .000,.000,.000];

blow [0.25,0.43, 0.99,0.12];

! scoop the wool up and over the head
```

```
startmove [(i*.13 + k*.03), 0.2, 3, rot(As,z,1.1)];
movetowards [surfp([0.5,0.0,0.25]), 0.6, rot(As,j,–1.0)];
movetowards [surfp([0.0,0.0,0.15]), 1.4, rot(As,j,–1.0)];
movetowards [(surfp([0.0,0.0,0.25])+[0.3,–0.3,0]),1.4];
movetowards [(surfp([0.0,0.0,0.25])+[0.3,–0.3,0]),1.4,Eframe([40,40,–40])];
movetowards [surfp([0.0,0.95,0.05]),0.2];

! shear along chin before scooping again

set shvel:0.12;
set force:6 & groll:2.5 & gpitch:5.0 & depmax:30 & xcurs:10;
set gproll:0.15 & gppitch:0.2 & gdroll:0.8 & gdpitch:1.0;
set glide:15 & approach:35 & heading: 0 & runin:0.05 & overshoot:0.02;
set pofsland:0.1 & blowstyle:7 & side:0;
set shyof:0 & shpof:0 & shrof:0 & Tswref:[ –.015,–.015,.000];
blow [0.25,0.30, 0.85,0.0];
```

Parameter interpolation

Each shearing blow is defined by a number of *t,u* coordinate pairs, each of which becomes a control point on a B-spline curve in three dimensions defining the blow path in Cartesian space. We use the control point weighting coefficients to interpolate shearing parameter values such as *toff*, *shpof*, *shvel*, etc. This allows the programmer to smoothly vary parameters and offsets along the path of a shearing blow.

```
interp shyof[ –0.3, 0.0, 0.0];
blow [0.25,0.30, 0.65,0.15, 0.85,0.0];
```

The appendix to this chapter provides a more detailed treatment of B-splines, parameter interpolation and trajectory computation.

Other commands

We have implemented facilities to manipulate surface models, take measurements using a vision system (chapter 10), control the manipulator (chapters 7 and 9) and record data collected during trials—all accessible through commands handled by CLARE.

Examples are:

cutter on;	turn on cutter motor
air off;	turn off compressed air
lock carriage;	lock carriage to overhead track
lights [0,0,1,0,0,1];	control lighting (for vision system)
p?;	list surface model patches held in store
slamp n3;	manipulate sheep to shearing position n3.

Some of the commands can be used to initiate operations which will operate in parallel with robot motions—other operations must be concluded first.

The interpreter also has simple conditional branching commands.

Thus the first part of the motion control system is the level at which the entire shearing system is integrated, through a common command structure.

Motion calculations

The calculations for motion control have been designed from the start with adaptation in mind. The aim is to keep the cutter next to the skin during shearing movements. To do this, we need to calculate the set point voltages for each actuator at a frequency of 100 Hz which is fast enough for smooth actuator motion.

Many robots incorporate the feedback control for each actuator in software too. We have avoided this by using hydraulic actuators with analogue servo controllers which can be decentralized on the robot structure. Adaptive control is more important than precise positioning so analogue circuits are accurate enough. The actuators are decoupled by design; this way we avoid complications with coupled control loops.

Unlike other robots which execute rather primitive movements, our motion control system devotes most of its effort calculating how and where the cutter needs to move. Only a small proportion of the time is spent solving the arm geometry and moving actuators.

The software runs in a single Hewlett Packard A900 processor which runs at approximately 3 MIPS[4] with a 1 MFLOPS floating point processor for computation. ORACLE was first controlled with a Hewlett Packard 21MX-E minicomputer (approx 0.7 MIPS, 0.2 MFLOPS). Some custom microprogramming was used to improve the speed. In 1983, this was replaced with the software compatible A900 super-mini computer—no custom microprogramming has been needed. All the code is written in Fortran 77, except for minimal operating system dependent code. Most of the system runs within the real-time environment provided by the proprietary RTE-A operating system.

As described earlier, the system comprises three major modules shown in figure 4.7 which run concurrently on different time-scales. The modules communicate through shared memory. CLARE operates in background in sudden bursts of activity and passes motion plans to TOM for execution.

TOM operates cyclically at each motion step; once each 80 milliseconds. TOM calculates the robot pose, how the pose is to change during the next motion step, and the actuator movements to achieve this. TOM also monitors the state of the hardware and software, and can initiate error recoveries when appropriate.

HEATHER runs at 5 millisecond intervals for sensing; every second time it updates the actuator positions, interpolating the movements calculated by TOM. The latter process incorporates corrections calculated from sensor inputs using the inverse Jacobian to transform these to joint space.

This multi-tasking structure is the key to our adaptation mechanisms. Figure 4.8 shows the functional arrangement more clearly, particularly how the inverse Jacobian transforms and surface curvature corrections are provided to HEATHER for fast path adaptation. Many robotics papers claim that inverse kinematics must be solved at high speed for adaptive control; we have shown that this is

[4] MIPS—million instructions per second; MFLOPS—million floating point arithmetic operations per second.

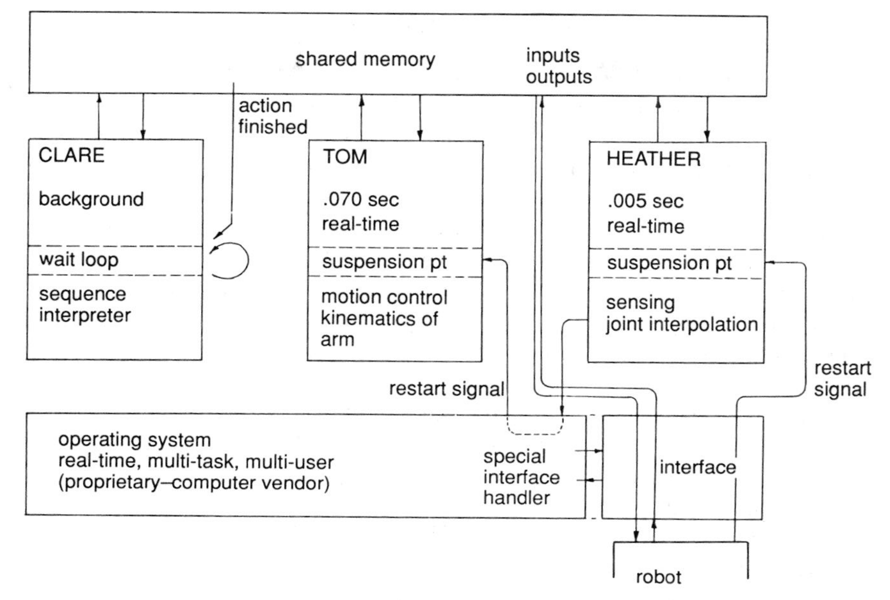

Figure 4.7 Motion control software modules—refer to text

not the case. We exploit the fact that the arm configuration changes relatively slowly, and the errors we introduce by updating the Jacobian slowly can be ignored. Our sensors correct the ones that matter.

It is also pleasing to note that comparable results have been obtained for full dynamic control. Sharkey *et al* (1989) have verified experimentally that updating the inertia and kinematic transforms slowly does not significantly affect adaptive force control with a computed torque control scheme.

Adaptation mechanisms

Figure 4.9 illustrates the major adaptation mechanisms which the motion control system implements. The top half shows continuous adaptations; the motion of the cutter is adjusted at each step to reduce errors progressively. However, there are practical limits to continuous adaptation mechanisms. For example, the cutter slows down if the drag force rises. If the drag force continues to increase, there is no point trying to shear forwards more slowly still; a recovery movement is needed to push wool out of the way and try again. We call this situation an operating error. The major part of the motion control system has a recursive structure to deal with operating errors, but this is discussed in later chapters. For the moment, I will focus on continuous adaptations:

(1) follower (HEATHER)
(2) height, pitch and roll error correction (HEATHER)
(3) cutter path steering (TOM)
(4) shearing speed adjustment (TOM)
(5) path manipulation and planning (CLARE)

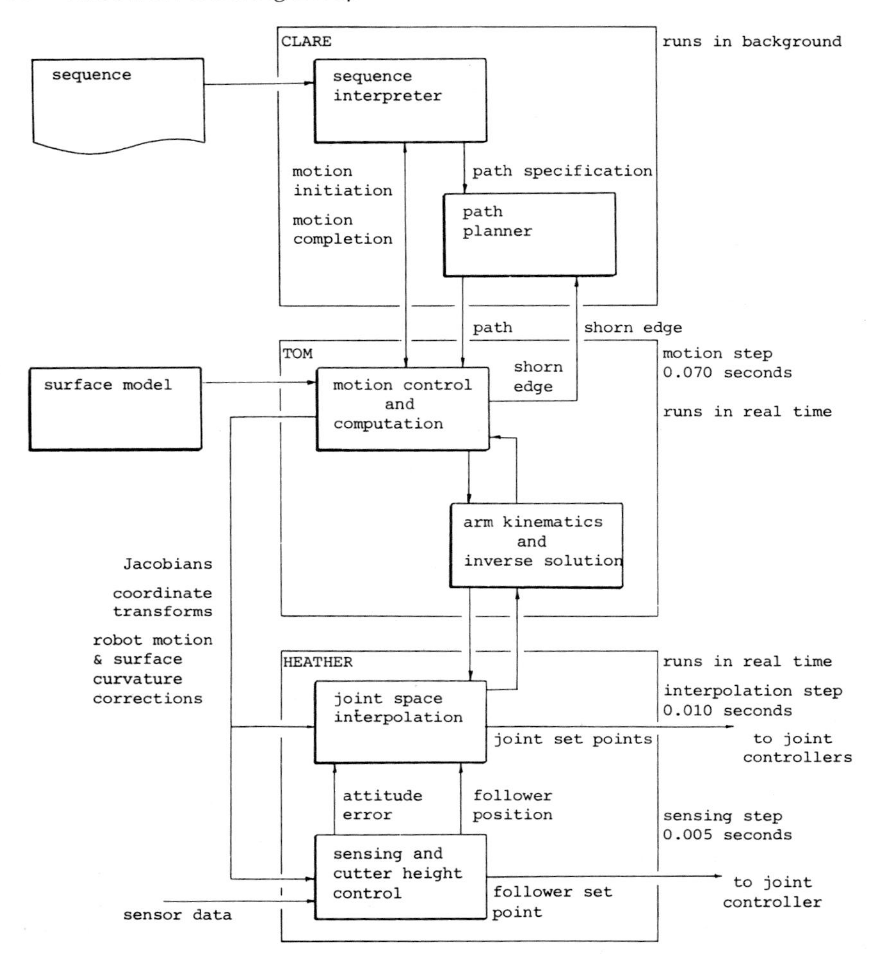

Figure 4.8 Software function for motion control—refer to text. Note that the motion step time interval was changed from 0.07 to 0.08 seconds in 1990.

Adaptation 1: follower

The surface follower mechanism was a central part of the vision of ORACLE described in chapter 2. In principle, it is similar to other active end-effector devices described in the robotics literature such as precision positioners (Taylor *et al* 1985). As our understanding has improved, both the mechanism and control have been rearranged several times.

The instabilities of the early trials with ORACLE were overcome partly by using surface model data and partly by improved sensor fusion methods. The particular problem of fusion is relating two very different forms of data from capacitance proximity sensing and resistance contact sensing with fast computation rates (figure 4.10).

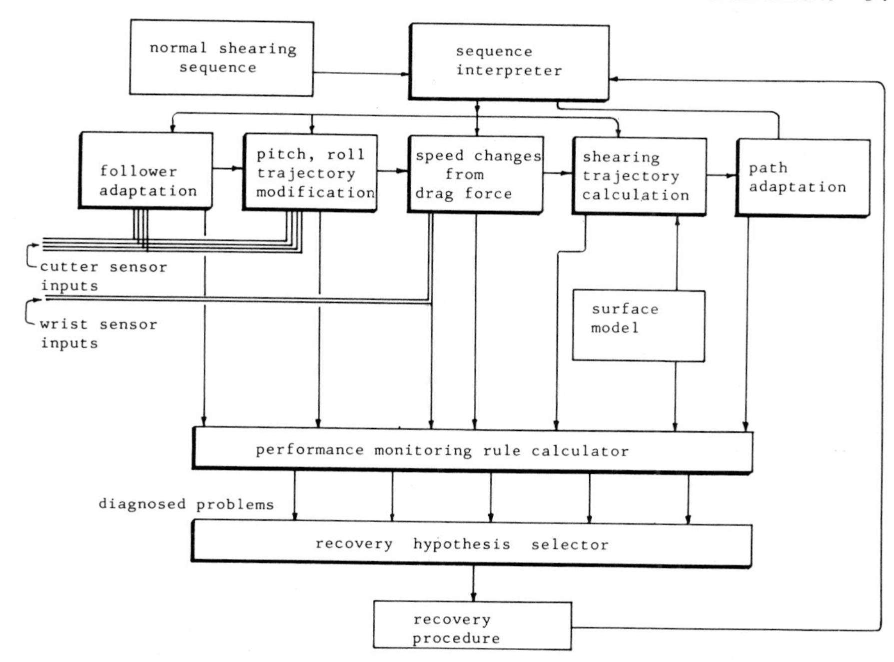

Figure 4.9 Adaptation mechanisms used during shearing. The upper group are continuous adaptations. The lower group activated only if operational limits for continuous adaptation are exceeded.

The resistance sensing provides an erratic signal (figure 4.11); there is no information unless the comb touches the skin, and even when it does, wool fibres, dead skin and other contaminants interrupt electrical contact. Moisture in the wool, particularly with high levels of potassium salts or suint,[5] results in greater base levels of conductivity through wool fibres, rather than by direct contact with the skin. Water or urine can give the same reading as a skin cut when the comb is in direct contact with the sheep's internal fluids. Combing forces can cause wool fibres to tangle round the end of a comb tooth, effectively tying the skin to the comb for a short time; this provides a confusing signal which is not affected by lifting the cutter!

On dry skin which has a low conductivity the robot has to press hard with the cutter to obtain sufficient electrical contact for the sensing to work. On moist skin only a very slight pressure is exerted. If the wool is conductive the cutter moves away from the surface of the sheep thinking that it is pressing too hard. If the cutter is not moving forward on the skin then wool fibres may come between the cutter and the skin allowing heavy pressure to be exerted on the sheep without the resistance sensing being able to detect this.

[5] see Glossary.

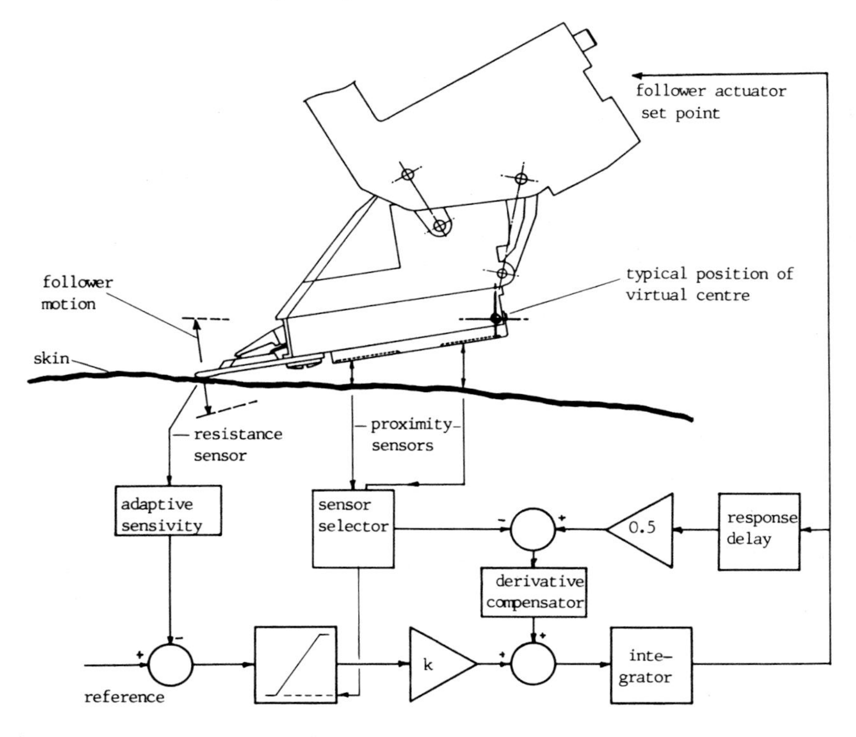

Figure 4.10 Position control scheme for follower actuator. The cutter assembly is articulated with a four-bar linkage mechanism which provides the motion illustrated. The locus of the comb tip is approximately straight. Three capacitance proximity sensors are used on the underside; the left and right sensors are just behind the comb and a third sensor is at the back end. The outer loop aims to obtain a steady signal from the resistance sensor, corresponding to moderate pressure on the skin. Constant skin deformation would be desirable. The lower saturation limit is set by the closest proximity sensor such that when the proximity reaches a minimum acceptable value, the signal passed to the integrator is zero or positive, lifting the comb. In this way, the follower will not seek down beyond a given minimum clearance at the proximity sensors. Without this, the base of the cutter can press too hard on the sheep. The inner loop aims to achieve velocity stabilization of the outer loop.

The principal advantage of the resistance sensing signal is its source—it is the only signal obtainable from the tips of the comb teeth.

The capacitance signal is more predictable. It is distorted by surface curvature near the sensors. The sensors need to be about 4 cm^2 in area so they can only be located behind the back of the comb, about 70 mm behind the front of the comb. Yet the distance indication is reliable up to 40 mm or more, and after calibration can be accurate to 1 mm.

The capacitance sensors are also affected by moisture and contaminants, only

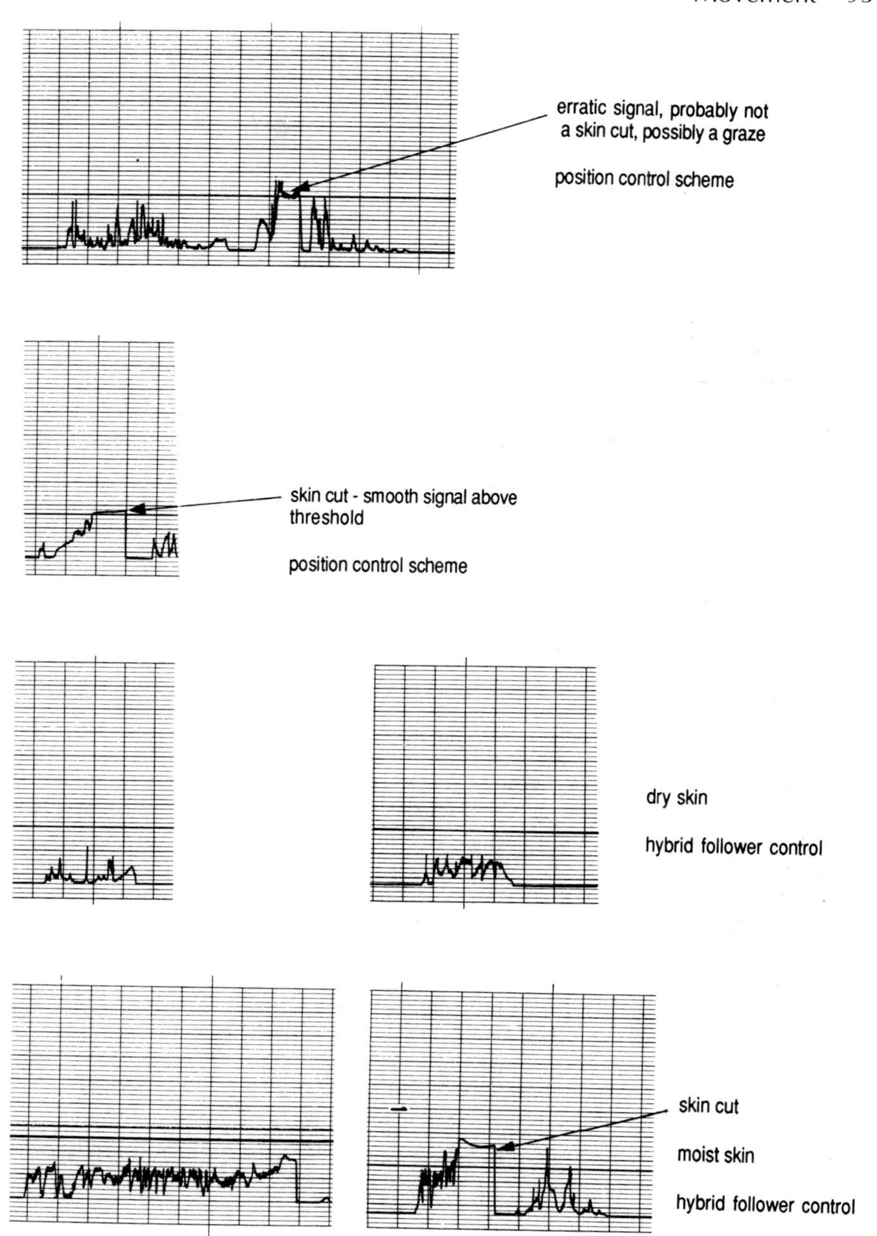

Figure 4.11 Typical resistance sensor traces taken from different sheep. The position control scheme is illustrated in figure 4.10; hybrid follower control is described later.

much more so than the resistance sensors. The most effective solution to this problem has been a pneumatically inflated shield which flattens contaminated wool against the skin. We also devised inductance sensors which could be used instead of capacitance sensors (chapter 6).

Apart from the fusion problem, the most difficult part of the design has been to deal with the limits on each signal: how to guarantee stable operation when operating near or beyond limiting values on one or more sensors or actuators. For example, the resistance contact signal can be lost for a time if the comb lifts off the skin. During this time, the follower position control switches to a different regime but must switch back smoothly when resistance contact is regained. This scheme was tested and refined between 1980 and 1987. By then we realized that there were fundamental difficulties with resistance sensing as our primary signal for keeping the cutter on the skin.

I yearned for an effective method to sense force in combination with distance and contact resistance. From expert shearers I learned to press the base of the comb firmly down while shearing. This flattens the skin under the comb allowing all the teeth across the leading edge of the comb to be in contact with the skin. If the teeth are not in contact with the skin then wool fibres ahead of those comb teeth lean over away from the comb. The comb teeth can no longer penetrate these more densely packed fibres and shearing breaks down. Thus, it is important to press firmly on to the skin to be able to use the full width of the shearing comb (figure 4.12). Under typical shearing conditions the resistance sensing control method results in very light contact forces and on rounded parts of the sheep only part of the comb width can be used. To shear well and efficiently, the comb contact force has to be controlled.

Yet there are some big obstacles to force control; the contact force is hard to measure, and robot force control has special problems.

The comb experiences large and variable bending moments from the closing force on the top of the cutter so one cannot measure forces on the comb directly. There are several potential sites for force measurement further back though. Several of these are described in chapter 6.

There are more general reasons to be careful with force sensing. Experiments conducted by Merino Wool Harvesting (p. 190) in Adelaide had shown that the combing forces can be large compared to typical contact forces applied to the cutter. As the comb moves forward the teeth penetrate between wool fibres; the tension force in the fibres pulls the comb teeth down on to the skin. In some circumstances, this effect can increase friction to the point where the comb snags the skin and the comb cannot be moved forwards without cutting the skin. A shearer often senses this and tips the comb back a little, relieving the pressure at the comb tips.

Merino Wool Harvesting went on to use force control but they experienced problems familiar to robotics researchers who have experimented with force control. The stability of the control loop is critically dependent on the compliance of the surface and the tool. They could not obtain stable performance

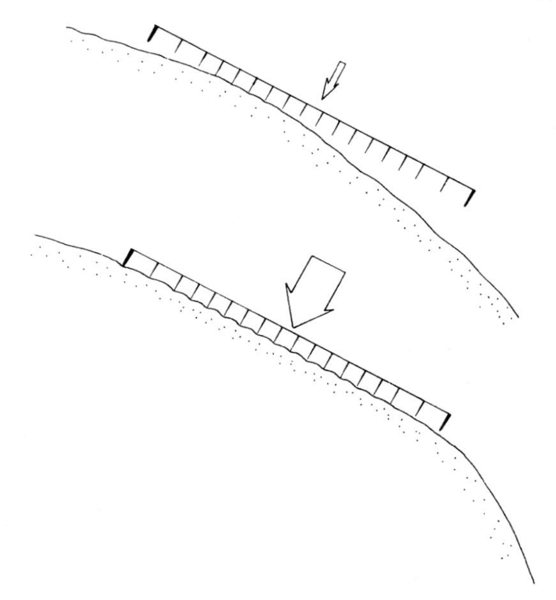

Figure 4.12 Effect of light comb pressure (above) and firm pressure (below)—refer to text.

under all the different operating conditions found on the sheep. Further, the design of the cutter was constrained by the need for delicate strain gauge beams for force measurement.

Hybrid follower

When we first considered methods for keeping the cutter on the skin in 1977 we briefly considered the use of spring force to keep the cutter in contact with the skin. However, to be effective this would require a relatively low spring compliance and therefore low natural frequency when considering the mass of the cutter mechanism. This would make it difficult to control high-speed robot movements close to the sheep because the cutter would tend to bounce uncontrollably on the end of the arm. For ten years of shearing trials we used a position control scheme with resistance sensing as its input.

Then at the beginning of 1989 I realized that the cutter could be held against the skin of the sheep by the equivalent of a hydraulic shock absorber or damper. With the addition of a servo valve to supply a fixed flow of oil to one end of the cylinder to maintain a fixed force between the cutter and the skin the force would be proportional to the pressure on the piston which would in turn be determined by the rate of oil flow through the holes in the piston (figure 4.13). In effect we have a 'constant force spring', with damping which suppresses unwanted bouncing. Furthermore, the servo valve would permit the oil flow to be determined partly by the position of the cutter so that a combination of active position control and passive open loop force control would be possible. I revisited the missed turning point referred to on p. 31 and called the new scheme a 'hybrid follower'.

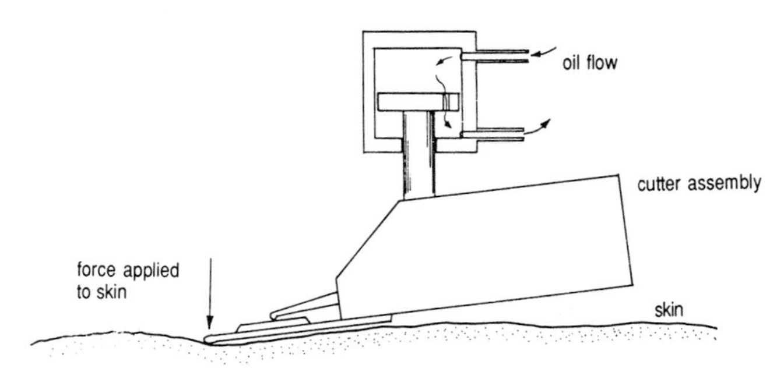

Figure 4.13 Operating principle of hybrid follower position control. Oil flowing through the hole in the piston results in a pressure causing a force to be applied to the skin by the comb. The magnitude of the force can be altered by changing the oil flow through the circuit.

The hybrid follower control method bypasses the stability problems associated with conventional closed loop force control by using a force generator. The force generator is further simplified by using open loop control avoiding force measurement; we use a force generator which is sufficiently accurate without having to monitor its output. This, of course, relies on achieving a low enough value of actuator friction but more of that later.

When one is working within an environment where one constantly relies on, and exploits the principles of feedback control, it is easy to overlook the possibilities of open loop control. This is a clear example where open loop control can be quite effective. Problems of dynamic stability, measurement noise, vibration interference and variations in the sheep surface compliance can all be neatly sidestepped.

In February 1989, we demonstrated completely automatic shearing of the whole sheep to the Corporation and their advisers for the first time. I had devised the hybrid follower idea during the frenetic preparations for the demonstration (we were running critically late), but I forgot to alert my colleagues. After the demonstration I announced my decision to step aside from directing the project to a surprised review committee; my colleagues were already aware of, and had accepted this. But when I went on to show how I intended replacing resistance sensing with hybrid force control, they were not so sure about my presence of mind!

Figure 4.14 shows mechanical details of the follower actuator. The bypass needle valve was originally installed for force sensing measurements in 1985. I thought that a small bypass flow would give rise to a position error proportional to contact force, but we dismissed the idea after measuring static friction levels.

Michael Ong developed some ideas for passive force control, but the subject was deferred when he left the team.

Figure 4.15 shows a schematic arrangement of the hybrid follower control scheme which is implemented by software and is the subject of a pending patent application. The oil flow signal comprises five components.

The first is determined by the force setting, and modified by a look-up table encoding the mechanical advantage of the four-bar support mechanism.

The second is derived from the measured velocity of the piston determined from the smoothed derivative of the potentiometer signal. Thus the servo valve provides just the right amount of oil flow to compensate for the velocity and thereby maintains a constant oil flow through the needle valve maintaining constant oil pressure across the piston.

The third component compensates for internal forces between the cutter and the follower assemblies caused by hose flexure.

The fourth component is a dither signal which imposes a slight oscillatory motion on the follower piston which helps to maintain an oil film between the seals and the rod of the piston shaft.

The fifth component is used to lift the cutter impulsively when a skin cut has been detected.

An extra component compensates for gravity; this is also not shown in figure 4.15. The approximate correction is calculated in proportion to $1.0 - k_3$ where k_3 is the vertical component of $\mathbf{k}$.

In certain shearing conditions, the negative force arising from the combing action can 'trip up' the cutter and lead to skin cuts. However, the large local contact force between the comb tips and the skin which precedes this can be detected by resistance sensing. A resistance sensing input with much reduced sensitivity overcomes this.

One of the significant problems which had to be solved was detecting skin cuts. In the past we simply monitored the level of the resistance sensing voltage. If the voltage exceeded a preset value (indicating a highly conductive path from the comb to the sheep) a skin cut was assumed to be the cause. With conductive wool sheep, this simple test was no longer appropriate. We found that crossbred sheep exhibit wool conductivity sufficient to give readings equivalent to a skin cut for much of the time.

We observed that skin cuts are associated with very smooth signals from resistance sensing; the second difference of the signal is a useful measure of smoothness:

$$s_k = \alpha\,|(V_k - V_{k-1}) - (V_{k-1} - V_{k-2})| + (1 - \alpha)\, s_{k-1} \qquad (4.1)$$

where s_k is the smoothed absolute second difference signal,
k is the sample number,
V_k is the resistance sensing voltage, and,
α is a smoothing factor, typically 0.1.

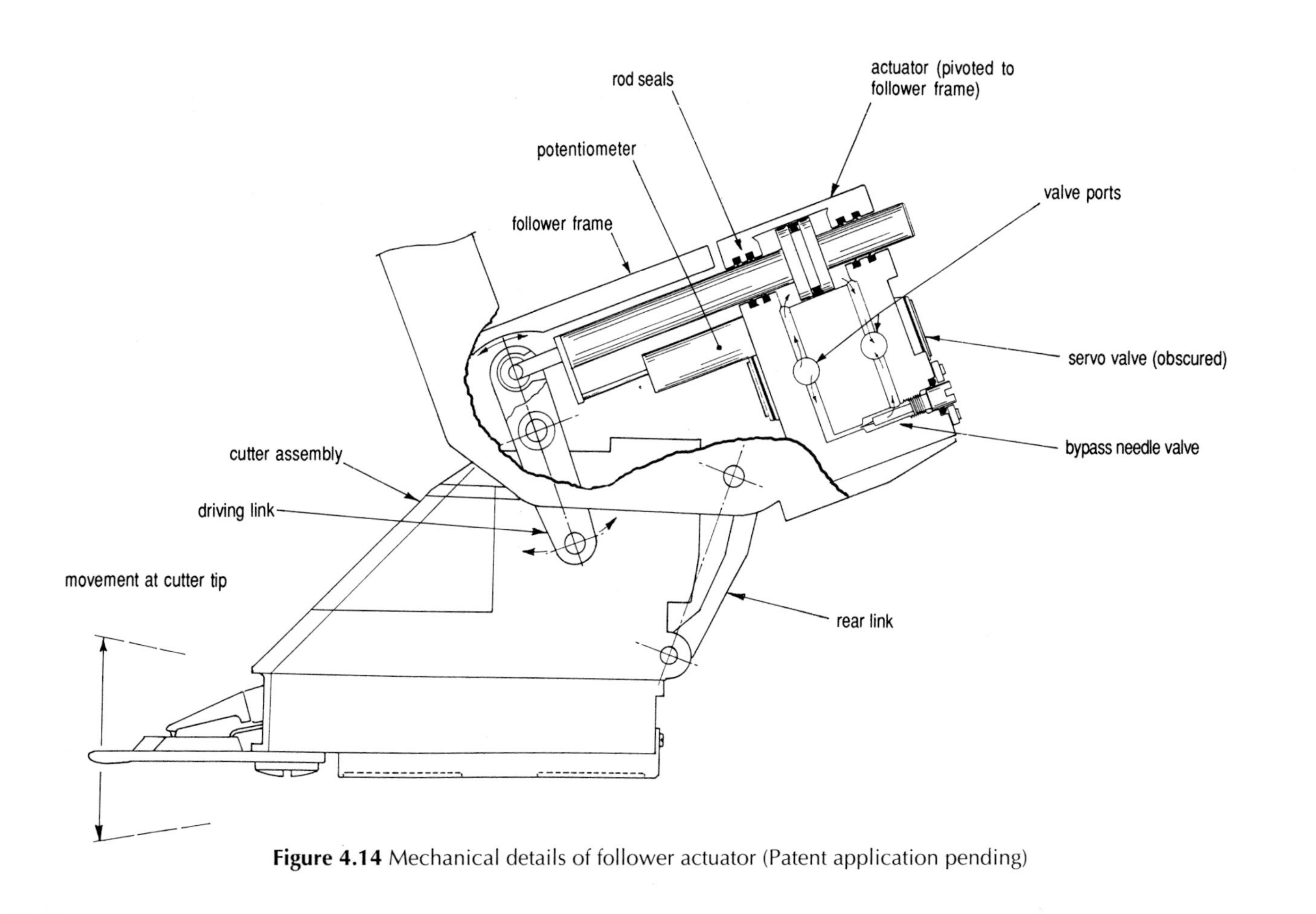

Figure 4.14 Mechanical details of follower actuator (Patent application pending)

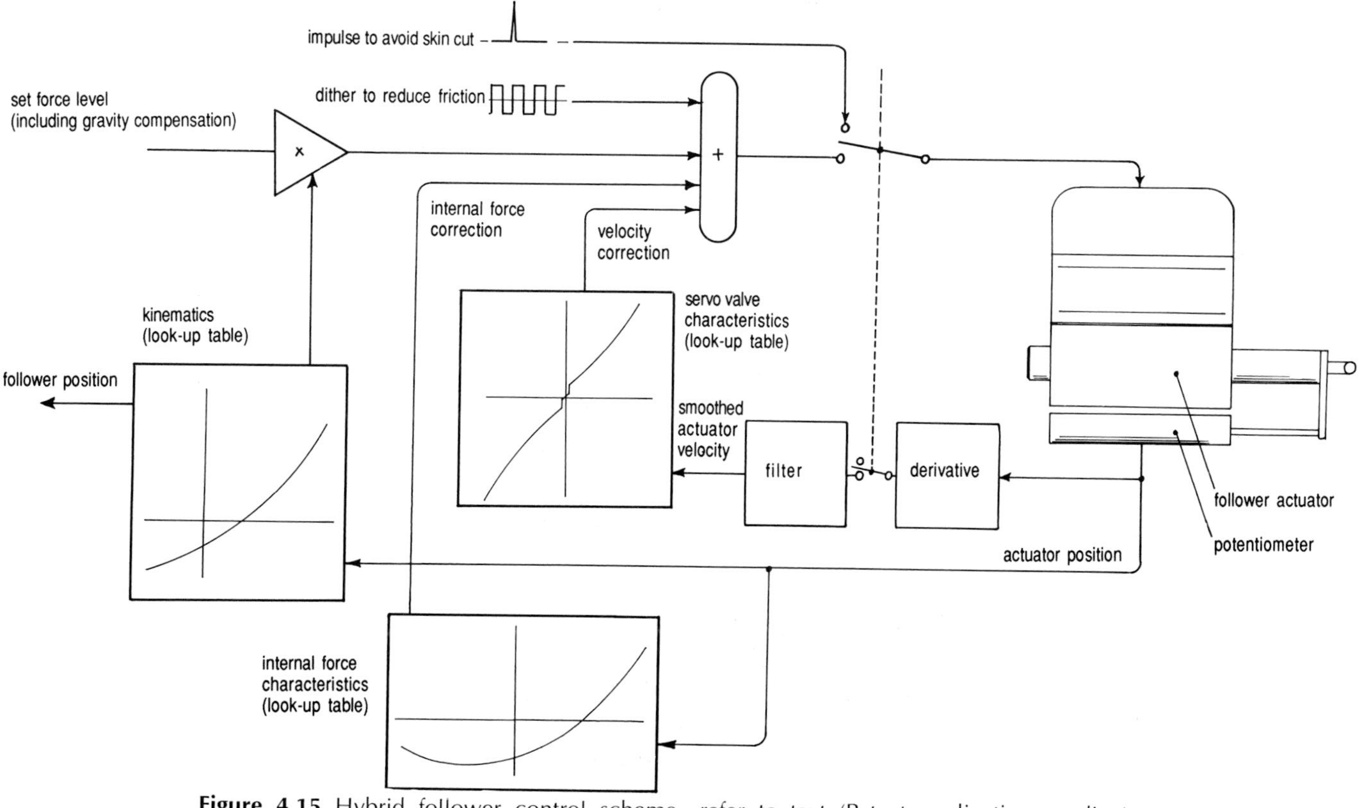

Figure 4.15 Hybrid follower control scheme—refer to text (Patent application pending).

When s_k is large then it can be deduced that the resistance sensing voltage is highly erratic and variable. This is typical of normal shearing conditions. If the second difference signal falls in value then the output of the resistance sensor is much smoother and this is typical of a skin cut. One explanation for this is that when one or more teeth of the comb have embedded themselves under the skin of the sheep there is a continuously available conductive path between the comb and the skin. However, under normal shearing conditions the electrical path between the comb and the skin is constantly changing and therefore, small but erratic changes in resistance sensing voltage occur all the time.

Friction in the actuator was also a problem. Initial tests showed that the friction was much less than we expected. A few weeks later the friction levels were much higher—far too high. Higher oil temperatures were found to be the cause and the hydraulic fluid was replaced with a higher viscosity oil with a low viscosity/temperature index. As a further measure, we removed one of the two sealing rings at each end of the actuator—the slight oil leakage on to the cutter which resulted from this was a help rather than a problem.

Naturally, when there is no contact with the sheep, s_k is zero; therefore we add the condition that V_k must be above a preset value and the robot must have been moving forwards for a skin cut to be presumed to have occurred.

Shearing results have justified my early confidence in the hybrid follower. As well as providing smoother and faster robot performance and more efficient shearing, it has been possible to shear damp and urinated wool. This would have been quite impossible with the resistance sensing control method used beforehand.

Adaptation 2: height, pitch and roll error correction

Compared to follower adaptation, roll and pitch adaptation is much simpler. The roll error is calculated by taking the difference between the left and right proximity sensors, and dividing by their effective separation distance. A surface twist correction term (derived from the surface model) and a roll offset term (set by the programmer) are added.

Once again, sensor limits need to be accounted for. If one sensor is too high to detect the surface, and the other is only just low enough, then the difference between them will not represent the roll error. It is then possible for the twist or roll offset corrections to swamp the measured roll error causing the cutter to roll over the wrong way! Careful coding is needed to avoid this in a reasonably efficient manner.

Pitch calculations are slightly more complex. First, there is a third proximity sensor at the back of the cutter body. This prevents the rear end being pushed into the sheep when shearing a concave region. The signal from the rear sensor gradually dominates the signal from the other proximity sensors as the distance measured by it is reduced.

One cannot simply take the difference between the contact resistance signal and the proximity signals to calculate a pitch error. The resistance signal is too intermittent and erratic. Instead, we integrate the resistance signal, apply saturation limits, and then take the difference. Then we have to decouple the displacement and rotary motions of the cutter arising from the approximate rotation about the virtual centre of the follower linkage.

Attitude control and joint space adaptation

The follower, however controlled to follow the skin, is in essence an active sensor so far as the rest of the arm is concerned. The relative follower deviation provides a signal to move the arm towards or away from the sheep. Coordinated motion of all the actuators is needed for this. Similarly, roll and pitch corrections need coordinated movements of the whole arm.

The follower deviation is converted into a Cartesian displacement by scaling a unit vector in the direction of follower motion calculated by the arm kinematics procedures, and passed to HEATHER with the arm Jacobian. Naturally, a gain factor is included such that only a small proportion of the deviation is corrected at each interpolation step. The result is combined into a pose change vector, the first three elements being the position change and the last three being an incremental rotation.

$$\delta\mathbf{pose} = \begin{vmatrix} f_x \\ f_y \\ f_z \\ \delta roll \\ \delta pitch \\ 0 \end{vmatrix} \tag{4.2}$$

The joint space correction is then given by

$$\delta\theta := \mathbf{J}^{-1} \begin{vmatrix} f_x \\ f_y \\ f_z \\ \delta roll \\ \delta pitch \\ 0 \end{vmatrix} \tag{4.3}$$

where $\mathbf{J}^{-1}$ is the inverse arm Jacobian,
f_x, f_y, f_z are components of the position change,
θ is the vector of arm joint positions, and
$\delta\mathbf{pose}$ is the change to the pose.

The joint space correction is then added to the interpolated planned joint motion.

We have obtained significant time savings by partitioning the Jacobian inverse such that orientation changes are treated separately from position changes.

When we developed this technique for SM, we found that great care was needed to ensure that joint motion limits were correctly accounted for. For example, a roll rotation of the cutter requires arm position changes to keep the comb centre in a fixed position. However, when actuator limits prevent further rolling of the cutter, the arm displacements must stop, otherwise the cutter suddenly moves sideways for no apparent reason! In the case of ORACLE, with its virtual wrist centre at the cutter tip, this was never a problem.

Filtering

Figure 4.16 shows a notch filter operation performed on the roll, pitch and follower deviation signals. This was introduced to suppress unwanted frequency components which tended to excite the hydraulic natural frequencies of the arm and wrist actuators. The notch filter was a frequency invariant (bilinear) discrete transform of (Gabel and Roberts 1980, p. 452; Phillips and Nagel 1984):

$$\frac{\left(1 + 2\xi_1 \dfrac{s}{\omega_n} + \dfrac{s^2}{\omega_n^2}\right)}{\left(1 + 2\xi_2 \dfrac{s}{\omega_n} + \dfrac{s^2}{\omega_n^2}\right)} \tag{4.4}$$

where s is the Laplace operator,
ω_n is the wrist actuator hydraulic natural frequency (10 Hz),
ξ_1 is the numerator damping factor (chosen to be 0.15), and,
ξ_2 is the denominator damping factor (chosen to be 0.8).

In effect this is a crude form of pole cancellation for cutter attitude control. It relies on a reasonably constant hydraulic natural frequency—the SM robot wrist design provides this. Texas Instruments (1985) provides a useful guide for applications of such filters.

The notch filters were replaced by uniform discrete low-pass filters on each of the arm joint position demand signals when we discovered that the planned motion signals could excite wrist vibrations as much as sensed attitude errors.

The position demand signals are linearly interpolated to provide updates often enough for smooth joint motion (figure 4.17). The joints are driven at constant velocity for the duration of each motion step, so there can be a sudden change in velocity at the end of each step. One can use higher-order interpolation schemes given some knowledge of acceleration and velocity. However, this approach is not feasible if unpredictable sensor adjustments are included.

When ORACLE was first commissioned, we noticed that coordinated joint motions computed for the arm appeared slightly uncoordinated. Each joint controller reacted with a different delay, though they all reached the correct final position. The error in each joint was approximately proportional to the joint velocity and the time constant for each servo as shown in figure 4.17.

Joint coordination was markedly improved by modifying the interpolated joint positions by a term proportional to each joint velocity and time constant. In effect, this is an open loop feed forward velocity compensation. The time constants for each joint were measured by injecting a constant velocity ramp

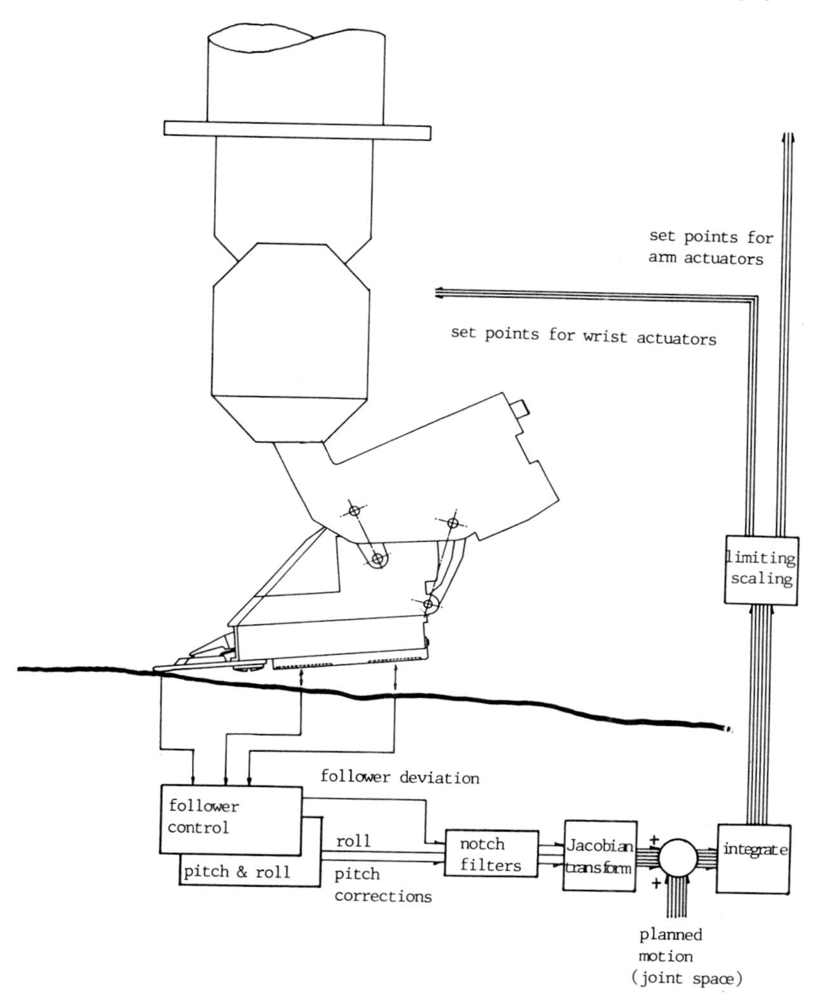

Figure 4.16 Joint space adaptation for height, roll and pitch errors. Follower adaptation provides a deviation signal requiring an arm position change. Pitch and roll corrections require a wrist angle change, which in turn requires an arm position change too.

signal and observing the position error. An on-line scheme could also be used, given appropriate statistical techniques.

Here we had to introduce a significant practical restriction. Each actuator has software limits imposed on its motion to avoid damaging collisions against the mechanical stops (figure 4.21). The feed forward correction introduced above could cause an excursion beyond the software limits, so we had to introduce a second limiting test to prevent the *compensated* set point from exceeding limits.

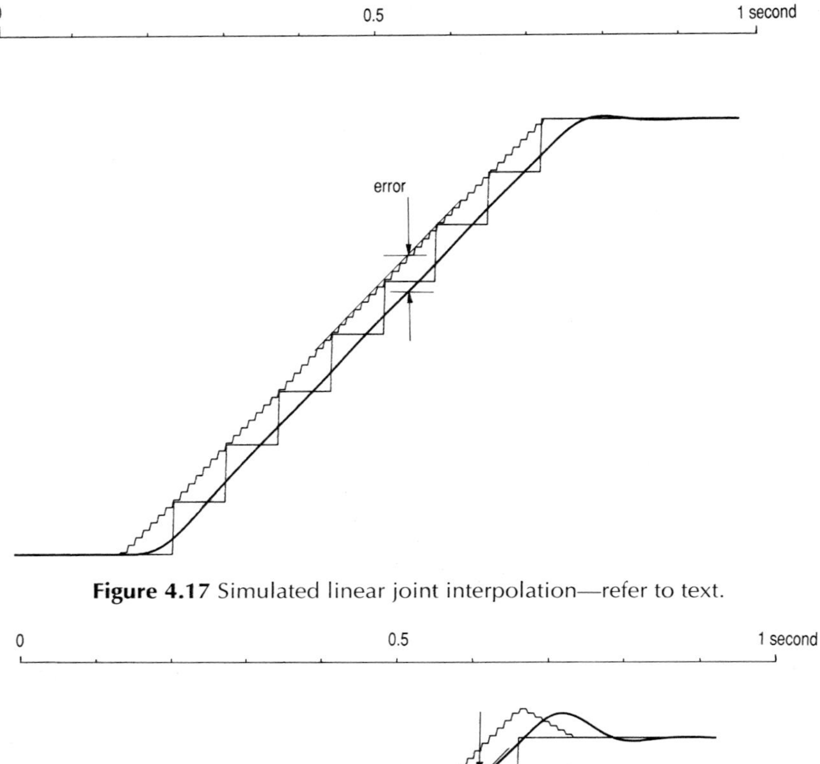

Figure 4.17 Simulated linear joint interpolation—refer to text.

Figure 4.18 Simulated averaged joint velocity for dynamic compensation.

The feed forward compensation caused further difficulties. Since there were velocity discontinuities at the end of each motion step, there were discontinuities in the corrected set points. We soon found that these discontinuities caused unwanted jerks and vibrations. In essence, the phase advance compensates only

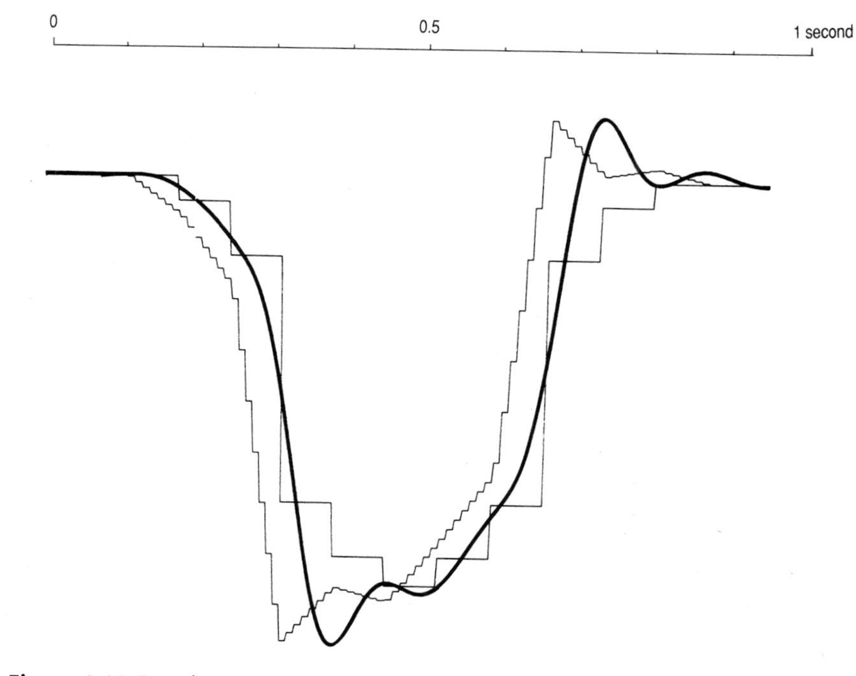

Figure 4.19 Rapid repositioning movement using averaged joint velocity compensation.

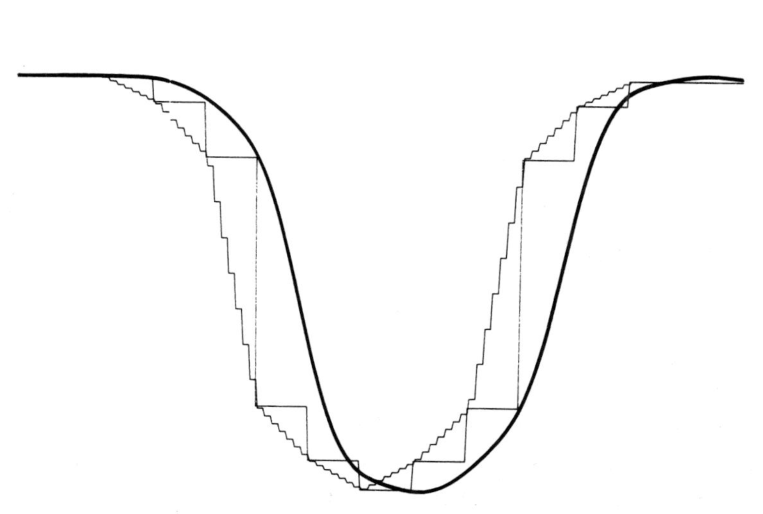

Figure 4.20 Rapid repositioning movement with a linear feed forward compensator and low-pass filter—refer to text.

one part of the actuator response, but not other parts. The parts which were not taken into account became more and more obvious. The worst effects of this were avoided by using an average joint velocity based on the position change over the previous 7 interpolation steps. Figure 4.18 shows simulated results using this scheme; actual results were close to predictions.

This scheme worked well for ORACLE and for the first two years of trials with SM. However, as the SM wrist actuator seals gradually polished the pistons and friction levels dropped, some unstable robot movements began to appear. Low frequency (7–15 Hz) vibrations appeared at certain wrist configurations particularly when the cutter was running. The cutter oscillates at about 50 Hz and sufficient vibrations are transmitted through the wrist to reduce friction levels significantly.

For some time, I suspected that the instabilities were associated with the attitude correction computations using the capacitance sensor signals. I have already referred to the use of notch filters to remove unwanted resonances. The problem gradually worsened until I felt I had to do something. Careful observations of videotaped shearing trials revealed that the instabilities could be induced even when the sensors were switched off.

I had always expected that dynamic coupling between actuators would cause some difficulties. The main SM arm actuators cannot interact with each other, but the wrist actuators can be affected by the arm actuators, and vice versa. I injected random noise into the arm actuators and observed the effects on the wrist actuators, but the coupling observed this way did not seem sufficient to cause the instabilities.

I came to realize that the difficulties were caused in part by frequency components arising from the motion step computation frequency (12–14 Hz). I was puzzled to find that there was no observable energy in the simulated output response from the actuators associated with this frequency when averaged over a long time span. However, it is clear in figure 4.19 that the actuator dynamics interact with the sampling frequency. Here we see the predicted result of a rapid repositioning movement on a single actuator. In practice, interactions between the arm and the wrist tend to cause more overshoot and oscillation.

I also realized that the feed forward compensation applied to the actuators made the problem worse.

As a radical experiment, I removed the feed forward compensation and passed all the actuator set points through a discrete filter equivalent to:

$$\frac{(1 + \tau s)}{\left(1 + 2\xi \frac{s}{\omega_n} + \frac{s^2}{\omega_n^2}\right)} \qquad (4.5)$$

where s is the Laplace operator,
τ is the time constant for each actuator,
ω_n is a low-pass filter cut-off frequency (5 Hz), and
ξ is the low-pass filter damping factor (chosen to be 0.8).

The numerator is a linear feed forward compensator to cancel position errors proportional to velocity. The denominator is a low-pass filter.

Figure 4.20 shows the simulated behaviour of a single actuator subjected to the same rapid reposition movement. There is no evidence of unwanted overshoot or oscillations, but the actuator response has been delayed by about 1 motion step. This would cause difficulties with the robot, but they could be overcome, I thought.

I was quite startled by the results when I tested the scheme on the robot. First, as expected, all the actuator movements were much smoother. Second, all of the irritating instabilities had disappeared. Third, the tracking performance on our dummy metal sheep seemed to be almost as good. Fourth, I could program much faster reposition movements without any apparent jerks, overshoot errors or erratic behaviour. The qualitative appearance of the robot's actions was immeasurably enhanced. There was no loss of joint coordination either since the filter for each actuator was set up with an appropriate value of τ. The time delay introduced by the denominator term was identical for every actuator.

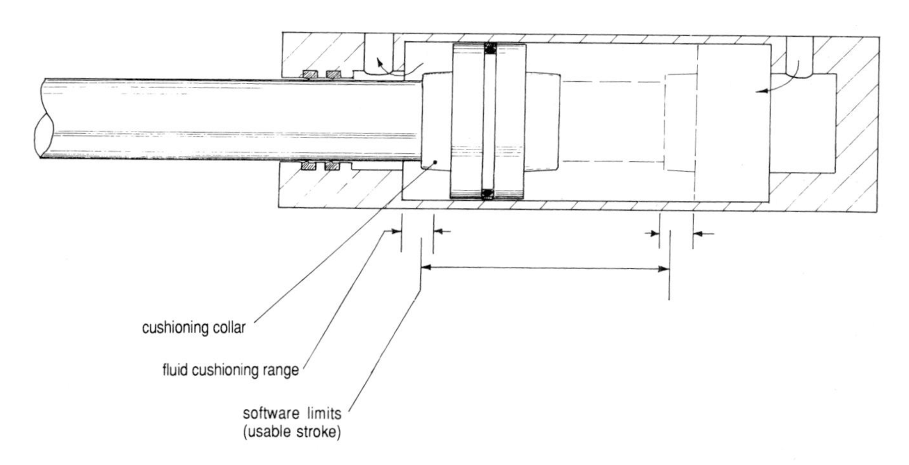

Figure 4.21 Joint motion limits. Each actuator can move within a range governed by mechanical limits. Collisions with mechanical limits can be damaging so software limits are imposed a safe distance from the mechanical limits. The software limits allow for errors such as overshoot. Fluid cushioning collars can also be fitted as a further precaution.

The time delay caused the expected complications and trial landings on our dummy sheep distinctly resembled those of Orville the albatross in Walt Disney's animated movie *The Rescuers*! The effect of the time delay at a landing speed of 200 mm/sec was to compute the cutter position about 15 mm below the surface at the instant of landing. Rapid corrections were needed and a few dents were left in the dummy sheep! Graceful landings were obtained by slowing the landing speed as soon as the surface was sensed by the capacitance sensors.

The low-pass filters appear to guarantee that the adaptation mechanisms will not have to respond to robot arm vibrations as well as sheep variations.

Adaptation 3: cutter path steering

The first two types of adaptation are implemented within HEATHER and operate at high computation rates (100–200 Hz). The next two types operate within TOM.

We calculate cutter movements in Cartesian space by a simple principle of path convergence; this operates in both shearing and reposition movements. Briefly, two components of movement are calculated:

(1) tangential to path at the projected position of the cutter on the planned path;

(2) a correction which tends to reduce the path-following error; the magnitude of the correction in any one motion step is limited so that graceful movements result.

This is based on the assumption that the cutter will rarely, if ever, exactly follow a planned path.

This method has proved to be robust in the presence of modelling errors, computation errors (resulting from the resolved motion rate control method), sheep movements, and actuator limits.

Shearing and repositioning paths are defined as B-spline curves in Cartesian space. An appendix provides details of B-spline curve calculations which form the basis of both 1. and 2. above.

Path following in three dimensions

The easiest way to explain path-following calculations is through diagrams. Figure 4.2 shows how the front edge of the comb is represented in the diagrams which follow.

The first method used for steering the cutter is shown in figure 4.22. It was only used for simulation experiments before ORACLE was constructed. It has obvious limitations for dealing with curved objects such as sheep (Trevelyan and Leslie 1979).

When shearing experiments started in 1979, we used the method shown in figures 4.23 and 4.24. Not only was the method simple, but it could be applied to any arbitrary surface within the bounding planes. The first blow followed one of the bounding planes; a special algorithm was developed for following the intersection of a plane with an unknown surface. The first demonstrations of successful back and side shearing in late 1979 used this technique.

The principal disadvantage of this method was the programming complexity. The shearing commands were encoded in Fortran with a subroutine which was linked in with the motion control software. It was difficult to change shearing programs.

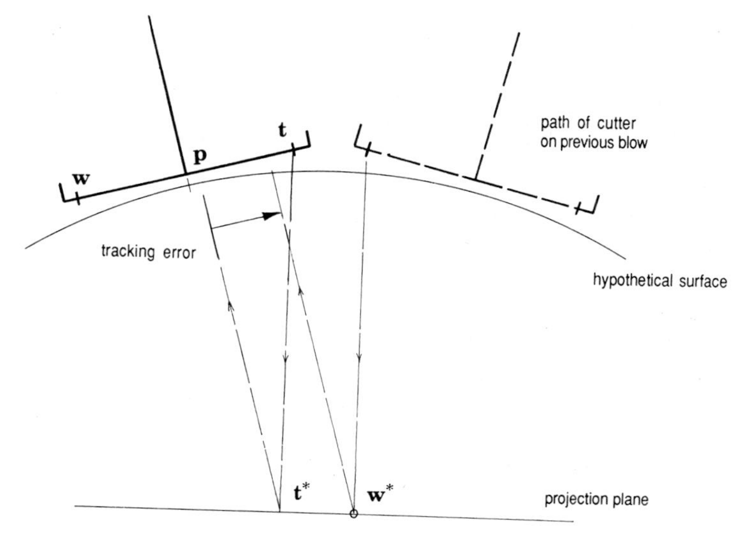

Figure 4.22 Plane projection steering (1977). The tracking error is calculated by projecting $\mathbf{w}^* - \mathbf{t}^*$ into the cutter plane ($\mathbf{t}^*$ is the projection of **t** on the projection plane, and $\mathbf{w}^*$ is the projection of **t** on the windrow line projected from **w** on the previous blow).

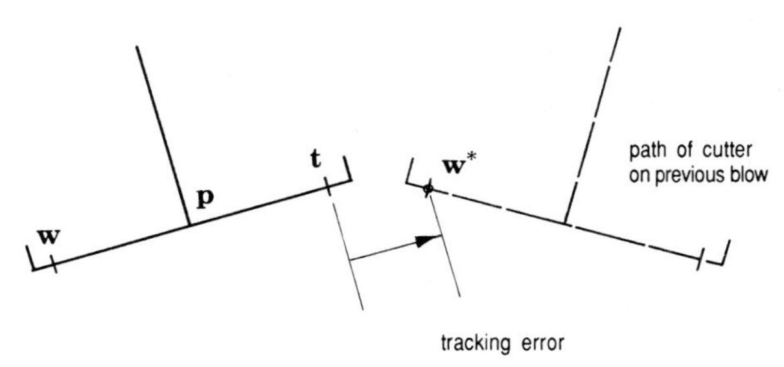

Figure 4.23 Three-dimensional steering (1978–80). The tracking error is calculated by projecting $\mathbf{w}^* - \mathbf{t}$ into the cutter plane ($\mathbf{w}^*$ is the projection of **t** on the windrow line retained from the previous blow).

The introduction of surface models led to major simplifications in programming and a new method of path following which avoided clumsy operations with bounding planes. Shearing areas could be defined in just two dimensions (on the model surface) (Trevelyan 1981).

Figure 4.25 illustrates the method, and shows how it failed when there were large differences between the model and the sheep. One reason for developing the deviation function method for on-line surface model adaptation (chapter 3) was to try to overcome this problem.

Figure 4.26 shows its successor which we still use. The surface tracking algorithm (chapter 3) made this method possible. And close inspection of the

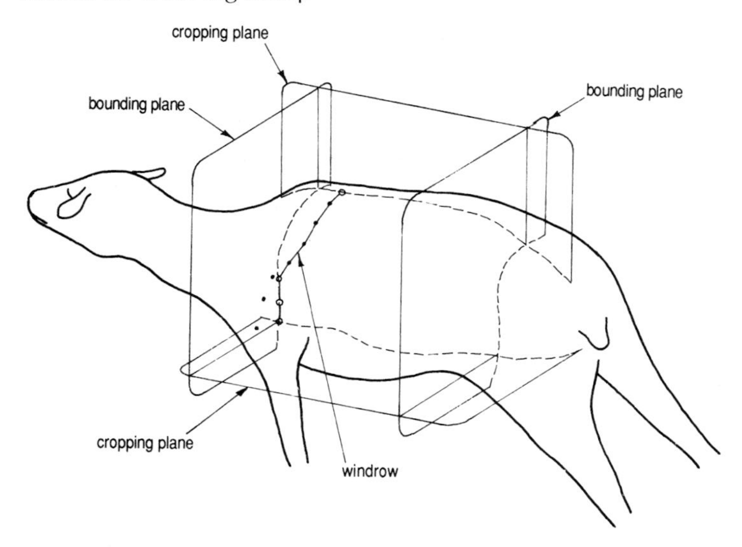

Figure 4.24 Use of bounding planes (1979). Points defining a windrow are constrained to lie to one side of a bounding plane. Cropping planes can be used to remove unwanted sections of a windrow. An incomplete windrow is extended to the cropping plane.

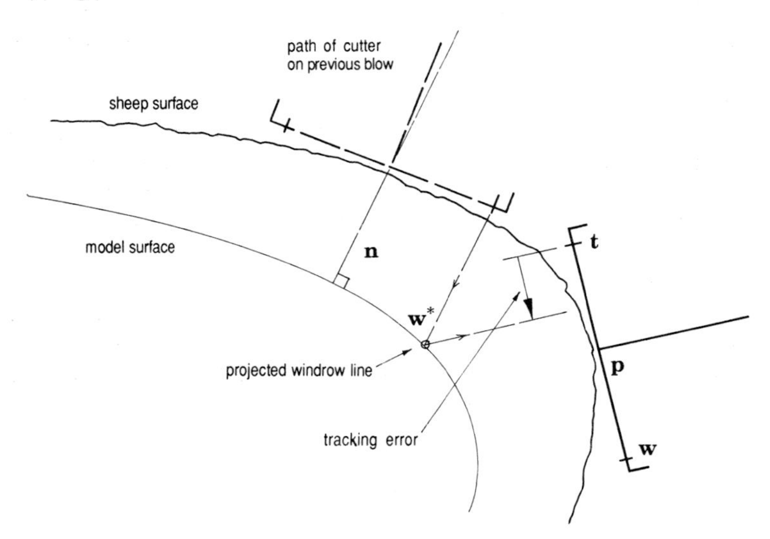

Figure 4.25 Three-dimensional steering (1980-83). The tracking error is calculated by projecting $\mathbf{w}^* - \mathbf{t}$ into the cutter plane ($\mathbf{w}^*$ is the projection of $\mathbf{t}$ on the windrow line projected from $\mathbf{w}$ on the previous blow). While this method often worked well, it failed in this case because of surface modelling errors. Note that the best estimate of the windrow line was obtained by projecting $\mathbf{w}$ parallel to $\mathbf{n}$. This introduced further error.

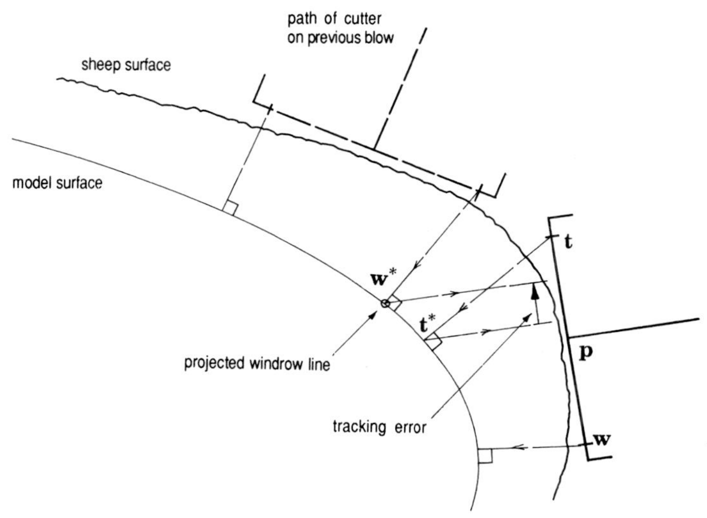

Figure 4.26 Surface model projection steering (1984-). The tracking error is calculated by projecting $\mathbf{w}^* - \mathbf{t}^*$ into the cutter plane ($\mathbf{t}^*$ is the projection of **t** on the projection plane, and $\mathbf{w}^*$ is the projection of **t** on the windrow line projected from **w** on the previous blow).

figure caption will reveal that the method is identical in principle to the first method in figure 4.22, though some time passed before we appreciated that! (Trevelyan 1989*a*)

A useful variant of this scheme aligns the cutter attitude with the surface normal computed at $\mathbf{w}^*$ (figure 4.26). This variant is used on sharply curved surface models where the deviation from the model is likely to be large compared with the radii of curvature which might cause surface tracking instabilities. In effect the surface tracking algorithm is applied as if **t** were located at $\mathbf{w}^*$.

Cutter steering

The tracking error provides the input to the steering calculation. An aiming vector **aim** is calculated by combining a component tangential to the planned path (or windrow being followed) and adding a component proportional to the tracking error. We set up a steering coordinate frame **Ass** as follows:

$$\mathbf{Ass} := |\,(\text{norm}((\mathbf{n} \times \mathbf{i}) \times \mathbf{n}) \quad \text{norm}(\mathbf{n} \times \mathbf{i}) \quad \mathbf{n}\,|. \tag{4.6}$$

This provides a stable frame pointing in the same direction as the cutter frame, but stabilized by the surface normal **n** so that steering calculations are not perturbed by cutter attitude changes. Note that the norm function normalizes a vector to unit magnitude.

We transform the aiming vector into steering coordinates to obtain components aim_1, aim_2, aim_3. aim_3 is immediately set to zero, constraining **aim** to be tangential to the surface, and we normalize the result.

The new aim_2 is now the sideslip component of the aiming vector. Sideways slip of the comb needs to be controlled; sometimes it is used for stretching the skin sideways, and other times it must be minimal. Therefore, we calculate a yaw (steering) angle change which will turn **i** towards **aim** during the next motion step. We also then restrict the value of aim_2 to a range:

$$shyof - sslf \leq aim_2 \leq shyof + sslf \tag{4.7}$$

where *shyof* is the shearing yaw offset set by the programmer, and
sslf is a side slip limit factor, which is normally preset but which can be changed by the programmer if necessary.

aim is renormalized again and scaled to obtain the desired shearing displacement for the next motion step, $\delta\mathbf{p}$.

Attitude corrections

The frame **Ass** is aligned with the surface normal. It follows that the first two components k_1 and k_2 of **k** (transformed into the **Ass** frame) indicate the cutter pitch and roll deviations from the surface model. These deviations are scaled by *–gppitch* and *–gproll* respectively to generate planned attitude changes to realign the cutter with the surface model. The first two components of the change in the surface normal since the previous motion step, Δn_1 and Δn_2, are scaled by *gdpitch* and *–gdroll* respectively to provide a feed forward attitude change which will anticipate the change in the surface normal over the coming motion step.

$$\Delta roll := +gproll\ .\ k_2 - gdroll\ .\ \Delta n_2 - shrof \tag{4.8}$$

$$\Delta pitch := -gppitch\ .\ k_1 + gdpitch\ .\ \Delta n_1 - shpof. \tag{4.9}$$

Since the surface model is different to the sheep, the attitude changes from the sensors will work against the proportional attitude corrections from the surface model. The extent to which the cutter follows the model or the sheep surface is determined by the relative values of the gain factors *gproll*, *groll* (for roll) and *gppitch*, *gpitch* (for pitch). Ideally, we could set *gppitch* and *gproll* to zero; then we would follow the sensor information, aided by the feed forward components from the surface model. In practice, the proportional corrections from the surface model are useful.

Attitude offsets

The cutter shears over a curved surface, yet the sensors and the comb lie more or less in a plane. Further, the proximity sensors used for controlling attitude lie some distance behind the comb and so they 'see' a different part of the surface. We have built-in offsets to compensate for surface curvature so that the attitude errors calculated from the sensor inputs will keep the *comb* at the right attitude.

Figure 4.27 shows why offsets are needed for pitch control; similar roll offsets are needed to compensate for surface twist.

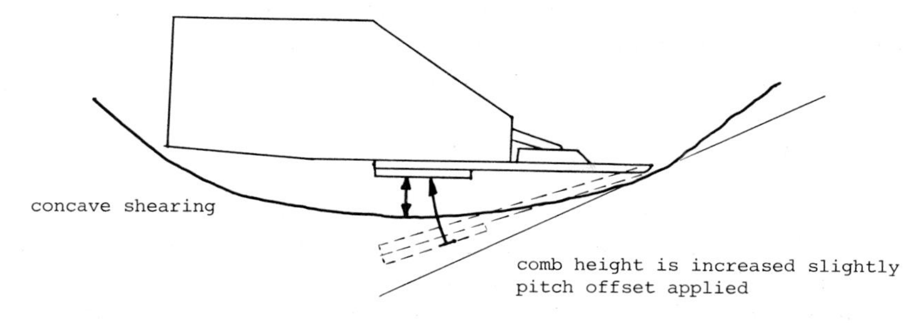

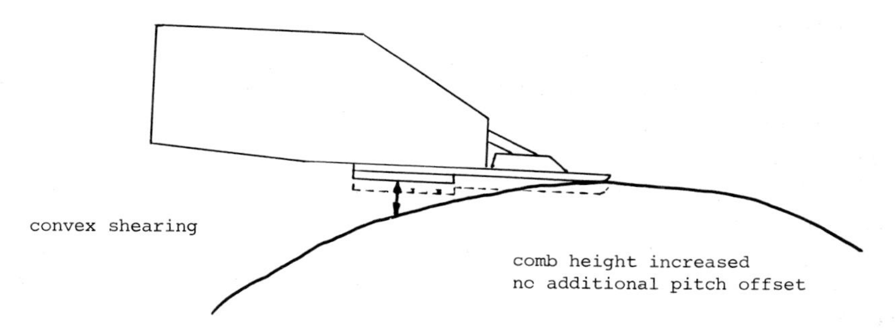

Figure 4.27 Offsets to allow for surface curvature. The diagram illustrates the effects for pitch offsets. Roll and height offsets are also required—refer to text.

As in the case of the Jacobian, the offsets are updated by TOM and are passed to HEATHER for use in roll and pitch error calculations.

We use a recursive calculation to estimate the difference in surface normal between the comb and the sensors:

$$\overset{*}{\mathbf{n}}_k = (1-\alpha)\overset{*}{\mathbf{n}}_k + \frac{\alpha}{|\delta \mathbf{p}|}\Delta \mathbf{n} \tag{4.10}$$

where α is given by $\dfrac{|\delta \mathbf{p}|}{d_s}$

d_s is the distance between the comb and the proximity sensors,
$\Delta \mathbf{n}$ is the change in the surface normal during the last motion step, and
$|\delta \mathbf{p}|$ is the distance moved by the cutter on the last motion step.

Then the estimated difference in surface normal between the comb and the sensors is given by

$$d_s \overset{*}{\mathbf{n}}_k. \tag{4.11}$$

The twist angle is given approximately by

$$d_s \overset{*}{(\mathbf{n}_k \cdot \mathbf{j})}. \tag{4.12}$$

The proximity sensor height offset for a convex surface is given approximately by

$$\frac{d_s^2}{2} \overset{*}{(\mathbf{n}_k \cdot \mathbf{i})}. \tag{4.13}$$

On a concave surface, we make some allowance for the back sensor which could otherwise hit the surface. Note that the correction is still positive—the cutter has to be pitched forward to keep the back end clear. The proximity sensor height offset is given by:

$$\frac{d_s^2}{2} \overset{*}{(\mathbf{n}_k \cdot \mathbf{i})} - \frac{d_{sb}^2}{2} \overset{*}{(\mathbf{n}_k \cdot \mathbf{i})}. \tag{4.14}$$

where d_{sb} is the distance from the comb to the rear of the cutter.

The pitch offset necessary to keep the rear of the cutter off the sheep on a concave surface is given by:

$$-\frac{d_{sb}^2}{2} (\mathbf{n}_k \cdot \mathbf{i}).^{*} \tag{4.15}$$

Adaptation 4: shearing speed adjustment

Although much simpler than the motion calculations described above, speed control is nevertheless a further form of adaptation.

The maximum forward shearing speed is programmed by the value of the parameter *shvel*. A further parameter *shaccn* governs the allowable forward acceleration along a shearing blow. This is typically between 0.5 and 1 m/sec^2.

The actual speed used for the next motion step can be reduced if certain other parameters exceed programmed limits:

(1) *drag* the drag force encountered during shearing movements is calculated using signals from the *W2* and *W3* wrist actuator differential pressure transducers[6] (only on the SM robot—there was no means for measuring shearing drag forces on ORACLE);
(2) *herr* height deviation from surface model;
(3) *aerr* attitude deviation from surface model;
(4) *raten* rate of change of surface normal vector.

The first three parameters are derived (albeit indirectly) from sensor inputs. The last arises from the surface model. In a strict sense of course the process of predicting the surface model which is described in chapter 3 is also a form of adaptation but for the present discussion the surface model is assumed to be static. This adaptation is not part of the motion control system.

A speed control factor is specified for each parameter; for example, a drag factor which is set to the drag level at which the speed should be reduced by about 40%.

[6] This information is incomplete; in theory we need to measure at least the three wrist actuator torques. In practice, however, sufficient information can be obtained this way.

We calculate the following quantity for each parameter (shown here for *drag*):

$$r_{\text{drag}} = \max\left(\left(\frac{drag}{\text{drag factor}}\right)^2 - 0.64, 0.0\right) \tag{4.16}$$

and then calculate the maximum speed by

$$v_{\max} = \frac{1.0}{r_{\text{drag}} + r_{\text{herr}} + r_{\text{aerr}} + r_{\text{raten}}}. \tag{4.17}$$

If the current cutter speed is less than the maximum, the speed is increased at the programmed acceleration *shaccn*. Otherwise, the calculated maximum speed is applied. This scheme provides the programmed speed at moderate values of the speed adaptation variables, with smooth speed reductions at higher values.

Adaptation 5: path manipulation and planning

We return to CLARE for the final level of adaptation.

The windrow is the path followed by the point **w** on the comb, projected on to the surface model. This path becomes the path for the point **t** to follow on the next blow. Before this happens, CLARE manipulates the windrow to keep it within the defined shearing area. Figure 4.28 shows how this was done up until 1988.

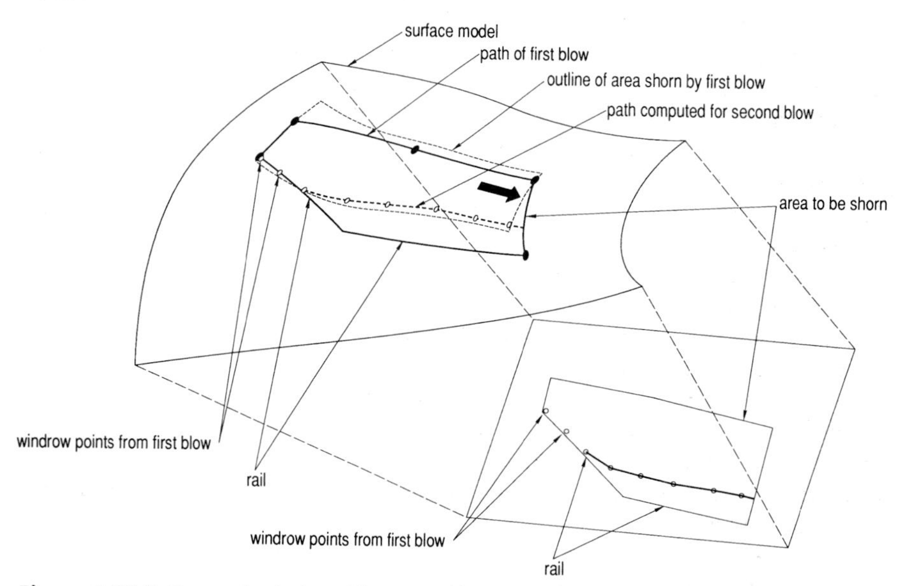

Figure 4.28 Path manipulation. The curved surface is mapped to a rectangular surface coordinate space. The shearing area is defined by the path for the first blow and a rail which is ideally the line of the last blow. Notice how the first blow has been constrained to follow the rail and the second blow commences where the windrow line diverges from the rail.

We exploit the opportunity to manipulate the path in two dimensional surface coordinates. The sections of the windrow lying outside the area are cropped off at the ends and clamped to the edge of the area in the middle.

The programming commands to shear an area of the sheep were (excluding parameter setting commands):

```
... repositioning commands ...
... parameter setting commands ...
blow [ t0,u0, t1,u1, ... ];   ! execute first blow
rail[ t0,u0, t1,u1, ... ];    ! define rail to align last blow –
                              ! t0,u0, t1,u1, etc have different values.
... more parameter setting commands ...
area [ m ];                   ! shear rest of area automatically,
                              ! maximum of m blows.
```

The **blow** command defines one side of the shearing area; more or less parallel blows will be used to shear the area. The **rail** command defines the edge of the shearing area which will be shorn last. The **area** command initiates a number of automatic blows and repositions to completely shear the area defined between the first blow and the rail.

Once the SLAMP manipulator was commissioned, the shearing pattern became more complex. We found that subtle parameter variations were required between blows and our original technique failed to allow for this. We also missed the freedom to plan individual repositions between blows. When the motion control system was redesigned with a recursive structure (chapter 11) we opted for a simpler approach to path manipulation with more explicit programming.

```
... repositioning commands ...
... parameter setting commands ...
blow [ t0,u0, t1,u1, ...]; ! execute first blow – t0,u0, t1,u1, . . .
set portion: getwindrow [ t3,u3, t4,u4, np ];
```

The **getwindrow** function returns the portion of the windrow lying between t_3, u_3 and t_4,u_4 subdivided to yield *np* points. (The exact location of the windrow is unknown of course—we retain the part between the projections of t_3, u_3 and t_4,u_4 on the windrow.) The windrow portion can be used as a parameter in a subsequent blow:

```
... repositioning commands ...
... parameter setting commands ...
blow [ ts,us, portion, te,ue];
set portion: getwindrow [ t3,u3, t4,u4, np ];
... etc...
```

Note how the facility to define vectors as a list of elements has been used to define the blow start point t_s, u_s and the blow end point t_e, u_e explicitly. This is a typical and important requirement.

This technique also provides more flexibility; windrows can be retrieved from different blows and combined for later blows (figure 4.29).

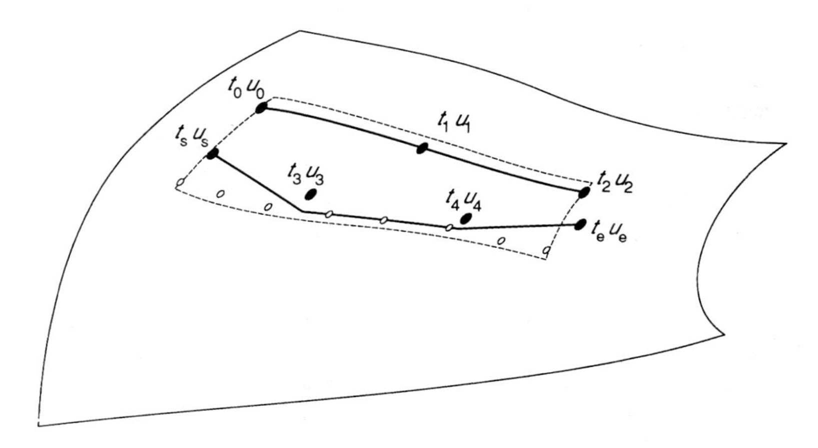

Figure 4.29 Modified and simplified windrow manipulation—refer to text.

Motion control timing

There are few papers in the robotics literature which discuss motion control timing beyond consideration of the computation time required for, say, a given inverse kinematics solution. The issue involves practical implementation which is beyond the scope of most articles dealing with theory.

Any robot which uses information derived from sensors to modify its behaviour is crucially dependent on timing for performance. It is not sufficient to use an inverse kinematics solution which takes, say, one millisecond if the total time taken to respond to a change in sensor output is of the order of 10 milliseconds or even 100. The stability and performance of a control system depends on the total computational delay and often it is the practical concerns of operating system overhead input/output response time and data communication bottlenecks which determine the performance more than the computation efficiency of the algorithms employed.

Recently there has been some focus on the use of transputers for robot control. These hardware devices look attractive for their built-in communications capabilities and the ease of implementing parallel processing algorithms. Sharkey *et al* (1989) discuss a transputer implementation of computed torque control of a PUMA 560 robot in a force control task. The communications timing problem was one of the most challenging they faced and the architecture of the controller had to be designed with that factor in mind rather than the computation problem. Computing capacity alone is not sufficient to provide adequate timing performance.

Computation times for a given algorithm can be shortened by computing slowly changing quantities less often. We exploit this technique by updating the arm kinematics (and the Jacobian transforms) about ten times less often than the actuator set points. Any minor errors incurred are corrected by responding to cutter sensors.

The SM robot uses a multiplexed communication system between the control computer and electronic circuits on the arm. Part of the robot performance specification called for a 2 millisecond maximum delay between resistance and capacitance sensor measurements and the arrival of an updated follower set point at the wrist electronics. The first step to achieve this was to arrange the multiplexing such that transmission of capacitance and resistance sensor signals to the computer immediately precedes the computation of the updated follower set point.

The multiplexing system is arranged such that completion of 45 input and output transfers triggers an interrupt which initiates the execution of the HEATHER software module controlling follower movement (figure 4.30). The completion interrupt needs to be generated at precise 5 millisecond intervals. The sequence of input/output transfers pauses after the first 18 channels until a pulse from a 5 millisecond crystal controlled clock allows the rest of the sequence to continue. The time-critical *inputs* to the system are arranged in multiplexor channels 19 through to 45 with the most critical being channels 41 to 45. Important *outputs* to the robot are arranged in channels 1 to 18.

The next important factor is the operating system overhead or interrupt response time. This is the time which passes between the hardware interrupt and entry to the HEATHER software. Depending on the hardware and operating system this can range from a few microseconds to as much as half a second. We used a commercial operating system—Hewlett Packard RTE-A—to provide a software development environment which allowed normal editing, compiling and debugging while operating the robot in real time. The typical response time to schedule a program within RTE-A can be as long as two milliseconds—clearly this method could not be used. However, RTE-A also provides facilities to embed user software within the operating system to obtain a faster response. The user software must save and restore control registers and initiate its own input/output transfers (referred to as housekeeping) but the operating system overhead can be almost eliminated.

We managed to reduce the housekeeping delay to about 100 microseconds. This time includes linkage from a skeleton control module within the operating system to the HEATHER software which is locked into a memory partition while the robot software is running. In essence, HEATHER is an extension of the operating system. This allows us to make changes to HEATHER without having to regenerate and reload the operating system.

HEATHER's execution time is either 0.6 or 1.3 milliseconds depending on

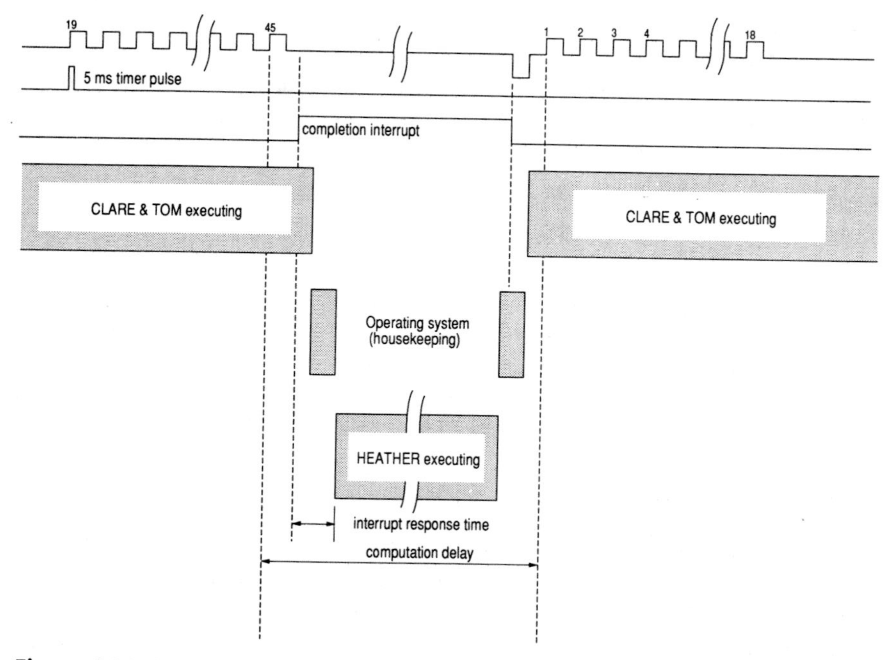

Figure 4.30 Arrangement of multiplexor timing and HEATHER execution—refer to text.

whether it interpolates the main actuator set points (every alternate cycle). These times include code for safety monitoring, data recording and error handling. The code within HEATHER is written in Fortran; the compiler is non-optimizing but it is possible with some discipline to write code which is almost as efficient as assembler. In this way we have been able to provide reasonable facilities for a research and development environment.

Motion step calculations

While timing is critical to reduce the computation delays in HEATHER it is also equally important for TOM in the context of the motion step cycle every 80 milliseconds. These calculations embrace Cartesian space motion control, surface model calculations and the inverse kinematics solution. Figure 4.31 illustrates the timing sequence.

We calculate the pose at the start of each motion step using direct forward kinematics vector algebra. Adaptation to sensed errors introduces a problem here; the interpolation sequence initiated by the last motion step is not yet completed. We therefore assume that no further sensor corrections will be added during the remaining interpolation steps. The actuator positions needed for the pose calculation are estimated by adding the sensor corrections which have taken place so far to the planned actuator positions at the end of the current sequence.

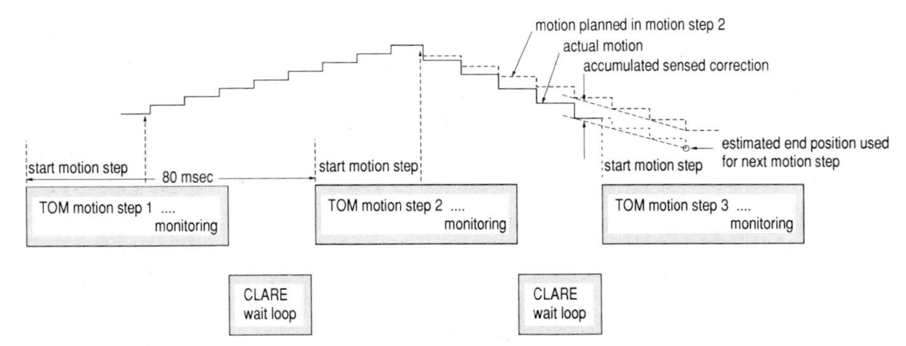

Figure 4.31 Motion step timing. The top stepped line shows the interpolated output for a typical actuator. HEATHER does the interpolation, but its execution is not shown to avoid cluttering the diagram. The actual motion varies from the planned motion because of the sensed corrections computed at each interpolation step.

The Jacobian transforms which are used elsewhere are by-products of the forward kinematics step.

Next we calculate the desired pose change in Cartesian space using the methods described in this chapter. We also need to update the projections of points on the comb (**t**,**w** and **s**) on the surface model as described in chapter 3.

Finally, we use the inverse kinematics to calculate desired changes in actuator positions.

We have used several forms of inverse kinematic solution. The ORACLE robot employed a closed form direct solution; the mechanical arrangement of the arm was constrained partly by a perceived need to avoid calculating transcendental functions (sine, cosine, tangent, etc.). The arm solution required only three square root calculations. The wrist, on the other hand, was solved using a direct formulation of the inverse Jacobian. This required some transcendental function calls. Note that singularities were avoided through design (see chapter 8).

For much of the time ORACLE was operated close to motion limits—particularly the tilt actuator. We did not find an entirely satisfactory way of avoiding sudden jerks under these conditions while using a direct solution method.

With a faster computer, we chose to experiment with a pseudo-inverse control exploiting the kinematic redundancy of the robot—it has 8 actuators in all. Peter Kovesi successfully implemented this technique which was used in the latter stages of its life (Trevelyan *et al* 1983).

The SM robot posed some intriguing inverse kinematic problems which are discussed more fully in chapter 8. The ET wrist does not have a fixed orientation centre and so a closed form solution is not easy to derive. We therefore implemented an iterative technique based on the resolved rate motion control method.

There were some intriguing differences between these techniques in the way corrections from the sensors were handled. In the case of the direct solution used for ORACLE we maintained two pose vectors. One was the actual pose and the other was a nominal pose assuming that the follower actuator was at its central position. The motion control system working in Cartesian space operated on the nominal pose position from which the inverse kinematics were calculated. However, the desired movement of the nominal pose included a component proportional to the deviation of the follower actuator so that this deviation would be progressively eliminated.

The resolved rate motion control technique permitted a simpler approach. Correction of the arm position for follower deviation takes place at the interpolation level (with a much more favourable computation time delay) and only one pose needs to be calculated. The motion control software in Cartesian space does not need to take follower deviations into account.

Notice that the follower axis for the ORACLE robot is not necessarily perpendicular to the skin (figure 2.10, p. 39). This means that a significant follower deviation can cause a significant lateral deviation of the cutter. This had to be allowed for in the Cartesian space calculations. We disposed of this added complication with some relief when we moved to the SM robot and relocated the follower axis to the end of the wrist.

Once the desired actuator changes have been calculated and checked for limits and excessive speed a message is transmitted to start the next interpolation sequence. Under steady conditions this message should arrive just as the previous sequence has been completed. A single-step queue was set up to allow for occasional disruptions so that the timing sequence does not have to be rigidly adhered to.

Once this message has been despatched TOM continues with hardware monitoring operations. Actuator errors and positions are monitored to detect servo failures. Several memory locations, particularly those between central or shared memory data blocks, are monitored to detect possible corruption of shared memory data. Actuator drift is monitored and correction terms are calculated to compensate for thermal drift. The shearing drag force is calculated—since this is not a time-critical parameter it can be handled within the monitoring routines.

Finally, data from the motion step calculations is recorded and TOM goes dormant until being reawakened in time for the next motion step.

On the useful shelf . . . collision avoidance and motion limit controls

In 1983 we completed the ARAMP manipulation platform (chapter 7) to hold sheep for shearing experiments. It was soon apparent that there was a danger of collisions between the ORACLE robot and many parts of the manipulator.

Peter Kovesi designed and implemented an on-line collision avoidance technique to prevent such collisions (Kovesi 1985). The robot end-effector and the obstacles are modelled by spheres or cylinders with spherical ends. As the robot approaches the vicinity of an obstacle, the component of velocity towards the obstacle is progressively removed, avoiding a collision.

While this technique worked well, we managed to overlook a most important factor; even with almost perfect guidance around obstacles, we still could not obtain good enough access to the sheep! The ORACLE cutter was quite small (figure 5.9, p. 139) but adding a commercial compressed air motor drive, a proprietary roll actuator, suitable bearings and a transfer gear box produced a rather large end-effector (figure 2.10, p. 39).

The SM robot cutter was much smaller, and we found that we could keep clear of the ARAMP obstacles. Once the SLAMP concept emerged, with no major obstacles for the robot to collide with, our obstacle avoidance software was placed, with some relief, on the 'useful shelf'.

Motion limits

Our shearing and repositioning trajectories often pass outside the workspace of the robot, at least for a short distance. This affects wrist orientation more frequently than the cutter position. There are no major practical consequences, provided the motion control software keeps the actual robot pose as close as possible to the desired pose (which cannot be reached of course). Under these conditions normal inverse kinematics calculations exhibit ill-conditioned behaviour—the calculated joint angles result in a pose which is rather different from the 'closest' pose in Cartesian space.

Piero Velletri developed ideas from Peter Kovesi and showed how this problem could be neatly overcome (Velletri 1989). In essence, the Jacobian matrix is used to calculate the position of an approximate obstacle plane (in six-dimensional space) corresponding to each motion limit. The velocity component towards the closest plane is progressively removed as the end-effector position approaches the plane. Even though the demanded end effector position passes through the plane, the actual robot pose remains at the closest attainable and there are no sudden velocity changes. Though the computation time and memory increase is small, we have now reached the limits of the A900 processor and so this useful gem is being kept on the 'useful shelf' until we can incorporate it into the motion control software.

Appendix

ON THE USE OF B-SPLINES FOR ADAPTIVE ROBOT PATH PLANNING

An edited paper by P. Kovesi, G. Walker and J. Trevelyan

This paper describes the use of B-splines for generating robot trajectories. B-spline trajectories have been chosen for their flexibility, intuitive specification, and ease of editing. In developing the use of B-splines for robot trajectories, techniques were devised for:

(1) changing an existing path to pass through a specified position in a specified direction;

(2) separation of velocity and position control;

(3) easy interpolation of parameters along trajectories.

An introduction to the B-spline

The shape of a B-spline curve is controlled by a series of control points, the curve order, and a knot vector. The control points can be considered as vertices of a polygon. The relationship between the defining polygon and the curve is governed by a set of weighting functions. The weighting or basis functions are local. Each control point affects the shape of the curve only over the range of parameter values where its associated basis function is non-zero. Conversely each point on the curve is affected by only a few of the control points. This range of influence is determined by the order of the curve. The curves are piecewise polynomial splines with degree one less than the curve order. For example, a third-order B-spline curve would be represented by a series of quadratic segments. Continuity is maintained up to the (order-2) derivative.

Parametrization along the curve is controlled by a knot vector. The knot vector is defined by a set of positive integers x_i, so that $x_i \leq x_{i+1}$ for all x_i. The first and last k elements are equal where k is the order of the B-spline. Duplicate intermediate knot elements can be introduced to alter the curve characteristics. We define a knot vector with no coincident intermediate knot elements and with unit spacing between elements.

The position **q** along a B-spline is defined by a parameter t, n+1 control points $\mathbf{q}_i$ and n+1 weighting functions $N_{i,k}(t)$

$$\mathbf{q}(t) = \sum_{i=0}^{n} \mathbf{q}_i N_{i,k}(t). \tag{4A.1}$$

The B-spline weighting functions can be defined recursively as (de Boor, 1972):

$$N_{i,k}(t) = (t - x_i)\frac{N_{i,k-1}(t)}{(x_{i+k-1} - x_i)} + (x_{i+k} - t)\frac{N_{i+1,k-1}(t)}{(x_{i+k} - x_{i+1})} \tag{4A.2}$$

$$N_{i,1}(t) = \begin{cases} 1 & \text{if } x_i < t < x_{i+1} \\ 0 & \text{otherwise} \end{cases}$$

where n is the numberof control points less 1
x_i are the elements of the knot vector
t is the B-spline parameter value (0 to t_{max})
t_{max} = (n-k+2) when the knot vector is defined as above.

For more information on B-splines the reader is referred to Rogers and Adams (1985) and Faux and Pratt (1979).

Changing a B-spline curve to pass through a specified position

One problem associated with the use of B-spline curves is that they do not pass through the defining control point positions. If one requires the curve to pass through some specified point and possibly also in a specified direction then special techniques for the placement of control points are needed.

A simple technique which can be used for a third-order curve exploits the fact that a third-order B-spline is tangential to the mid-points of the defining polygon. Hence by specifying appropriate positions of control points on either side of the desired path point the desired trajectory can be achieved. To achieve a straight path segment then several points would have to be placed in a line. As the curve order increases so does the number of control points that must be added.

In some cases increasing the number of control points is not desirable. An example of this would be when editing an existing robot movement where adding extra control points would require the trajectory to be completely respecified and replanned.

To avoid this problem an alternative technique has been developed that calculates the minimum change required in existing control point positions in order to force a B-spline curve to pass through some specified point in space. This technique can be applied to curves of any order, and defined by any number of control points.

Some point $\mathbf{q}(t)$ on the original curve has been forced on to a new point $\mathbf{q}^*(t)$ by moving the control point $\mathbf{q}_i$ to some new position $\mathbf{q}^*_i$. The assumption is made that the parameter value t at the position $\mathbf{q}$ and the position $\mathbf{q}^*$ is the same.

The positions $\mathbf{q}$ and $\mathbf{q}^*$ can be written in terms of the control point positions.

$$\mathbf{q}^*(t) = \sum_{i=0}^{n} \mathbf{q}_i^* \, N_{i,k}(t) \tag{4A.3}$$

$$\mathbf{q}(t) = \sum_{i=0}^{n} \mathbf{q}_i \, N_{i,k}(t) \tag{4A.4}$$

$$\Delta\mathbf{q}(t) = \sum_{i=0}^{n} \Delta\mathbf{q}_i N_{i,k}(t). \qquad (4.A4)–(4.A3)$$

Now consider the function g where:

$$g(\Delta\mathbf{q}_0, \Delta\mathbf{q}_1, \Delta\mathbf{q}_n) = \sum_{i=0}^{n} \Delta\mathbf{q}_i N_{i,k}(t) - \Delta\mathbf{q}(t) = 0. \tag{4A.5}$$

Attempting to minimize $\Sigma(\Delta\mathbf{q}_i)^2$ supplies an additional constraint.

A solution to the above can now be obtained by using the method of Lagrange multipliers.

We attempt to minimize:

$$H(\Delta \mathbf{q}_0, \ldots \Delta \mathbf{q}_n, \lambda) = \sum_{i=0}^{n} (\Delta \mathbf{q}_i)^2 + \lambda[\sum_{i=0}^{n} \Delta \mathbf{q}_i N_{i,k}(t) - \Delta \mathbf{q}(t)]. \quad (4A.6)$$

Solving for the $\Delta \mathbf{q}_{i\text{'s}}$:

$$\Delta P_i = \frac{-N_{i,k}(t)\,\Delta \mathbf{q}(t)}{\sum_{j=0}^{n} (N_{j,k}(t))^2}. \quad (4A.7)$$

An expression for $\Delta \mathbf{q}(t)$ can be developed:

$$\begin{aligned}\Delta \mathbf{q}(t) &= \mathbf{q}^*(t) - \mathbf{q}(t) \\ &= \mathbf{q}^*(t) - \sum_{j=0}^{n} \mathbf{q}_j N_{j,k}(t).\end{aligned} \quad (4A.8)$$

Using this the solution for the new control point positions can now be written:

$$\mathbf{q}_i^* = \Delta \mathbf{q}_i + \mathbf{q}_i \quad (4A.9)$$

$$\mathbf{q}_i^* = \frac{N_{i,k}(t)[\sum_{j=0}^{n} \mathbf{q}_j N_{j,k}(t) - q^*(t)]}{\sum_{j=0}^{n} (N_{j,k}(t))^2} + \mathbf{q}_i. \quad (4A.10)$$

It is clear that only the control points with non-zero weighting functions will experience position changes. The 'range of influence' that the change will have on a curve is therefore a function of the curve order.

Changing a B-spline curve to have a given tangent vector at some specified position

A variant of the position changing algorithm can also be used to calculate the change required in control point positions in order to force a B-spline curve to have a tangent vector of specified direction at some specified point in space.

The tangent vector of a B-spline curve at some parameter value t can be written as (De Boor, 1972):

$$\mathbf{q}'(t) = (k-1)\sum_{i=0}^{n} \mathbf{q}_i^{(1)} N_{i,k-1}(t) \quad (4A.11)$$

where $\mathbf{q}_i^{(1)} = (\mathbf{q}_i - \mathbf{q}_{i-1}) / (x_{i+k-1} - x_i)$. (4A.12)

This expression for the B-spline slope can be rewritten as:

$$\mathbf{q}'(t) = (k-1) \sum_{i=0}^{n} \left\{ \frac{N_{i,k-1}(t)}{(x_{i+k-1} - x_i)} + \frac{N_{i+1,k-1}(t)}{(x_{i+k} - x_{i+1})} \right\} \mathbf{q}_i. \quad (4A.13)$$

By letting

$$\mathbf{M}_i(t) = \left\{ \frac{N_{i,k-1}(t)}{(x_{i+k-1} - x_i)} + \frac{N_{i+1,k-1}(t)}{(x_{i+k} - x_{i+1})} \right\} (k\text{-}1) \qquad (4A.14)$$

we now have:

$$\mathbf{q}'(t) = \sum_{i=0}^{n} M_i(t) A_i. \qquad (4A.15)$$

This is of the same form as the B-spline position formulation. The solution for the new control point positions is therefore:

$$\mathbf{q}_i^* = \frac{M_i(t)[\sum_{j=0}^{n} \mathbf{q}_j M_j(t) - q'^*(t)]}{\sum_{j=0}^{n} (M_j(t))^2} + \mathbf{q}_i. \qquad (4A.16)$$

where $\mathbf{q}'^*(t)$ is the desired slope at the parameter value t.

As before only the control points with non-zero basis functions will be moved. There is no position change for the point on the curve corresponding to the parameter value t.

Parameter variation along B-spline trajectories

The need for a robust technique for velocity control and smooth interpolation of parameters along a shearing blow has led to the idea of a generalized control point. Rather than have only position associated with a B-spline control point we now allow any number of steering or sensing parameters.

$$\mathbf{q}(x,y,z,p_0,\ldots,p_n)\,\Big|_t = \sum_{i=0}^{n} \mathbf{q}_i(x,y,z,p_0,\ldots,p_n)\, N_{i,k}(t) \qquad (4A.17)$$

where $\mathbf{q}_i(x, y, z, p_0, \ldots, p_n)$ is the i'th control point.
x, y, z are the Cartesian position of the control point.
$p_0, \ldots, p_n$ are parameter values associated with these control point positions, e.g. velocity, acceleration, steering parameters.

The B-spline weighting functions are used to interpolate both position and any other parameters associated with the generalized control point.

Using such a technique has several advantages.

(1) It provides a simple technique for smoothly interpolating parameters along a trajectory.

(2) Specification of parameters along a path becomes intuitive and straightforward.

On-line tracking of B-spline trajectories

The task of the real-time robot motion control program is to calculate, at each computation step, the change required in the robot end-effector position to achieve the desired velocity and position. This process must include appropriate corrections so as to eliminate tracking errors from the previous calculation step.

The following algorithm has been used.

Notation:

t_{last}	–	parameter value at the end of the last step
δt_{last}	–	parameter increment used at the last step
δt	–	parameter increment this step
$\mathbf{v}(t)$	–	velocity at t
$\mathbf{p}_{\text{last}}$	–	end-effector position at the end of the last step
$\delta\mathbf{p}$	–	change in end-effector position this step
$\mathbf{e}$	–	tracking error from the last step
$\mathrm{d}t$	–	the duration of the computation step (80 msecs).

The current parameter value is incremented by δt_{last}.

$$t = t_{\text{last}} + \delta t_{\text{last}}$$

The velocity $\mathbf{v}(t)$ and position $\mathbf{q}(t)$ are calculated at this value using the B-spline weighting functions as described in the previous section.

The distance *aDis* between $\mathbf{q}(t)$ and $\mathbf{q}(t_{\text{last}})$ is calculated along with the desired distance to be moved this step, *dDis*.

$$aDis = |\mathbf{q}(t) - \mathbf{q}(t_{\text{last}})|$$
$$dDis = \mathbf{v}(t)*\mathrm{dt}.$$

A new value of t which will produce the desired movement is then estimated.

$$\delta t = \delta t_{\text{last}} * \frac{dDis}{aDis} \tag{4A.18}$$

$$t = t_{\text{last}} + \delta t. \tag{4A.19}$$

A correction for the tracking error must be made (the magnitude and direction of the correction allowed is limited). The required change in robot position this step can then be calculated.

$$\mathbf{e} = \mathbf{p}_{\text{last}} - \mathbf{q}(t_{\text{last}}) \tag{4A.20}$$

$$\delta\mathbf{p} = \mathbf{q}(t) - \mathbf{q}(t_{\text{last}}) - \mathbf{e} \tag{4A.21}$$

$$\mathbf{p} = \mathbf{p}_{\text{last}} + \delta\mathbf{p} \tag{4A.22}$$

Care must be taken to ensure that the above algorithm remains stable. This is done by limiting the maximum value of (*dDis*/*aDis*) and the magnitude of the tracking error correction.

The velocity profiles which result from the use of the above algorithm and the generalized control point are very smooth.

Experimental results

B-spline trajectories have been successfully used for the control of reposition movements in a number of shearing trials with the SM robot. The velocity control technique has been recently implemented and the resultant smooth velocity and acceleration profiles have enabled robot repositions to be executed at higher speeds than was previously possible. Velocity during the landing phase of a reposition can be achieved independently of the velocity for the rest of the reposition move resulting in smoother and more reliable transitions to and from shearing blows.

Rotations

An orientation change is specified by a new frame $\mathbf{A}_{new}$ to which the wrist must be rotated. We find an axis $\hat{\mathbf{a}}$ and angle ϕ which will rotate **As**, the current cutter frame, to $\mathbf{A}_{new}$. $\mathbf{A}_{new}$ is considered to consist of three column vectors $\mathbf{i}_{new}$, $\mathbf{j}_{new}$ and $\mathbf{k}_{new}$ just as **As** consists of **i**, **j**, and **k**. We have favoured this technique over alternative techniques in the literature in which the wrist actuator angles are simply interpolated because we usually want to turn the cutter in a Cartesian plane more or less tangential to the sheep surface nearby. Wrist angle interpolation did not work.

In theory $(\mathbf{i}_{new} - \mathbf{i}) \times (\mathbf{j}_{new} - \mathbf{j})$ can be used to calculate $\hat{\mathbf{a}}$ but the result may be ill conditioned; $\mathbf{i}_{new}$ may be coincident with, or opposite to **i**. To avoid this problem, we calculate $(\mathbf{j}_{new} - \mathbf{j}) \times (\mathbf{k}_{new} - \mathbf{k})$, $(\mathbf{i}_{new} - \mathbf{i}) \times (\mathbf{j}_{new} - \mathbf{j})$, and $(\mathbf{k}_{new} - \mathbf{k}) \times (\mathbf{i}_{new} - \mathbf{i})$ and use the cross product which gives the greatest resulting magnitude. For example, if $|(\mathbf{i}_{new} - \mathbf{i}) \times (\mathbf{j}_{new} - \mathbf{j})|$ is the greatest of the three,

$$\hat{\mathbf{a}} := \mathbf{norm}((\mathbf{i}_{new} - \mathbf{i}) \times (\mathbf{j}_{new} - \mathbf{j})).$$

If the largest magnitude is small (< 0.01) then it is because $\mathbf{A}_{new}$ and **As** are very nearly coincident or very nearly opposite to each other. The latter case is treated as an error; there is only a small chance that this will occur in practice and it can easily be avoided by programming an intermediate orientation movement. The former case can be handled by a special computational method which we call twitching.

The angle of rotation is calculated by:

$$\phi := \text{sign}(\ \text{atan2}(\ |\mathbf{e} \times \mathbf{f}|, \mathbf{e} \cdot \mathbf{f}\ ),\ (\mathbf{e} \times \mathbf{f}) \cdot \hat{\mathbf{a}}\ ) \qquad (4A.23)$$

where $\mathbf{e} = \hat{\mathbf{a}} \times \mathbf{j}_{new}$ and $\mathbf{f} = \hat{\mathbf{a}} \times \mathbf{j}$.

If $\hat{\mathbf{a}} \cdot \mathbf{j}$ is greater than 0.9 (meaning that this too may be illconditioned and **a** may be coincident with **j**) then **i** is used instead to calculate **e** and **f**.

Frame rotation

Quarternions are used for rotation of vectors. To rotate a vector **v** about axis $\hat{\mathbf{a}}$ through angle f, we use

$$\mathbf{v} := \sin\phi\ (\hat{\mathbf{a}} \times \mathbf{v}) + \cos\phi\ \mathbf{v} + (1 - \cos\phi)(\mathbf{v} \cdot \hat{\mathbf{a}})\ \hat{\mathbf{a}}. \qquad (4A.24)$$

We rotate a frame by treating each column as a unit vector and using eqn (4A.24) to rotate the vector.

Multiple turns of wrist

The *W1* axis can turn more than 360° and therefore a new angle for the cutter may be provided by more than one *W1* position. It is important for the progam-

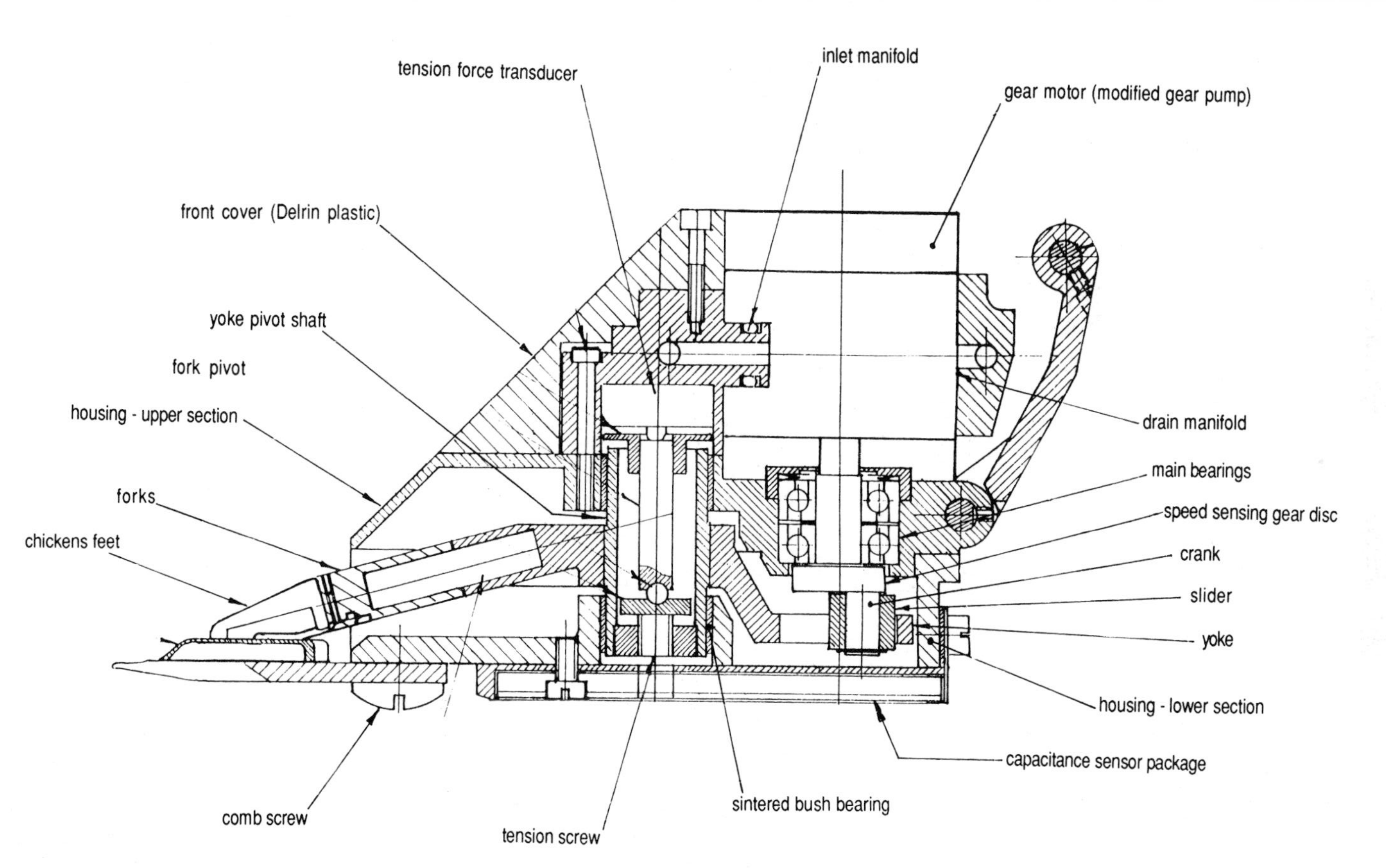

Figure 5.10 SM robot cutter mechanism 1985. Refer to text for details.

tensioning adjustment screw. An electronic tension force transducer was also installed and we soon found that the tension force was remarkably stable; there was no need for further adjustment provided it was checked before each sheep was shorn. The hydraulic tensioning device was never installed. But there were other problems.

From earlier tests, we knew the motor would be tricky to restart under load. The motor had first to be allowed to drain, and then the full oil flow had to be applied suddenly to start it turning. Once turning, the dynamic lubrication effects reduced friction, and the oil flow could be eased back to obtain the desired running speed. All this was reasonably easy to set up by using software to control the servo valve.

It was soon apparent that the cutter needed more power. David had purchased a slightly larger gear pump as a precaution, and soon had it ready for testing. We were all dismayed to find it would not start at all. Patient investigation by an honours student revealed that the main output shaft was being deflected by the start-up oil pressure and there was insufficient torque to overcome friction (Hochstadt 1986). Meanwhile, the original motor was performing adequately, even after conversion to wide shearing combs.[5]

David realized that a major redesign was needed but other priorities delayed this until 1990. Once again a cautious approach was necessary and the work is still to be completed. Yet it is often the case that an enforced delay leads to insights which might have been otherwise missed. David has now adopted a direct reciprocating drive and has devised an arrangement which is twice as powerful, yet half the size of its predecessor. Figure 5.11 shows a size comparison between the manual handpiece, the original SM cutter and the new 'interim cutter' arrangement.

Much of the size reduction has been used to raise the underside of the rear end of the cutter so that for the first time, we will be able to shear with the comb almost flat on the skin. We now know enough to realize that this is essential for fast shearing on Merino sheep.

Future developments

Because the cutter is such a critical part of any shearing robot, it will be worth undertaking a costly and lengthy development phase to refine the design. The

[5] The standard comb had 10 teeth and the cutter had three. Wider combs were popular in New Zealand. In 1982 and 1983 there was a bitter dispute among Australian shearers over the allowable width of the comb. Since the 1920's the shearer's union had enforced a ban on wider combs, initially at the bidding of farmers. By the 1980's New Zealand shearers had mastered the use of wide combs on Australian sheep and were popular with woolgrowers as they ignored some restrictive union rules. The union attempted to outlaw the wider combs by withdrawing their labour. When New Zealand and Western Australian shearers moved in to the Eastern States sheds there was violence. The dispute was resolved in the Arbitration Commission; the union lost and wide combs were accepted. Ironically, all the combs were manufactured in Australia.

Because of the dispute, we had some difficulties obtaining wide combs for shearing experiments. Nor did we want to antagonize the shearers' union unnecessarily. After 1986 we adopted wide shearing gear for shearing trials.

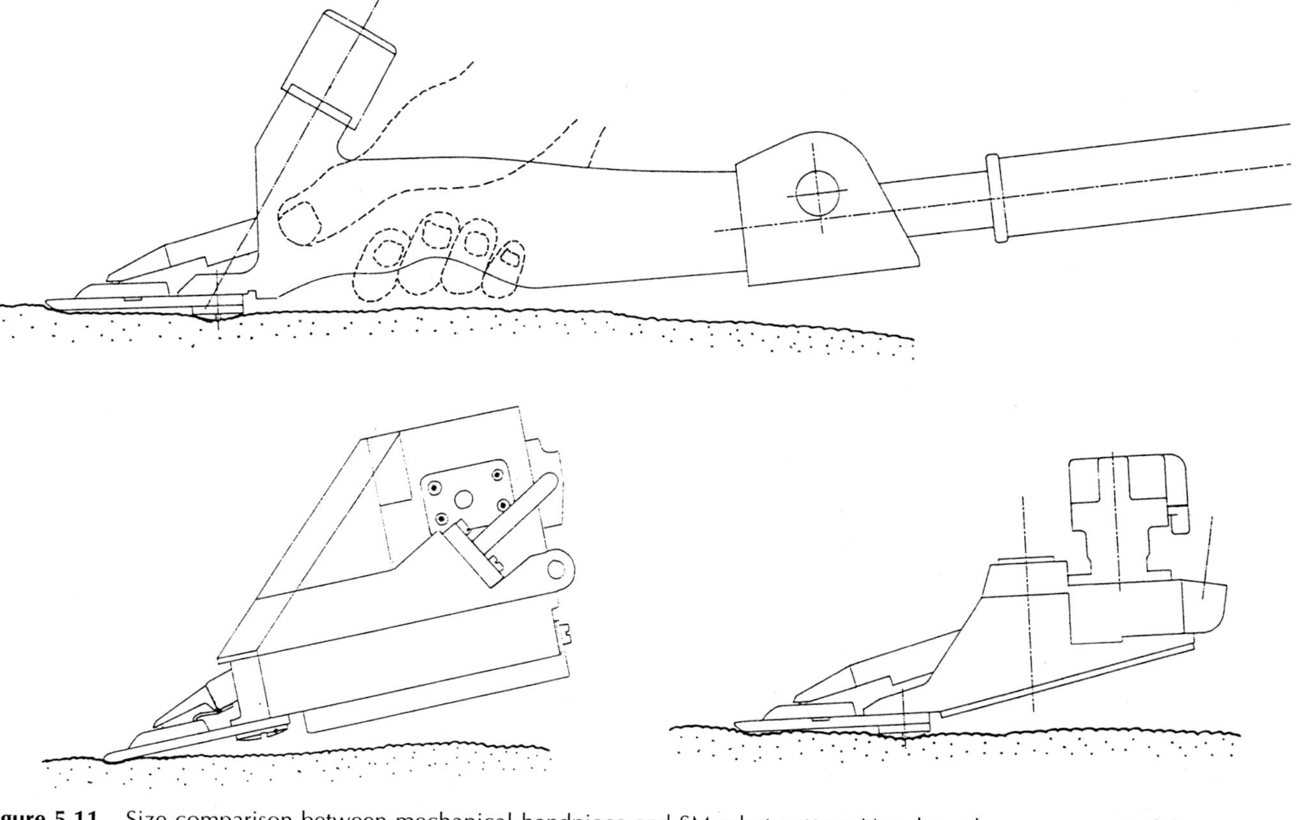

Figure 5.11 Size comparison between mechanical handpiece and SM robot cutters. Note how the arrangement of the sensor package on the underside of the first SM cutter dictates a comb angle of up to 10°. The thinner sensor and sloping underside of the interim cutter design permit a flat comb angle.

cutter vibrates and runs in a hot, dusty, dirty and greasy environment laden with wool fibres. We have only focussed our efforts on cutter design to the extent necessary for our on-going research. It would be a mistake to presume that our work is anything but a useful starting point.

Our work with cutters underscores the importance of process knowledge for robotics researchers. We have to know about cutters to make a sheep-shearing robot work, and we were unable to avoid the necessity of acquiring 'deep' knowledge and experience. To work with robots, it is necessary to be expert with both the robots and the application. Since this is rarely achieved by individuals, teamwork and collaboration skills are essential prerequisites.

We have not been concerned with automatic comb and cutter changing in our work; however, we have collected numerous ideas to make this a feasible proposition. New developments in comb and cutter materials and surface coatings may provide very long-lasting tools which do not need to be replaced so often; this would be a great simplification.

6
Sensors

Sensors are a crucial part of a sheep-shearing robot, and there are tighter constraints on sensor design than on any other aspect of the machinery.

Wool's virtues make it opaque to almost all forms of sensing. The density of its interlocking fibres defeated the wheels of the PATS machine; sound absorption defeated ultrasonic ranging; thermal properties defeated infra-red imaging; electrical properties defeated microwave ranging. Inorganic salts in the sheep's natural lanolin protection almost defeated capacitance sensing; its texture confuses machine vision; and high moisture content distorts capacitance and contact resistance sensing signals.

The cutter needs to be as simple as possible, and no sensing device can project into the path of the wool flowing around the cutter. Any device mounted on the cutter must be resistant to heat, vibration, oil, grease, dust, chafed particles of wool, grass seeds and cleaning fluids. It must be robust enough to survive encounters with horns and other boney projections.

Four forms of skin and touch sensing device have been used on the cutter so far; force sensing pads, resistance contact sensing, capacitance sensing and inductance sensing. Sensors used adjacent to the cutter include ultrasonic radar and drag force sensing. Remote sensing devices have included vibration sensors, machine vision and passive infra-red imaging.

It might be surprising then that there has been less research on sensors than on any other aspect of the machinery. We have developed robust, simple and reliable touch and proximity sensing devices because there are few suitable alternatives. Most of our effort has been focussed on processing the information from the sensors. Adding more sensors could increase the reliability of sensed information, but only at the expense of increased processing time. The increase in processing time is likely to be proportional to the number of sensors squared because each combination of sensor signals has to be examined.

This chapter explores some of the shearing-specific sensors we have developed. Several conventional sensors are used too, such as vision, and strain gauges for pressure and force measurements. Vision is explored in chapter 10; the image processing problem is more interesting and deserves a chapter in its own right.

Resistance contact sensing

In figure 6.1 we see the basic resistance sensing circuit. We have built many variants during the years of research—as many as five different approaches in the first six months of shearing trials.

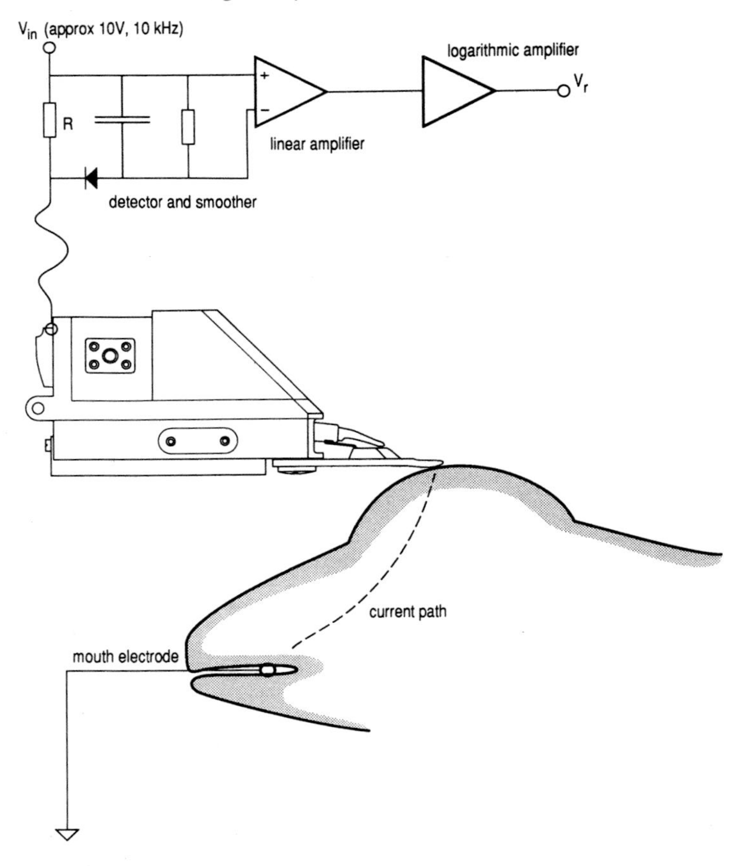

The earliest circuits used direct current. The sheep was isolated from the robot and a voltage of about 10V was applied to the animal through a mouth electrode. This was little more than a piece of stainless steel wire held in the mouth by elastic around the head.

We experimented on ourselves too—but not with the same mouth electrodes! We observed polarization effects—the resistance measured would increase with time and was polarity dependent. This was no surprise to experts in physiology who use skin resistance measurements to observe neural activity. When we discussed our problems with them, they showed us special silver electrodes, connected to the skin with a special jelly. They used AC circuits too; their aim was to eliminate the contact resistance variations which we wanted to observe. Nevertheless, we adopted an AC circuit with frequencies between 1kHz and 10kHz (Greenhalgh *et al*, patent, 1981).

The resistance levels varied enormously. On normal sheep skin, beneath a full growth of fleece wool, a typical resistance level between a shearing comb and the skin was about 2000 ohms. Yet on dry skin, covered by thin scraggy

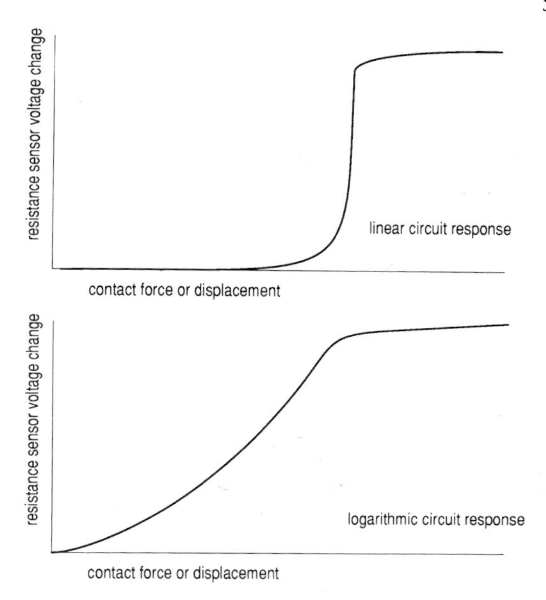

Figure 6.1 Resistance contact sensing principles. The graphs show notional characteristics only for the purpose of illustrating the effect of the logarithmic amplifier. The actual characteristics are erratic and time-dependent.

fleece, the level could be as high as 500,000 ohms. Human skin, being exposed to the air, tended towards the higher end of the range. If we cut the skin, the resistance dropped to about 500 ohms, but the levels varied from one sheep to another, and during different shearing conditions.

Hand shearing experiments often revealed low resistance values, even lower than skin cut readings. However, there was nearly always significant noise—erratic signal variation—when the skin was not being cut. We have supposed that this is caused by cutter teeth 'dancing' on the skin. The closing force between the comb and the cutter is sufficient to deflect the comb teeth, but the deflection changes as the comb sweeps across different teeth from side to side. Thus each comb tooth oscillates vertically while the cutter is running. The whole cutter mechanism also vibrates sideways due to the inertia of the moving parts. All this contributes to contact noise. However, when the skin is cut, at least part of the comb is in consistent contact with body fluids and much of the fluctuation disappears, fortunately providing a second means of sensing skin cuts.

We incorporated a logarithmic amplifier to expand the useful range of the sensor output signal.

Once we commissioned the ARAMP manipulator (chapter 7), we encountered electrical isolation problems. Although the metal structure was isolated from the floor and the robot by plastic spacers, sensor wires ran throughout, and one wire of each pair was connected to ground. Gradually and unavoidably, urine and other excretions penetrated into the most inaccessible places, and reduced the isolation to the stage where we were forced to reverse our sensing circuit. We isolated the robot cutter instead and grounded the sheep.

It was not easy to include an electrical isolation barrier. To maintain a mechanically stiff connection, we used a thin plastic spacer between metal components and under connecting bolt washers. On the SM robot, the barrier was inserted between the wrist yoke and the W1 actuator shaft (figure 8.31, p. 245). Occasionally, metal particles seemed to lodge on the gap causing intermittent resistance sensing failures. The outer case of the capacitance sensor package was also grounded and had to be isolated from the cutter. This too was a vulnerable spot and metal wear particles sometimes accumulated here, bridging the gap intermittently.

After a few weeks of operation, more inexplicable intermittent 'spikes' appeared on the resistance sensing when the robot was in motion. We had initially associated these with metal particles. Each time we noticed them we cleaned the wrist and cutter with solvent, without eliminating the problem. After some days of frustration, we found that the spikes were caused by sudden contact changes in the bearings between elements of the wrist and cutter mechanism. Initially, the bearing surfaces were rough enough to provide constant metal-to-metal contact. As the robot was used, the surface imperfections were rolled or polished away, and a complete oil film would form, temporarily breaking electrical contact. All the wrist elements were isolated from the rest of the robot by the spacer in the yoke. Yet each element had its own capacitance, and even though we were using a comparatively low frequency (10kHz) this was sufficient to cause spikes when the elements made or broke contact with each other. The problem was solved by connecting each of the major mechanical parts with bonding wires.

Refinements

Many refinements were proposed to solve some of the intrinsic problems of resistance sensing, particularly to be able to distinguish the location of any skin contact on the comb. All of these required some modifications to the comb. And as long as the comb remained part of the severance mechanism and was subject to wear and resharpening, the idea of a non-standard comb was likely to be an expensive choice.

Two possibilities remain.

If the life of the comb could be extended by surface treatment or new materials, non-standard equipment would be less of a problem. It might then be feasible to integrate sensing electrodes into the comb structure such that they could connect with suitable electrical contacts on the comb mounting.

At the present time, there are no suitable surface treatments.

For many years, Merino Wool Harvesting experimented with D-Gun carbide coatings developed by Union Carbide. The top (rubbing) surface of the comb is bombarded with high-velocity carbide particles in a vacuum chamber. It was claimed that the resulting surface continually sharpened the cutter as well. Comb lifetimes of hundreds of sheep were claimed, but we are not aware of extended field trials supporting these claims. There were problems with coating adhesion, and distortion of the comb during bombardment. The roughened surface finish at the comb tips can interfere with combing though masking can alleviate this problem.

More recently, several laboratories are experimenting with ion implantation coatings, laser surface hardening and titanium nitride coatings.

There have been several attempts to produce ceramic cutters, none so far successful. Combs are much more slender and complex, and it is unlikely that ceramic combs could be feasible before cutters can be produced satisfactorily.

The second possibility is to incorporate a comb which is separate from the severance mechanism and hence does not have to be regularly replaced. The rotary cutter (figure 5.8, p. 138) incorporates this idea, although the makers are still using metal combs at the moment. Plastic combs might be a feasible alternative.

Wool and skin properties

The resistance sensor is affected by both skin and wool characteristics; also to a secondary extent by the internal body resistance of the sheep and the connection between the sheep's body and the mouth electrode.

One method we used to relate skin deformation to resistance variation was to vibrate the cutter up and down using the follower actuator. In this way, we could estimate the slope of the notional curve relating skin deformation to voltage change under dynamic conditions. It was not a helpful experiment except to indicate how much the slope varied from one part of the sheep to another. Skin covered by a thick layer of fleece was moist and soft and the rise in voltage was steep. Skin exposed to the air, or covered by short fleece, was dry and there was less rise in voltage. In these areas, there was a big difference between sheep; some were dry and scaly and others were moist and soft; just as with humans. Some parts of the sheep are subjected to abrasion and wear, for example the brisket between the front legs, and these can be covered by a thick layer of non-conductive dead skin, or can be moist and wet if the sheep has been lying on damp ground. The conductivity of freshly shorn skin falls noticeably in a few minutes, presumably due to drying.

Later in the programme, an on-line slope estimator was included in the resistance sensing position control scheme (figure 4.11, p. 93). At each sensing step, we calculated the first differences in the resistance sensing voltage and the commanded follower displacement. We took the absolute value of the differences and passed them through first-order smoothing filters with a time constant of about 0.35 sec. The ratio of the filter outputs provided an estimate of the slope. We used this to modify the sensitivity parameter relating position error to the sensor voltage output.

Ultimately we judged the usefulness of the resistance sensor from the shearing performance of the robot. And this was inextricably linked to our understanding of shearing skill and the nature of the shearing comb. It is the comb teeth which provide the electrical contact with the sheep's skin and the way the teeth run over the skin determines the signal received. As none of these considerations affects the nature of the sensing device, I have deferred the rest of this discussion to chapter 11 in which I explore shearing skills.

Wool conductivity can interfere with resistance sensing under extreme conditions. However, the effect is much more pronounced on the capacitance sensors.

From time to time we have experimented with phase measurements in the resistance sensor, but we have no useful results yet.

Capacitance sensors

David Henshaw devised the first capacitance sensors and tested them on his shearing machine in 1978. They were far from satisfactory and were soon discarded. We were fortunate to have an enigmatic electronics genius close at hand. Duncan Steven was a friend of Jim Blair, and had a habit of designing electronic circuits on a scrap of paper which was pulped the next time his shirt was washed. As a lecturer in the Department of Electronic Engineering, he wished that he could have insisted that all his students be able to handle a soldering iron *before* they started his course.

Duncan designed a simple voltage divider arrangement using germanium diodes for their low capacity. For simplicity the circuit was driven by a 455 kHz oscillator from a conventional radio (figure 6.2). It provided a good measure of distance from zero to 12mm. The design was rugged and required only a single miniature coaxial cable carrying AC power to the sensor; the output appeared as a DC offset on the AC power supply. We suspected that the device would be extremely sensitive to EMI (electromagnetic interference or radio interference) but it was not. The sheep provided good shielding.

The sensor was very sensitive to temperature changes, due to the germanium diodes and within a few months, we were able to obtain silicon-based Schottky diodes which were much more stable (Baranski *et al*, patent, 1982).

The design was inherently robust, and our prototype sensors survived many scrapes against our dummy tin sheep and sheep supports.

Effects of wool properties

The major weakness was a sensitivity to wool contamination. This first became apparent towards the end of 1979, but it was not until 1981 that we were really confronted with the problem. One particular sheep appeared with wool which was highly reactive to the sensors. Named after its ear tag Red 8, it was to become famous in another respect too.

We constructed a special probe consisting of several capacitance sensors directed down and sideways to measure the electrical properties of the wool.

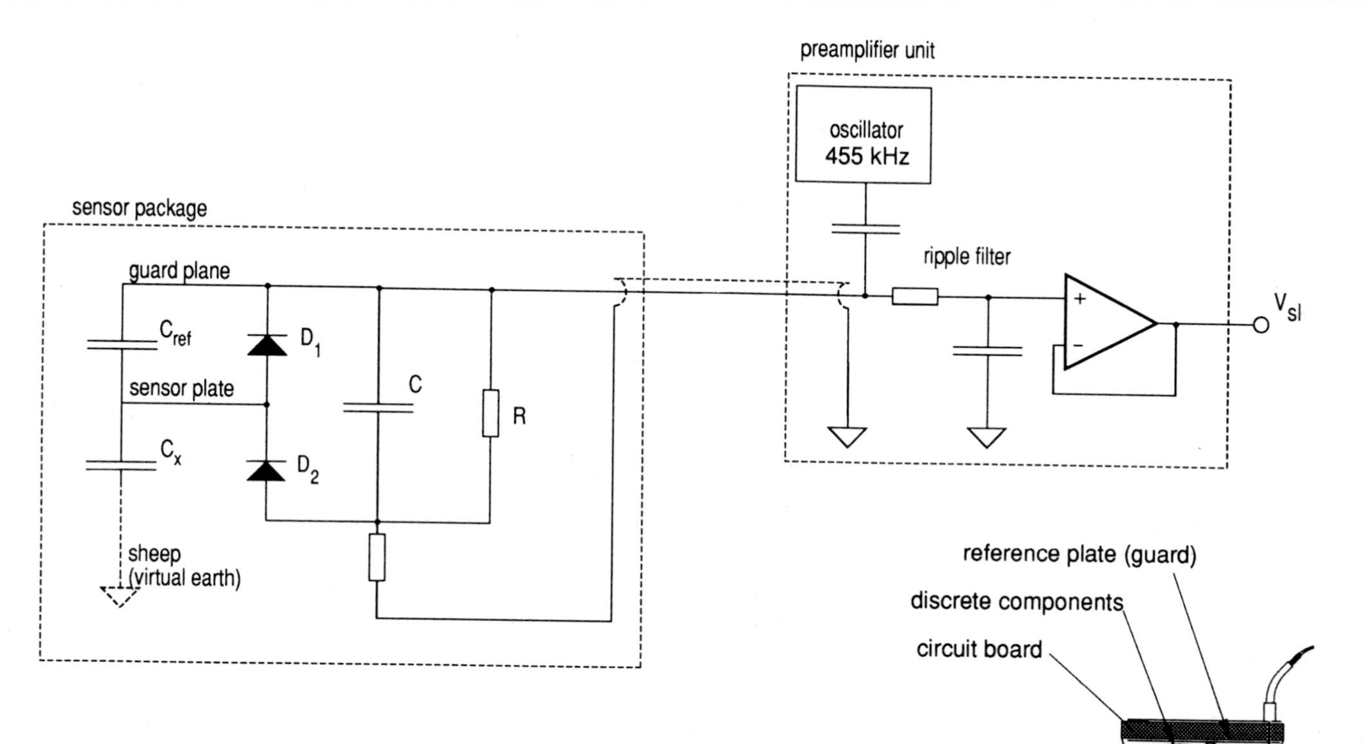

Figure 6.2 Capacitance sensor (1978). The sensor was connected to the oscillator and amplifier unit by a single miniature coaxial cable, approximately 1 metre long. AC power was transmitted to the sensor and the signal returned as a DC offset. Low-capacitance diodes are essential. The device operates with a range of up to 12mm. The frequency was determined by readily available RF components.

But before it was finished, Red 8 was shorn by mistake and the new coat of wool failed to show the problem.

Red 8 was always noticeably nervous, agitated, almost neurotic, particularly when taken on the 2 km truck ride from the animal house to the laboratory pen. None of the usual ovine acquiescence for him.

After growing a new fleece we brought him back for more tests, but he had rather different intentions. No sooner had he been transferred to the little laboratory pen, then he started jumping and trying to escape. Seconds later, he was out and bounding down the laneway towards freedom, across the car park and the street beyond, having little regard for normal traffic regulations.

Three robot engineers in white coats set off in desperate pursuit and, following advice from passers-by, converged on a dilapidated weatherboard house on the far side of the street. The house had been a practice clinic for psychiatry students, but was overdue for demolition. Conferring, they decided that one of them would creep round each side of the house, while the third kept watch at the front door, swinging open in the breeze. The first two met up in the back yard, finding it empty. They checked the fences and peered into the neighbouring gardens; the fence was too high. So they followed up the only escape route—the open back door of the house.

Creeping inside, they checked the rooms until at last they found the errant patient—crouching on the one remaining piece of furniture—a psychiatrist's couch!

Reactive wool

Red 8's wool came to be known as 'Reactive Wool'—wool to which the capacitance sensors reacted strongly. The electrical effect was caused by high conductivity to high-frequency AC. The DC conductivity was often insignificant. The result was that the sensors measured the distance to the wool rather than the skin.

Whereas fleece wool is usually a brilliant white under the surface grime, reactive wool is usually discoloured from grey through to brown and yellow. The problem usually appears in warm humid weather, typically in spring or early autumn. The strength can vary from one day to the next, and a sheep with a reactive fleece does not necessarily grow another similar fleece the next year.

The first quantitative observation was that reactive wool absorbed more water; up to 30% by weight compared with up to 10% for normal wool. A high suint content was also typical (inorganic salts in wool grease).

It was soon clear that we did not have sufficient physiological expertise to understand the nature of the problem. I chose to research countermeasures because an as yet unknown number of sheep would be affected. Once we understood the extent of the problem, we could select the most appropriate countermeasures.

We thought that sweating would contribute to the problem. However, we learned from a sheep physiologist that sheep do not use sweating to keep cool.

If anything, the moisture from sweating *warms* the sheep when the wool absorbs the moisture.[1] We also learned that sheep sweat in bursts between 10 and 20 minutes apart. They keep cool by panting.

We thought that wool reactivity was affected by stress. Sheep were affected by transport, accommodation, feed; in short almost every aspect of our experimental situation could affect the condition of the sheep. This was a relief because we could defer serious investigations until field trials.

The first countermeasure was inductance sensing and we showed that this was at least five times less sensitive to reactive wool. However, for reasons which will become clear, we have continued to use capacitance sensors for our shearing trials.

We also devised an effective mechanical shield for the capacitance sensors. Two versions were used. The first consisted of a flexible plastic shield which pressed wool on to the skin under the sensors. This way, the distance measured to the skin was almost correct. If the wool was stubble, the error was very small indeed. The shield was not as effective as I would have liked. Wool to the side of the cutter could still affect the sensor, and when the cutter turned, or moved backwards close to the sheep for repositioning, wool could become caught between the shield and the sensor (figure 6.3).

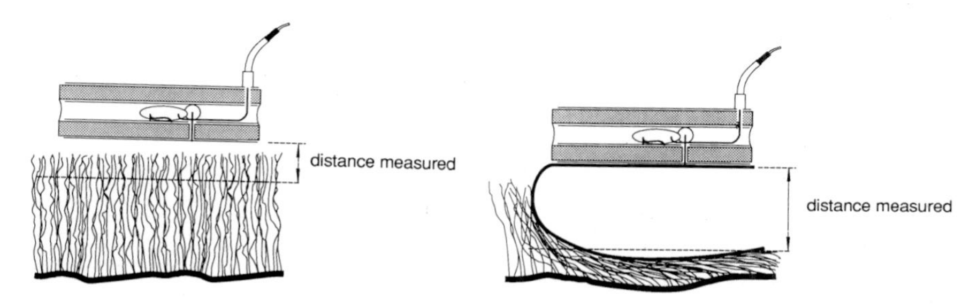

Figure 6.3 Effect of reactive wool and mechanical shield—refer to text.

I devised a pneumatic shield which was much more successful but could pose some interesting production problems. The shield is inflated by a low-pressure air mover using the Coanda effect; a small volume of high-pressure air can be used to drive a large volume of low-pressure air. It has the advantage that the inflation pressure is more or less constant, and is not greatly affected by volume changes. Quite large leaks can be tolerated (figure 6.4).

The air inflates a U-shaped sausage which is contained within a teflon coated outer bag. The outer bag defines the inflated shape and the teflon coating reduces wool drag (Trevelyan, patent, 1985).

[1] The most striking instance of this which I have observed was on leaving a long flight to Singapore wearing a woollen jumper which had been desiccated by the dry cabin air. As it absorbed the humidity in the air the wool released a distinguishable dry warmth. However, I soon realised I could do without that!

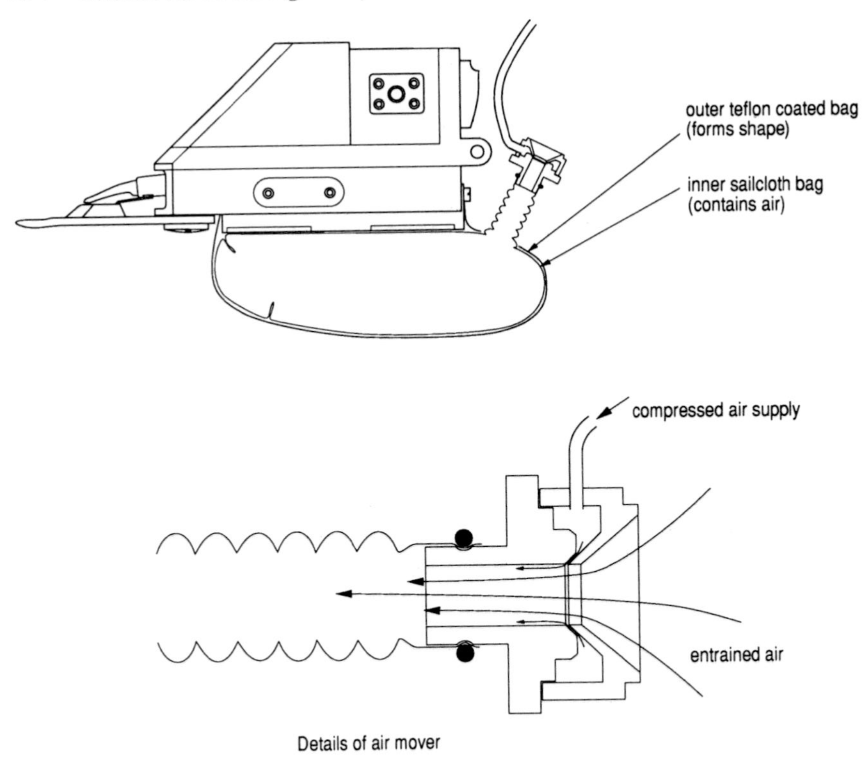

Figure 6.4 Air bag shield.

The bag inflates sideways as well as underneath the sensor. This way, wool to the side is fended off.

The air bag has worked well, except when the cutter is turned in the wool. Then a large amount of wool is forced against the side of the cutter and the air pressure in the bag is not sufficient to flatten the bulk of the wool. However, we can arrange the shearing pattern to avoid this problem.

More recently, the hybrid follower control scheme has greatly lessened the impact of conductive and reactive wool. Improvements in cutter design allowing a flatter shearing attitude have relaxed attitude error requirements. We have yet to confirm results with extended shearing trials, but we expect that most reactive wool sheep can be shorn now using capacitance sensors. In short, the reactive wool problem seems largely to have been overcome, but we cannot be sure without extensive field trials.

Circuit refinements

Jan Baranski joined the team in 1982 to develop the inductance sensor described later. Once he had pushed this as far as limited resources permitted, he turned to the capacitance sensors, looking for refinements. The original design used only passive components but five years later, some active components had emerged as alternative possibilities.

The sensitivity and miniaturization of analogue components had advanced as rapidly as the complexity of digital components and at the beginning of 1985 Jan designed a package combining three independent sensors, a self-contained oscillator and detector elements, the J3 design (Baranski 1986*b*, Baranski, patent, 1987*b*).

Figure 6.5 depicts the circuit arrangement which was designed around a single integrated circuit containing four high-input impedance operational amplifiers. One amplifier was used for a square wave oscillator; each of the others amplified the output of a capacitive voltage divider to generate a DC signal from a diode detector. The design was a convenient arrangement; screened wires were sufficient to carry the DC power supply and signals to the wrist electronics. The package was very robust; although the nylon mounting screws sheared off twice during accidental collisions, the same sensor was used for six years without a failure.

The circuit was constructed on top of a double-sided circuit board; the sensor plates were on the underside of the board. The board thickness formed a reference capacitance between the copper coatings. The circuit was potted inside a thin metal box which was mounted within an insulating cover on the underside of the cutter assembly. The thickness of the sensor package was 8mm—more than I had wished. A thinner arrangement has since been devised but is still at the experimental stage.

Passive infra-red imaging

In 1980 we tested the use of infra-red imaging to detect the boundary between shorn and unshorn wool. We hired an AGA Thermovision camera for a day and the results looked promising, even if the camera price tag seemed daunting.

We have not tested the idea further. First, when the air temperature is close to body heat during the summer shearing season there will be no temperature difference to observe. Second, since the wool is such a good insulator, the temperature of the wool about half way between the skin and the outside is still almost the same as body heat. The shorn wool would have to stand exposed to cool air for a short time before a usable temperature difference could be obtained. Finally, the cost of a passive infra-red sensor has seemed a major obstacle.

Perhaps a decade later it would be worth a second look.

Vibration contact sensor

Michael Crooke joined us in 1981 to develop new sensing devices and design electronic aspects of the ARAMP manipulator. The main cause for concern was

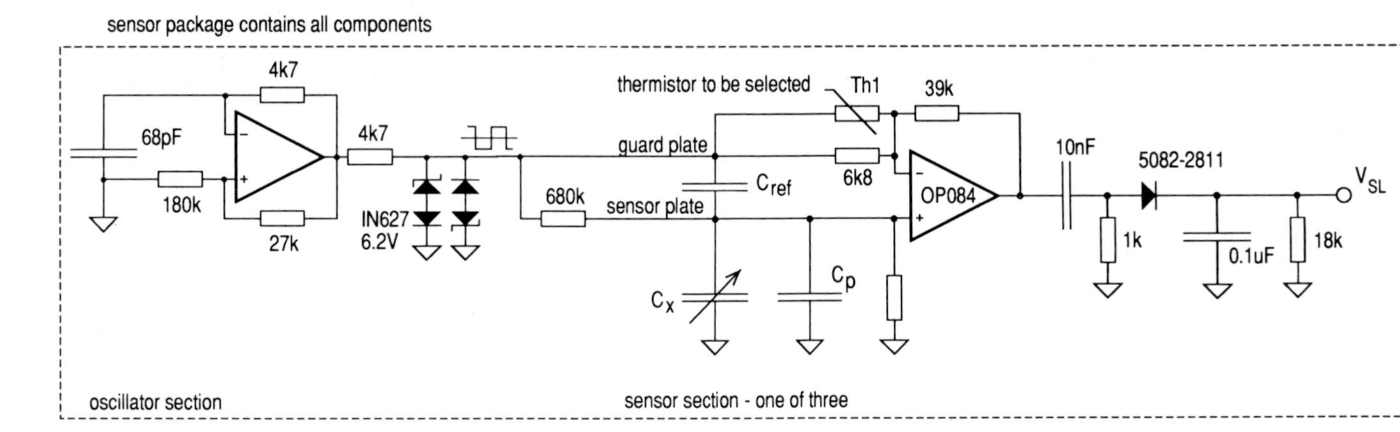

Figure 6.5 J3 capacitance sensor (1985). The circuit diagram illustrates only one sensing channel; two others were built into the same package. The output characteristic is given by

$$V_{sl} = K_x \, V_{osc} \, \frac{C_p + C_x}{C_p + C_x + C_{ref}} - V_p$$

where V_{osc} is the oscillator output,
K_x is a constant,
C_p is the parasitic capacitance to ground,
C_x is the capacitance to the sheep,
C_{ref} is the reference capacitance across the circuit board, and,
V_p is a further parasitic allowance.

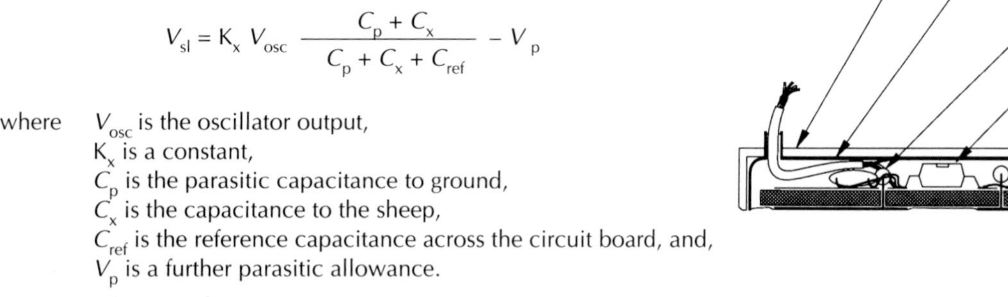

reactive wool which affected the capacitance sensors. However, if the conductivity of the wool was sufficiently high, the resistance contact sensor was also affected. Michael and I were discussing this problem when it occurred to us that the vibration of the cutter mechanism could be used for sensing contact with the sheep. When the cutter was pressed on to the sheep, the cutter vibrations were transmitted to the sheep's body.

Michael devised a cradle for the sheep which could be used to measure this vibration. The cradle was made from a lightweight steel plate mounted on a foam mattress. He mounted an accelerometer on the steel plate. Later, the accelerometer was mounted on a cantilever beam attached to the plate as figure 6.6 shows; the beam was tuned to the cutter frequency (Trevelyan and Crooke, patent, 1982).

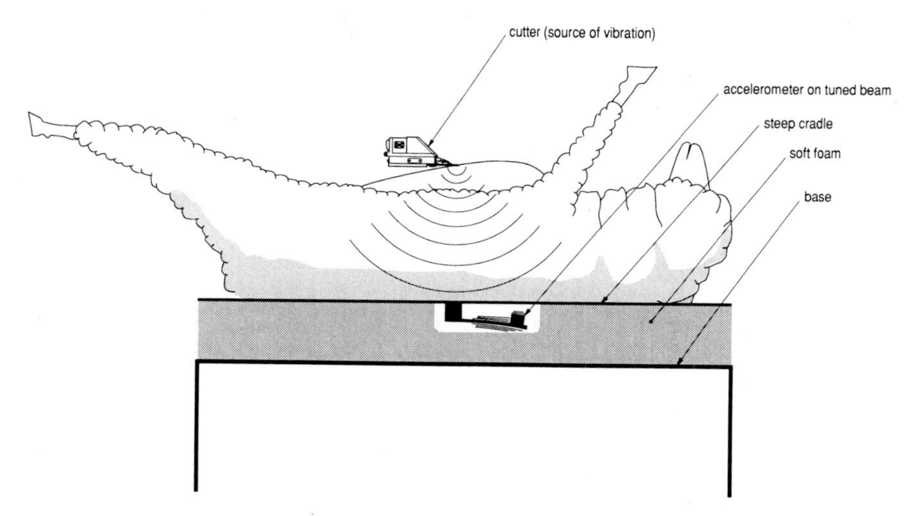

Figure 6.6 Vibration contact sensor.

The sensor worked, particularly when shearing the belly which was most likely to be affected by conductive wool. If the sheep urinated during shearing, the belly wool became very conductive. There were problems of course.

Vibration was transmitted to the sheep through the wool as well, particularly with dense fleece.

Vibration transmission from the extremities of the sheep was poor; presumably vibration sensors would need to be attached to leg clamps and other parts of the cradle.

Another problem was vibrations from within the sheep. We soon discovered that the sheep's heartbeat was easy to detect. Until we adopted the tuned beam, the heartbeat dominated the accelerometer output. But what if the sheep were to bleat during shearing? Michael failed to persuade a sheep to do this for him so when he thought no one was looking, he mounted the cradle himself and did his best to imitate a sheep.

He was overheard. In the office above, we wondered if a sheep had escaped since there were not supposed to be any sheep around that day. Unfortunately, no one was carrying a camera to record a sight for sore eyes; Michael lying spreadeagled on the cradle doing his best to sing 'baaa baaa black sheep'.

The results of the experiment were never recorded.

The cutter speed varied so we proposed variable-frequency electronic signal processing to replace the tuned beam.

The ARAMP design was well advanced by then and there was little prospect of changing it to accommodate the vibration sensor. By the time the next manipulator was designed, Michael had long since left the team and the idea was never revived. It looked exciting at the time but the complications seemed open-ended. The frequency response of the sensor was poor and the complications for cradle design were unwelcome. Finally, there was no prospect of distinguishing skin cuts; the resistance sensor was still the only means of doing this.

Inductance sensing

Inductance sensing had long been considered a possibility, but the relative ease of capacitance sensing held us back until the need became obvious. The principle is the same as that used by metal detectors except that the aim is to detect the skin of the sheep. The sheep skin conducts electricity so an alternating magnetic field from a nearby coil will set up alternating eddy currents. These change the characteristics of the coil.

Theory

An alternating magnetic field normal to a material penetrates to a notional depth of

$$\delta = \frac{2}{\omega\mu\gamma} \tag{6.1}$$

where ω is the frequency,
μ is the permeability, and,
γ is the conductivity of the material.

Eddy currents form in a circular path parallel to the surface. If the field emanates from a coil close to the material, the eddy currents affect the coil inductance.

The inductance sensor coil consists of a single circular turn and thus the eddy current is largely similar to the shape of the coil, forming in effect a second coil in the material.

The mutual inductance is given by (Baranski, 1983, 1986*a*, patent 1987*a*):

$$M = \frac{2\mu_0 r}{k}\left((1 - \frac{k^2}{2})\ J(k) - E(k)\right) \tag{6.2}$$

where $k = \dfrac{4r^2}{4r^2 + d^2}$

μ_0 is the effective permeability,
r is the coil radius
d is the distance between the coils, nominally $h + \delta$, and,
h is the distance from the surface.
J(k), E(k) are second-order elliptic integrals.

Assuming that $h \ll r$,

$$\Delta L = -\frac{1}{\mu_0 \gamma \delta}\left(\frac{\partial M}{\partial h}\right) - M. \tag{6.3}$$

Two sensors are needed—one on the right side and one on the left to measure roll and pitch errors. A third sensor is needed towards the rear.

The coil is printed on the underside of a piece of circuit board; the upper side is grounded and a radial Faraday shield is placed under the coil.

The ground plane affects the inductance as given in eqn (6.3) so the resonant frequency had to be found experimentally. The trimmer capacitor is altered to obtain a suitable operating frequency, typically 830 MHz.

As the sensor approaches the skin, the resonant frequency shifts, changing the voltage across the circuit. The sensor was found to work over a distance of up to 60mm from the skin.

Electrical conductivity, if present in the wool, exists largely along the fibres rather than between them. By its nature, the inductance sensor reacts to skin conductivity in the plane of the skin. Therefore, the sensor is not affected by wool conductivity as much as the capacitance sensors. Several experiments confirmed this result.*

Jan discovered many practical problems in his research. The first sensors were more or less self-contained with miniature oscillators. Frequency stability was important. The resonant coil frequency was affected not only by temperature changes, but also by changes in the moisture content of the plastic materials used for the coil formers and insulation. Jan used two coils, working on the well-established principles of a voltage divider. In the end though, this approach was defeated by coupling between adjacent sensors through the sheep. The coupling was only apparent when the sheep was close to both sensors, but the effect was to reverse the sensor output characteristic!

Jan simplified the sensors by using a remote oscillator. A splitter element was needed to distribute the RF power to each sensor via a detector box. Separate boxes were needed to obtain correct impedance matching. The prototype arrangement worked, but it was too bulky to be fitted to a robot for shearing trials (figure 6.7). We also found that any flexing of the short cables between the splitter, the detectors and the sensors distorted the sensor outputs. We requested additional funds to design a special purpose sensor incorporating the splitter and detectors in a single package but the Corporation never

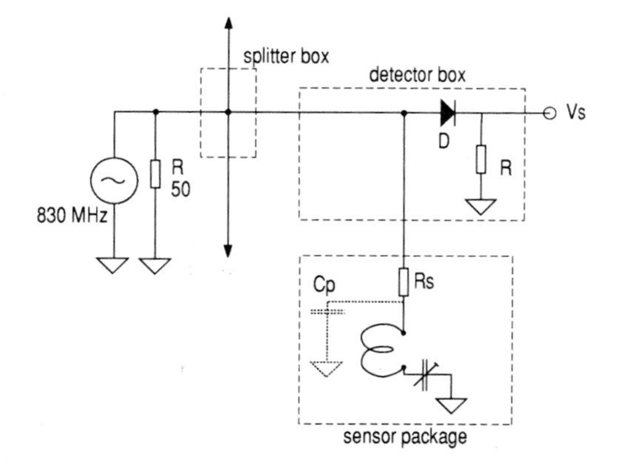

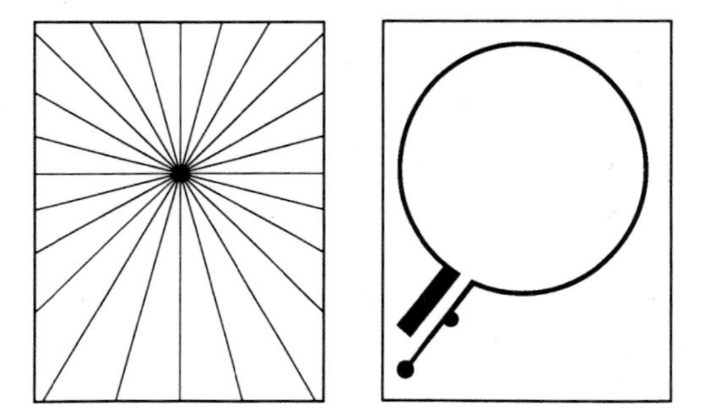

Figure 6.7 Inductance sensor (principles). The rectangles illustrate the circuit board patterns for the Faraday shield (left) and the single turn coil (right).

authorized funding. Improvements to capacitance sensors have allowed us to proceed with shearing trials in the meantime and presumably the science of microwave design and component design has also improved. We believe that such a device could be made when it is needed. But it would be comforting to be able to confirm that.

Ultrasonic ranging

Ultrasonic sensing of the skin ahead of the cutter was conceived at the outset of our research. The Corporation commissioned the Electrical Engineering department at the University of Adelaide to first investigate ultrasonic sensing and then develop a prototype for testing. This took several years and in the meantime our impatience for a forward sensor led us to use surface models to simulate a sensor. The surface models turned out to be very useful in any case and eventually served as an alternative.

The major problem with ultrasonic sensing is the sound absorbent properties of wool. The energy loss along the signal path about 240mm long (figure 6.8) was measured, with difficulty, as 120—130dB—comparable with the loss in a conventional radar. Early research was aimed at energy focussing devices and an ingenious ultrasonic lens was invented. Phased arrays were also evaluated and discarded for reasons which will shortly become apparent. The final choice was a spherical transmitter/receiver array coupled with digital signal processing hardware which allowed a single pulse to be spread in time, and then reconstituted after reception. The individual electrostatic transducer elements were specially designed; transmitting voltages of up to 1 kV were required testing the bond strength between the aluminium film coating and the special plastic film to its ultimate limits. Without individual amplifiers for each transmitting element, a phased array was inconceivable with such high operating voltages.

The performance was in the end too marginal. The major technical difficulties were specular main beam surface reflections, grass seed clutter in the middle layers of the fleece (where the attenuation on the transmit—receive path was so much less) and the highly variable fleece propagation characteristics. Some types of fleeces have cracks, like crevasses in glaciers, allowing sound to penetrate to the skin almost without loss. As the sensor moved over the fleece, the propagation loss for the skin echo would vary enormously. Other fleeces were dense without any apparent openings. Reverberation of the transmitted signal in the sensor mounting, specular side lobe reflections and cutter noise were also troublesome problems to resolve.

Despite the failure to achieve a useful result, the project provided many useful ideas, among them the ultrasonic lens (figure 6.9; Bognor and Nissinck 1981; Bryant *et al* 1983).

Force sensing

Resistance contact sensing is an indirect means of force sensing which has several advantages, yet the variations in skin and wool properties outweigh these. There have been several attempts to sense the contact forces between the comb and the sheep and to use these signals for controlling the cutter position.

Michael Ong proposed an ingenious method for measuring the skin contact forces on the outermost comb teeth. Figure 6.10 shows a comb with slots cut by spark erosion or laser cutting. The sections of the outer teeth isolated by the slot lie beyond the sweep of the cutter and are not subjected to the closing forces. However, contact with the skin causes them to bend; the bending can be measured by a spring loaded beam safely inside the cutter housing. There is no need to mount strain gauges directly on the comb.

The major difficulty with this method is to prevent wool fibres entering the slots. Michael proposed filling them with soft silicone rubber. He built a prototype, but it was never used in shearing trials.

Michael also proposed a strain gauge bridge on the trailing link of the SM follower mechanism. He asked David Elford to prepare a special link with a

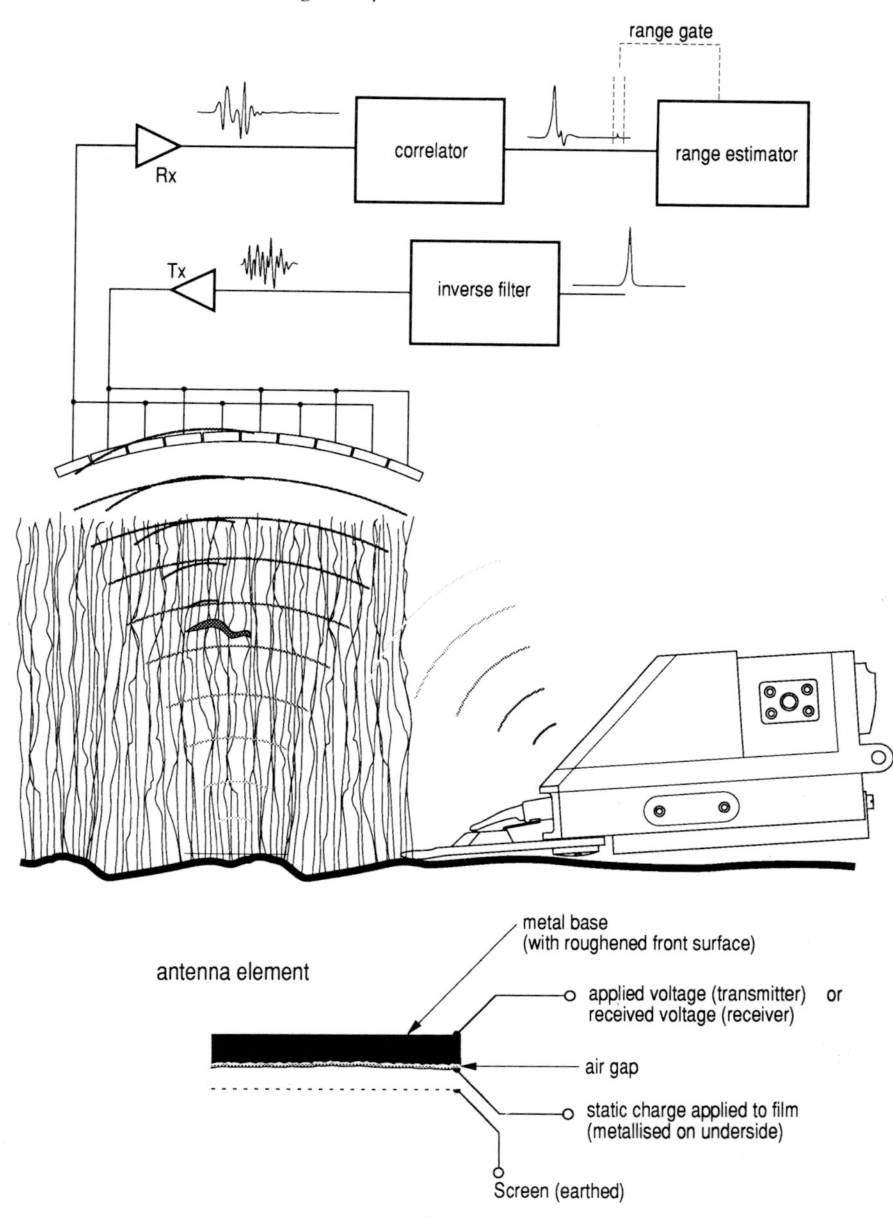

Figure 6.8 Ultrasonic range sensor (principles). The correlator and filter circuits were implemented with several TMS32010 digital signal processing integrated circuits, some custom designed circuits and microprocessors. The diagram illustrates one fundamental problem. The echo from the grass seed clutter high in the wool is much stronger (by orders of magnitude) than the skin echo which travels through more sound absorbent wool.

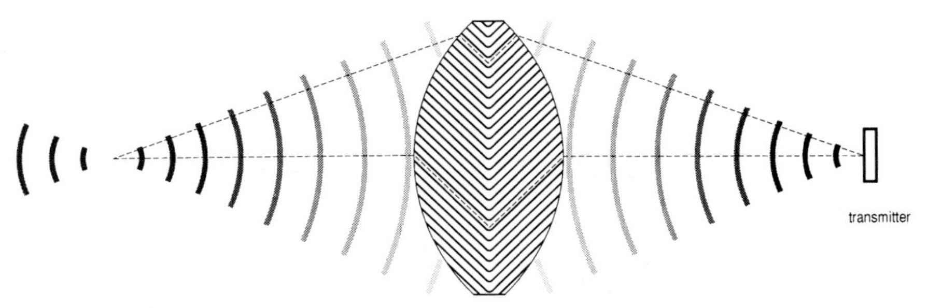

Figure 6.9 A lens suitable for focussing an ultrasonic beam.

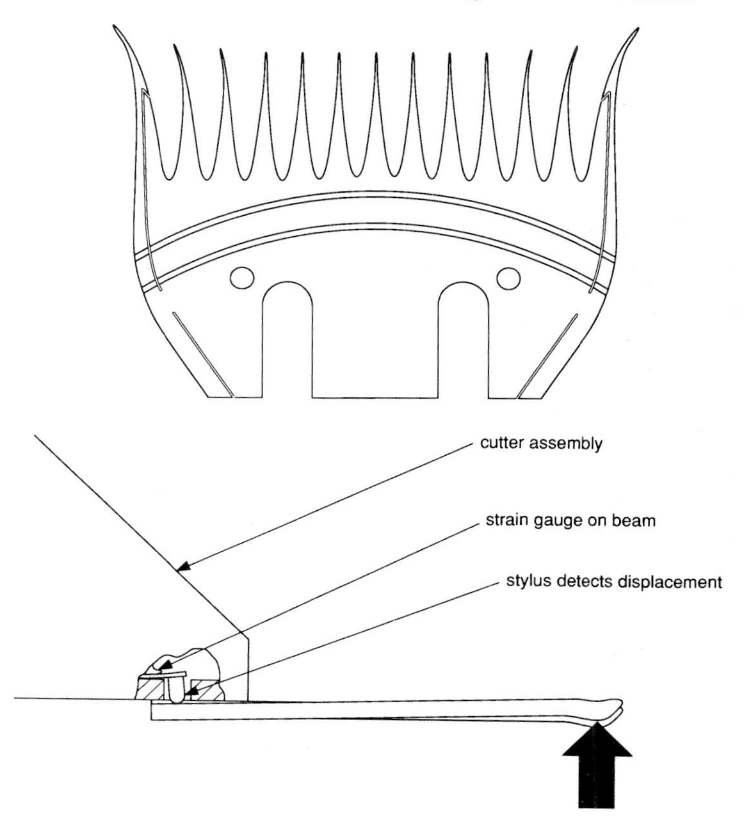

Figure 6.10 Contact force sensing comb proposal (1985). The comb is cut so that the outer teeth will bend slightly in response to skin force. The outer bending sections are away from the sweep of the oscillating cutter. This scheme was never used for shearing.

thinned section where the strain would be concentrated. David was reluctant to weaken the link, and was also concerned about additional wiring near the cutter. The weakened link was fitted shortly before Michael left the team, but the strain gauges were never fitted to the link. Instead they were bonded to a special set of wooden salad servers with the handles thinned as much as we dared; several metres of fine sensor wire were then installed on the salad servers and the set was presented to Michael on his departure! A few weeks later, a sheep heaved against the cutter at the wrong moment, destroying the thinned rear link. The original one was promptly replaced.

Michael proposed several other methods of force sensing (Ong 1985), but the only one which was implemented was measurement of shearing drag forces by measuring the differential hydraulic pressure on the wrist actuator cylinders. Although these measurements were affected by inertial effects, gravity and friction, the results were consistent enough to use for adapting shearing speed and detecting snags.

I veered away from force sensing because of the control problems which appear when one tries to make use of force signals. Inevitably, the stability is sensitive to the compliance of the sheep surface which varies from the hard boney legs and brisket to the softness of the abdomen. Merino Wool Harvesting attempted to use force sensing elements in their cutter mechanism to measure and control the skin contact force. They found they had to adopt a dynamically balanced cutter mechanism to reduce fatigue damage to their delicate strain measurement beams. A sensitive accelerometer was needed to compensate for gravitation and they had stability problems which were never resolved.

7
Holding the sheep—part 1

The first question most people ask me about robot shearing is 'How do you keep the sheep still?' Next comes 'How do you catch the sheep?' Finally 'How do you turn it over to shear the other side?'

On the surface we ignored these questions for the first three years because we started with 'How can a robot shear a sheep?' We built ORACLE to provide answers, and by 1981 we had them. We had found that ORACLE could shear any part of the sheep, given a reasonable surface model, and that the shearing speeds were fast enough to suggest, for the first time, that robot shearing might be a practical economic proposition.

At a deeper level, we had begun to answer the first questions too. Before we started building ORACLE, Roy Leslie had rephrased the first question in terms of keeping the sheep still in a 'good shearing position'. A human shearer holds the sheep in a series of positions; each one has been evolved to present one part of the sheep at a time in the best possible shearing position. The skin is firm, gently rounded and the skin is stretched to make shearing as easy as possible. Manipulation of the sheep into shearing positions was the main issue.

At the same time Roy appreciated what every woolgrower knows; a sheep which is firmly but comfortably restrained will not struggle. It seems to be one of the most intelligent behaviours exhibited by sheep. They keep still and relaxed, conserving their energy, until they feel a chance of escape. Other animals struggle furiously to the point of exhaustion. Sheep are not regarded as intelligent animals. But then there is a grudging respect from farmers illustrated by the following riddle. 'What animal is more stupid than a sheep?' Answer: 'A farmer who runs round after them trying to make money from them!'

Observation of sheep in the cradles at CSIRO and Muresk had confirmed that they remain still most of the time except when they are uncomfortable. Sheep struggling did not seem to be a problem. You will recall from our vision of ORACLE 'the robot would have to pause when the sheep struggled momentarily, before resuming'. On the odd occasion when a leg slipped out of a clamp, then vigorous struggling would follow as the sheep sensed the machinery undoing itself.

When the first phase of shearing trials with ORACLE was completed by early 1980, we had shown an adequate competence at shearing the back and side of the sheep supported in the original upright cradle. Roy, Stewart and I planned a new research programme for shearing each part of the sheep. The animal would be supported by fixed cradles, each one holding the sheep in a favoured shearing position. The issue of manipulating the sheep from one position to the next would come later. We needed all our resources focussed on the issue of whether the robot could shear all parts of the sheep, and how it was going to do it.

Chairs and tables

We devised three cradles.

The first was an inclined table for side and back shearing (figure 7.1). Here we were again shearing the back and side of the sheep, but this time the sheep was better supported. The upright cradle allowed the sheep to sway from side to side, and the side bars supporting the middle of the sheep prevented shearing far enough down the side to reach the beginning of the belly. The legs, particularly the front legs, needed to be more relaxed.

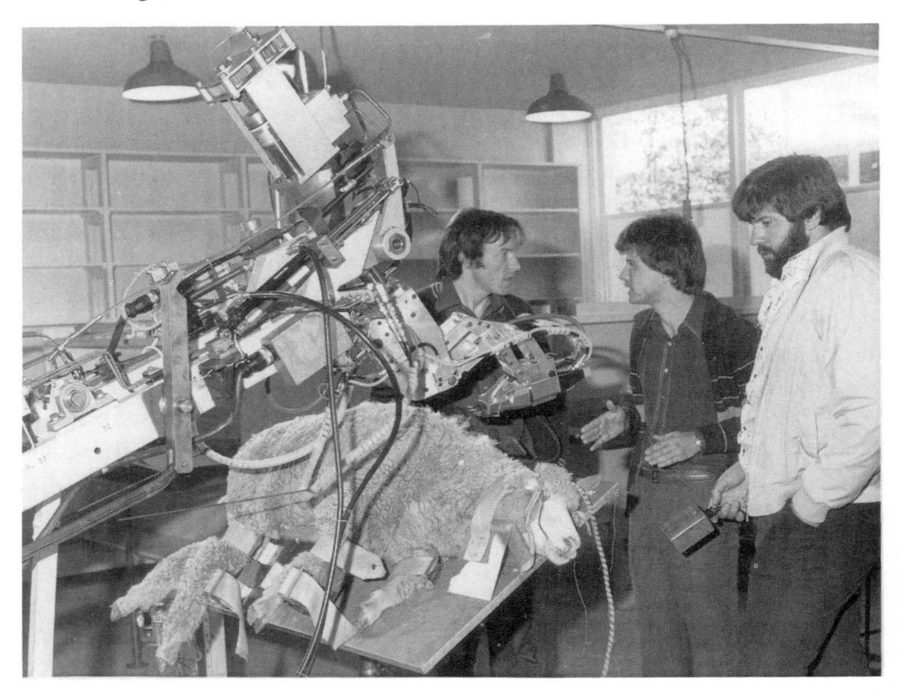

Figure 7.1 Shearing trial mid 1980. From left: author, Stewart Key, Roy Leslie. The sheep is supported on a side shearing table. The mouth electrode forms part of the resistance sensing circuit.

The second cradle was for belly shearing. Here the sheep was lying on its back in a trough with the legs held down (figure 7.2). We exploited some of the known 'pressure points' on the sheep. By pressing down firmly on the back legs, just above the knee joint (called the stifle joint on a sheep) the back legs were effectively straightened and immobilized.

In this position, the normally floppy abdomen is taut; even if the tummy muscles relax, the belly is well supported and constrained. There is excellent

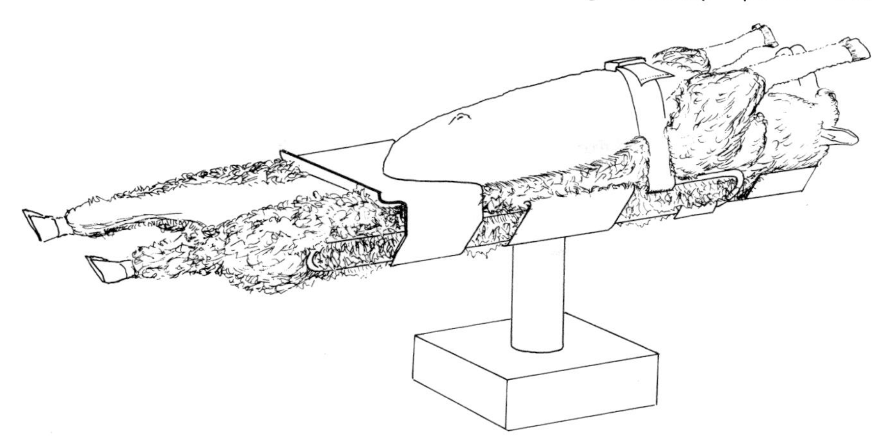

Figure 7.2 Belly shearing cradle. Note the guard plate which covers the teats and presses on the stifle joint on the rear legs, keeping them extended and stiff.

access for the robot, and the extremities of the rear legs could also be shorn if we wished.

The last cradle was affectionately known as the 'baa baa's chair'. Here I feel bound to include a delightful poem on a parallel theme—the 'Bush Barber' by Dixie Solly.

The sheep sat in the chair with its neck folded over an arm on one side, exposing the one side of the neck for easy shearing (figure 7.3). The neck of the Merino sheep has large folds of skin down the sides and along the bottom; these can only be shorn after the skin has been stretched by bending the neck. Even then, the dewlap along the underside of the neck unfolds as a long strip of free skin with no flesh inside it.

We sheared the neck of the first sheep in the chair by hand so we could clearly see how the robot would behave on the delicate dewlap. We crouched down to watch as the cutter moved gently over the arch of the neck; the pressure on the skin seemed imperceptible. We stood back, thrilled as each piece of stubble came off—the delicate skin unmarked. The resistance sensing allowed the comb to follow the skin flap so lightly that it hardly moved. Three woolly sheep followed, each with near perfect results.

From elation to frustration. The next three sheep we tried to fit into the chair had short necks. No matter how we tried, they simply would not fit; their necks would just not bend far enough over the side arm. After this experience, we modified the side shearing table so that shorter necks could be bent over a neck rest.

The trial programme was soon finished, and we felt comfortable with excellent shearing results. From our experimental results we predicted an eventual shearing time of two minutes for about 85% of the wool using the shearing patterns shown in figure 7.4 (Key 1981). We expected that the remaining wool

The Bush Barber

Reproduced by permission

He was wandering past a barber's shop,
When a notice he did spy,
It read "Wanted Gents Hairdresser,"
"Only Good Man Need Apply."

He said "I've never shorn much hair,"
Up on the shearing run,
But when I gets some practice why,
I'll be a bloody gun!"

"The award's four bucks a haircut,
I should make a stack of dough,
I'll show these city shearers how,
A nor-west gun can go."

He walked into the barber's shop,
Asked for the "leading hand,"
He said he wasn't doing much,
And he wouldn't mind a stand.

The boss, one "Mr Joseph,"
Said he'd give him one day's trial,
And whether he was hired or fired,
Depended on his style.

"Well if it's style you're after,
You've got your man indeed,
And not only am I stylish,
I'm blessed with lightning speed!"

I'll turn out some new hair cuts,
That you city blokes ain't seen,
I'll perform the rousies mohawk,
That I learnt in Wilson's team.

He was at the shop next morning,
In some very flash attire,
A pair of pig skin shearing boots,
Tied up with fencing wire.

A faded navy singlet,
With a breastplate tied down low,
His trousers were the finest cut,
Created by Milro's.

There were some uneasy glances,
And some bloody anxious stares,
As he threaded up his needle,
And hung it by his chair.

He loaded up his handpiece,
Then set to work with zeal,
He grabbed a long haired hippie and,
He made him fairly squeal.

He struggled gamely in the chair,
Then his eyes went dull,
As the nor west barber,
Bounced the hand piece off his skull!

His technique was deplorable,
The execution worse,
He shore a sort of tally hi,
And mohawk in reverse.

He choked them down when ere they frowned,
He kneed them in the gut,
He used his comb and cutter tin,
To do his basin cut.

He made free use of tension nut,
And elbows in the face,
And all the while maintained his style,
And set a cracking pace.

A baldish chap came thru the door,
He grabbed him in a flash,
He held him down and shore him,
With one sweeping back hand slash.

And as he let the "flier go,"
He kicked him in the rear,
"Oh I'll make me bloody fortune,"
If they let me use pulled gear!"

But old habits they die slowly,
And he got as rough as guts,
As he went for a tally,
In a shower of second cuts.

He nearly scalped some long haired lout,
He gouged another's ear,
And the patrons of Joe's Hair Salon,
Drew back in mortal fear.

They say he only did a day,
And now he's back up north,
And if you're talking hair styles,
Then this fellow will spring forth.

He'll tell a few "hair raising" tales,
His blood shot eyes will gleam,
Of the day he shore two hundred,
In the barber's shearing team.

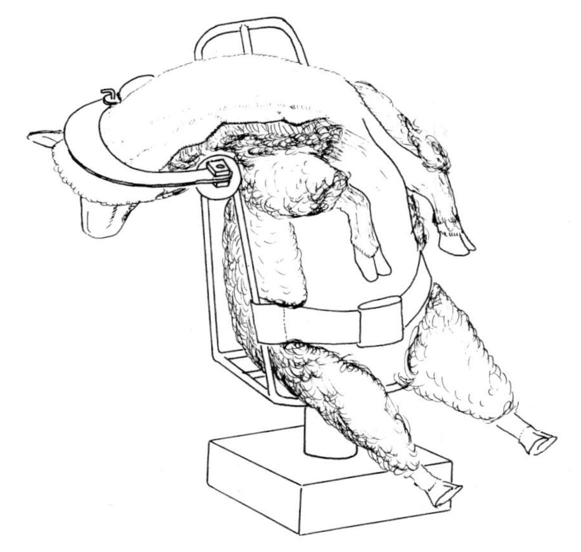

Figure 7.3 The Baa Baa's chair used for neck shearing experiments.

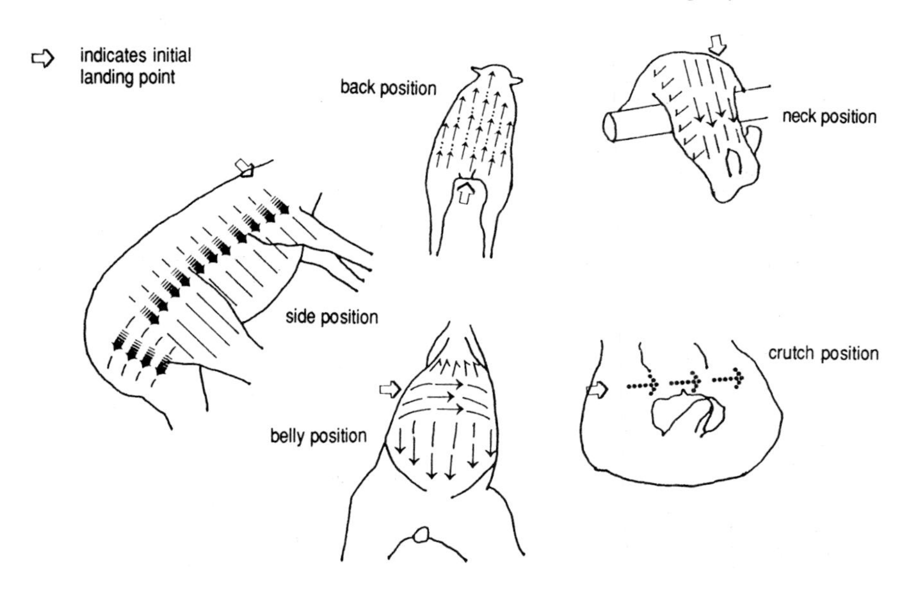

Figure 7.4 Shearing positions and shearing patterns—mid 1980.

would take about a minute to shear. These times were based on faster repositioning between blows than we could manage with ORACLE. Nevertheless it was hard for many people to understand how we could make such predictions when it took so long to set up laboratory tests and we encountered considerable scepticism, even from informed sources (Williams and Phillips 1981).

Cunderdin jumpers

There are about 220,000,000 sheep shorn each year in Australia. Which makes it hard to believe that finding sheep for shearing tests was one of our biggest headaches! With wool more valuable than the sheep's body, most sheep at the saleyards are cleanly shorn. Farmers were often unwilling to lend us good sheep, especially with the risk of unwanted diseases or pests accompanying them on their return.

We borrowed ten sheep from the Cunderdin Agricultural College for trials—sheep were difficult to obtain at the time and we were glad of any good ones we could get. The Cunderdin flock were due for shearing any day, and so had full twelve month fleeces.

Within two days of arriving, we were alerted to their unusual athletic abilities. They resided in the university animal house along with sheep being used for other research programmes on nutrition, reproduction, etc. We found our sheep evenly distributed among the pens reserved for the other animals, and when we attempted to shepherd them back to their own pen, they simply jumped the metre high fences, a feat few other sheep had ever attempted. We found we had to erect special 2 metre high fencing around their pen to keep them in.

Two weeks after their arrival, we had to send them back, shorn. Ian Hamilton, our mechanical technician, arrived early in the morning to find a sheep straddling the top of the fence, with a hind leg caught in the wire on the wrong side. He helped it down, counted out the sheep into our truck and released them three hours later into their home paddocks. A day or two later, we received an anguished telephone call from the Animal Science department alleging that we had taken a precious experimental animal to Cunderdin. Ian could not understand. He had double checked the count, knowing how they could jump.

We went to the animal house, and were shown a recently shorn sheep without horns in the pen belonging to the missing animal which had had horns (but had also been recently shorn). We could only conclude that our jumpers had introduced the otherwise slothful university sheep to the excitement of show jumping. One of ours had jumped out, over the two metre fence, *and* the missing animal had jumped in!

Now we faced a new problem. If the Cunderdin flock had been shorn, it would be rather difficult to find the prized sheep, even though it had horns. Ian left for Cunderdin immediately, and found to his relief that shearing had been delayed. The ten shorn sheep were easily spotted, and the missing animal was soon retrieved.

Not long after, the Corporation was persuaded to provide money for our own flock.

Project review

The first major review of our work was scheduled for March 1981. In fact, since we started, our funding has always been approved for a maximum of only twelve months. Each February or March the Wool Harvesting Research Advisory Committee (WHRAC) met to review progress and debate the levels for future funding. Nevertheless, the March 1981 review was a little different. Our successful trials with ORACLE, and major difficulties with the biological defleecing programme, had brought an air of optimism and expectancy. The Corporation was known to have set aside a major sum to expand research on robotic shearing.

About this time, I noticed a change in my role in the project. From the start of ORACLE, I had taken a leading role at the technical level. At first I was officially only working one day per week on the project, but both my interest and actual time involvement were far greater. I was employed by the University three days per week as a lecturer, and worked on a number of small consulting contracts at the same time.

Jim Blair, Prof. A-W, Roy Leslie and I handled the non-technical issues on a joint basis. However, Jim became ever more involved with his research and consulting work on the dynamics of long iron ore carrying trains. In late 1979 and early 1980, Jim went to Cambridge on study leave, and Prof A-W took a less active role in the project. This left Roy and I very much in charge. Naturally Stewart Key took a more active role as well. I was really most fortunate that Roy, and later Stewart, shouldered the burden of administration, orders, invoices, accounts and the details of budgeting. It was this factor that gave me the time and freedom to shape our visions for the future of the project.

Roy of course had started the project—for the first 18 months, he was the only person with any substantial technical involvement apart from an honours student, Frank Wittwer. Frank had laid the groundwork for the software sheep development by showing that sheep surface profiles could be described by parametric curves.

At the end of 1980, Jim Blair found that working on his industrial research contracts was becoming too difficult in the university environment. Apart from the time pressure, he found the many administrative restrictions and discouragements a constant source of frustration and irritation. He asked the university if they would employ him on a half-time basis, but there was no agreement on superannuation or entitlements. So Jim decided to form his own company, ACE-T, in partnership with Prof. A-W and Duncan Steven.

He offered Roy a partnership in the company. Roy had a difficult choice to make. If he stayed with the sheep-shearing project, he faced annual reviews and uncertainty about funding each year, relatively low pay, and the prospect of working 'in my shadow' for the foreseeable future. With Jim, he would have a growing capital asset, and the prospects for building his own career.

I was relatively naive about these issues at that time, and I can only say that I was quite surprised by Roy's decision to join ACE-T. I had not been sufficiently sensitive to Roy's situation to recognize his shift in perception early enough. Once he resolved to leave, he brought renewed vigour back with him. The enervating doubt and indecision which had sapped his energy for several months had gone. He turned his energy to designing a hand-operated manipulator platform (HOMP) so we would be able to show a sequence of shearing for the March review. This would show how the sheep could be restrained and turned over by a mechanical cradle. Automating it later would be straightforward.

By the end of 1980 we were keenly conscious of the growing interest in our work from the Corporation. Experiments with chemical shearing were abandoned when it became clear that a wool-dissolving chemical would damage all the fleece, not just the part closest to the skin. Furthermore, the wool did not really dissolve; it turned to a sticky mess which clung to both the remaining wool and the skin.

The main alternative to robot shearing was also in trouble. The search for a chemical which could induce a sheep to shed its fleece through biological actions without side effects had been fruitless. Several promising alternatives had been tested yet all showed unacceptable side effects. The issue of long-term residues in the meat had not even been considered. The search for a safe defleecing chemical was proving to be very difficult indeed.

With growing confidence, we started preparations for a major demonstration for the March review. New software was needed to link all the shearing sequences together for the new HOMP cradle. Yet the new shearing software was plagued by bugs. These were only resolved by the end of the first week in March, just days before what Paul Hudson had specified to be 'an impressive'

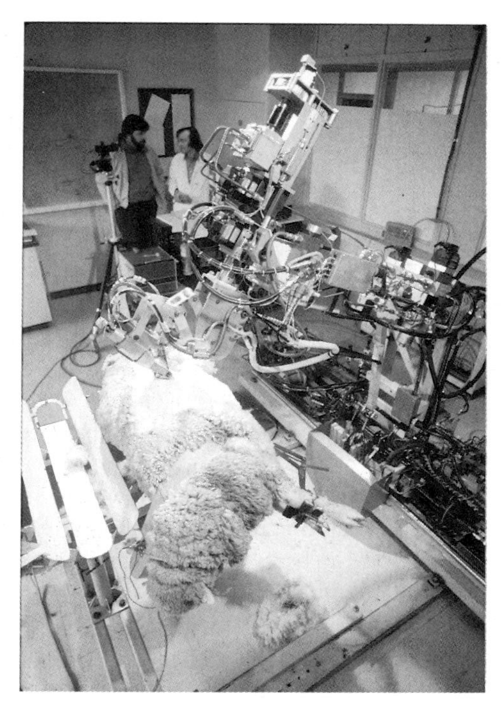

Figure 7.5 Sheep shorn in HOMP manipulation cradle—end 1981.

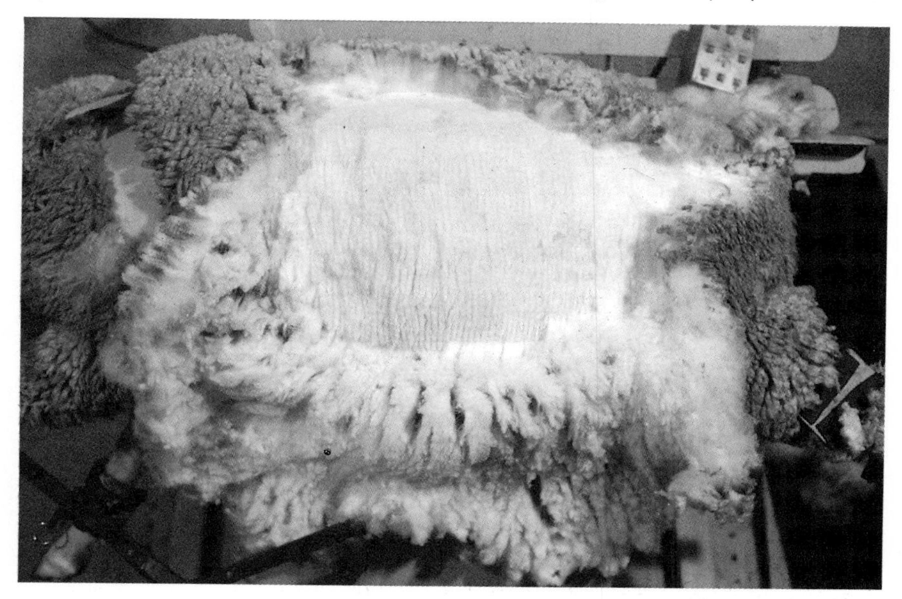

Figure 7.6 Typical side shearing result—end 1981.

demonstration. (Paul Hudson took over Alan Richardson's role in 1978, coordinating wool harvesting research for the Corporation.) Since the performance measurement trials had finished the previous September, we had only a handful of successful runs. Roy and Ian Hamilton worked against time to finish HOMP, but the paint was still drying when Paul Hudson arrived with the review committee on March 21st.

We split the visiting party into two groups after they had been formally welcomed. While one group watched ORACLE shearing, the other was given a demonstration of sheep manipulation and hand shearing in HOMP which had not yet been linked to the robot.

The shearing was my responsibility and the poor results, particularly for the first and most influential group, are a painful memory. The weather was hot and humid, conditions we now know as the worst for reactive wool. Inorganic salts, which are normally bound up in the wool grease, absorb moisture and raise the conductivity of the wool by a factor of a thousand or more. This confused the sensor readings and the robot tried to shear away from the skin, leaving a ragged untidy result. Not at all impressive, and a disappointment for all of us. The second sheep went much more smoothly, but that came too late.

I was tied up with ORACLE while the first group were watching Roy demonstrate HOMP. I can only presume from the eventual outcome that they were not particularly impressed by this part of their visit either.

When we finally sat down with the committee to discuss future options, it was clear that they were keen to proceed with the project, albeit cautiously.

They asked if we could develop an automatic manipulator to work with ORACLE while steadily improving the standard of shearing.

Prof. A-W eloquently explained how this course could be counterproductive. ORACLE after all was only designed as a test rig—an experimental device—to find out how to build a real shearing robot. Although it worked, we now knew how to build a much better robot. Any manipulator designed to work with ORACLE would be fettered and constrained by its limitations, just as ORACLE itself had been constrained by political limitations imposed by CSIRO. But another factor had emerged from a most unexpected quarter.

Behind any successful research project you will always find political skill combined with scientific know-how. For the biological defleecing project, this expertise lay with Trevor Scott, chief of the CSIRO Division of Animal Production at Prospect near Sydney. Trevor had the breakthrough he needed to ensure his project's survival. With a string of disappointments through 1979 and 1980, biological defleecing was attracting 'the axe'. The breakthrough came in mid 1980, but Trevor played it to maximum advantage. Days before the March review, he announced the discovery of Epidermal Growth Factor (EGF)—a naturally occurring compound which causes sheep to shed their fleece. And it was a gentle, safe biological compound, quite unlike the synthetic chemicals and drugs which had been tried before. At long last the search for a safe defleecing compound was over. EGF had been found to be the factor which inhibited the growth of wool on sheep embryos before birth. It could be obtained from the salivary glands of mice for experimental purposes; advances in genetic engineering would provide a way of manufacturing the compound in commercial quantities.

With this in mind, the Corporation knew that they still had a two-way bet. Both our programme and CSIRO's biological defleecing were continued with increased funding. Yet with no interest in a new robot of any kind, we were disappointed. It seemed that all the momentum had been lost, dissipated. Money was set aside for what we saw as a time-consuming but relatively mundane task of automating a sheep manipulator 'to demonstrate fully automated shearing and manipulation'. Any thought of commercial development was to be set aside until we had demonstrated shearing of at least 95% of the wool under automatic control.

After Roy left us at the end of March, our spirits sank even lower. With Rob Greenhalgh away in Britain on holidays, we were reduced to Stewart Key and Ian Hamilton full time, and myself part time.

Looking back, we could hardly have done better. In research, as with any human venture, success depends on having good people. Two months later we would have the people we needed; we jumped from 4 persons to 16. It is nine years later as I write this, and it took seven of those years to achieve the 95% objective. Were all those years really necessary? I don't think so, but who can really tell?

Automating manipulation

We received three interesting applications to our advertisement for a mechanical design engineer. Two had useful but not entirely relevant experience. The other applicant had designed fruit canning machinery and some interesting automatic machines, but expressed a strong interest in the project in his letter. David Elford brought an intriguing photograph album to his interview. For behind the impressive size and speed of the canning machinery lay some amazing mechanisms. If you like peaches, you will appreciate that peeling a clingstone peach cleanly is no easy task. David had built machines to turn the peaches the right way up, peel them, split them in half and cleanly remove the stone, at the rate of hundreds per minute.

Stewart and I knew that we needed a person of exceptional design skill; there was no doubt that David had the skills we needed. Yet David had been accustomed to managing his own organizations and had held some very senior positions in large companies. He was also quite a bit older than I. How would he settle in to a small group of young research staff? I decided that we could handle these problems without too much effort, but that we could not do without David's skills.

A week later, the university staffing office informed us that we could not offer the position to someone who had not qualified as an engineer. (David's design skills were in such demand that he never had time to complete his formal engineering degree course.) We bypassed the staffing office and hired David as a consultant—he has been with us since.

Shortly after David joined us we appointed another mechanical engineer, Zbigniew Lambert, an electronics engineer, Michael Crooke (whose family still manufacture shepherd's crooks), and a mathematician, Robyn Owens, to work on surface model statistics. With two extra mechanical technicians, a part time secretary, and three students, the team size reached 16 persons.

David soon concluded that the HOMP design was no basis for automation—a completely new design concept was needed. For several weeks he patiently reviewed countless inventions for handling sheep. He soon narrowed these to three alternatives (figure 7.7):

(1) a cradle turning on a horizontal axis with sections which folded back to reveal parts of the sheep within;
(2) twin horizontal rollers with or without a belt between them, similar to the cradle devised by Norman Lewis for the PATS shearing machine;
(3) a cradle with pivoting sides similar to HOMP.

All three designs would have to rotate on a central vertical axis and translate in the *y* axis direction relative to ORACLE to move each part of the sheep within reach.

Option (1) was rejected because too much space was needed to fold back the opening sections, and restraining the legs and head would be a problem. With vertical rotation and translation motions added, the machine would be very large and expensive.

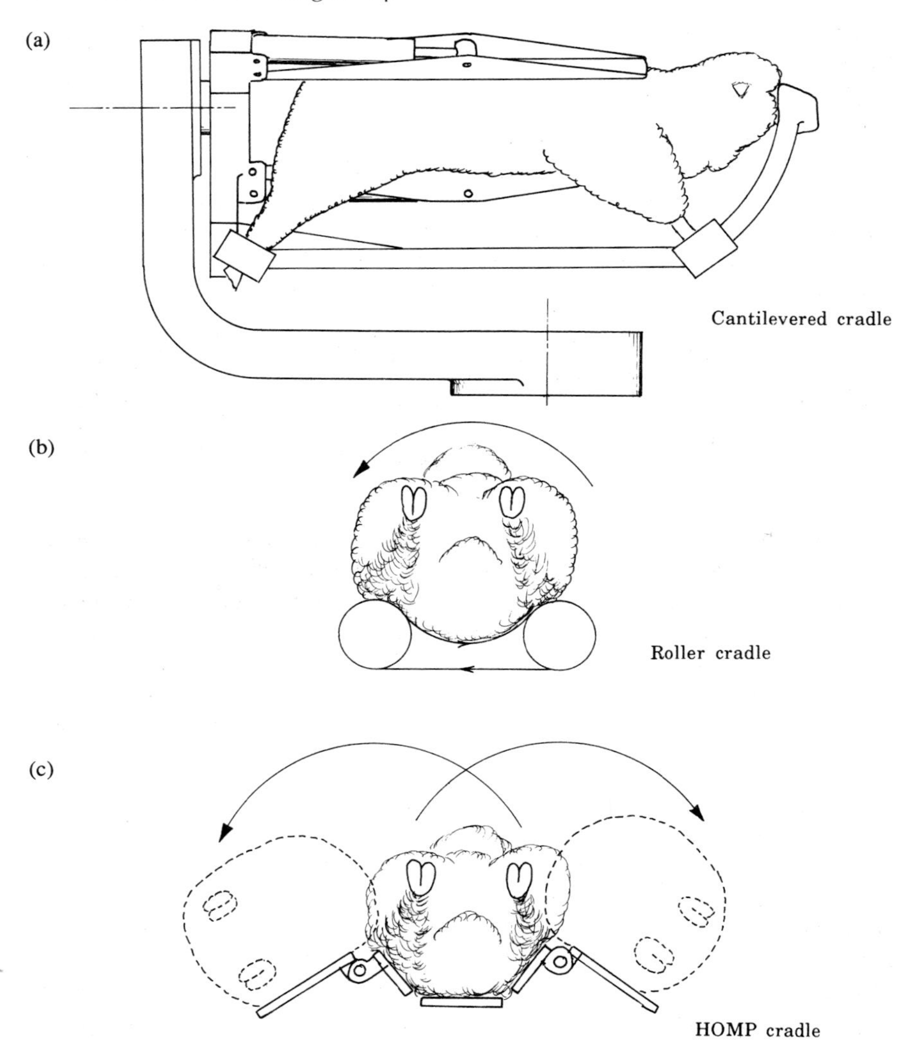

Figure 7.7 Alternative manipulator configurations evaluated: (a) Cantilevered cradle; (b) Roller cradle; (c) HOMP cradle.

Option (2) was rejected because we remembered how unpredictably sheep turned over in the PATS cradle. Predictable location of the sheep was a cornerstone of our surface modelling strategy. Eventually we were to return to this idea.

Among the more unlikely proposals given brief consideration was one in a letter from Tasmania suggesting that the sheep be pulled through a hole containing knives to shear off the wool . . .

'It begins by observing a mouse pushing its pliable body through a hole that is barely big enough. The mouse adjusted its body to the shape of the hole so to speak. I think I learned something from that.

'My sheep to be shorn first walks on to a moving floor that has a ribbed section where its legs go down and are then firmly held.

The next movement is pulled steeply upwards by the two front legs pushing to head well down and now it faces that "hole" mentioned before. That hole represents a method of shaping the body so that it becomes accessible for the massive, semi-circular, pliable shearing-knives-battery awaiting the correct shape. Can you see the picture?

I repeat it: The sheep is practically up-standing, in fact, gently leaning at a sharp angle and held by the front legs, the body resting on the moving part that draws the whole sheep up and into the "hole" . . .

. . . First the "hole" really consists of a number of plates, sheets of small sizes, vertically and rigidly held together, spring loaded or sensitized in some way, to follow the exact shape of the topside of the sheep. They are bobbing up and down, calculated pressure, and tell the knives what the shape is at a particular point....The computer should make it feasible and practicable to adjust the deadly knives so quickly out of the paths of obstructions, that no damage will be done.'

With some reluctance, and a growing concern at the complexities of the task, David returned to the HOMP concept which he had initially rejected. We christened the new concept ARAMP—the 'Automated Restraint And Manipulation Platform' (Key and Elford 1983). Stewart Key started on the control software and Michael Crooke, Rob Greenhalgh and I specified the electronics.

The ARAMP manipulator

The manipulator which David and Zbigniew created in their drawing office was designed to meet a simple objective: to show that sheep could be automatically manipulated into shearing positions, and shorn by the ORACLE robot. The objective might have been simple, but the task was most certainly not. The useful workspace of ORACLE was quite small in comparison to its size and the manipulator had to be mounted on a mobile platform. Two servo controlled movements were provided to turn the manipulator through 270° and to move it towards or away from the robot though a range of 600 mm to place the sheep within reach of the robot (figure 7.8).

The sheep's body had to be kept at more or less the same height when it was rolled on to its side, again to keep it within reach of the robot. Thus two different roll axes were required, spaced about 300 mm apart.

The legs had first to be immobilized by restraint bars, and then grasped by clamps which had to be transferred to each side as the sheep was rolled. This was accomplished by designing a separate leg manipulation beam, containing all its own actuators and mechanisms. The beam is secured to either of the side panels of the cradle by hydraulic key locks.

All the hydraulic control valves and electronics had to be contained within a compact pedestal under the cradle. Over 40 actuated movements were needed,

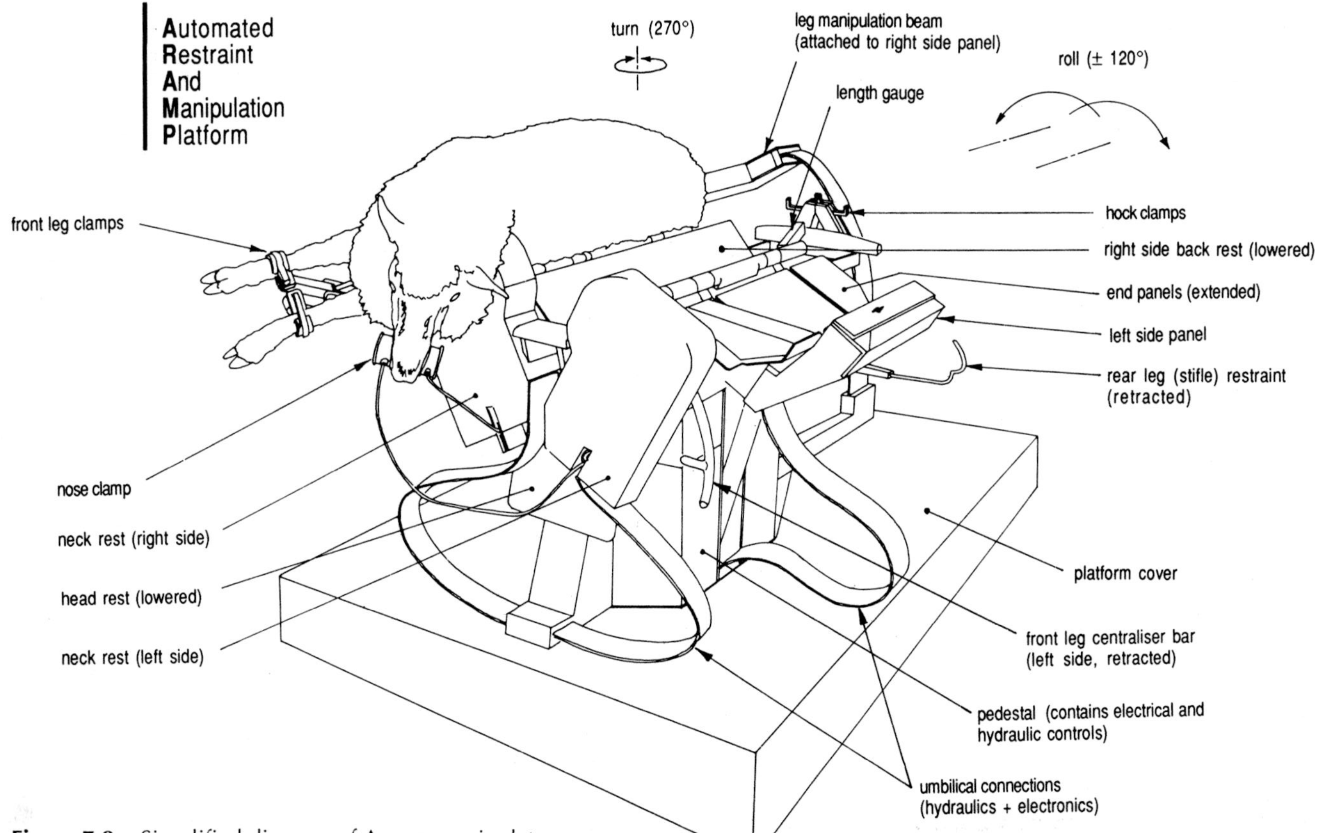

Figure 7.8 Simplified diagram of ARAMP manipulator

and it was not feasible to provide hydraulic connections to so many actuators from a remote control unit.

With all this complexity, it would have been easy to lose sight of the overall objective—a practical mobile shearing plant with reasonably simple equipment. Figure 7.9 shows one of the concepts which we developed for analysing the economics of robotic shearing. With ARAMP we were committed to complexity to provide a manipulator for ORACLE to demonstrate automatic shearing and manipulation. The next robot and manipulator would be quite different.

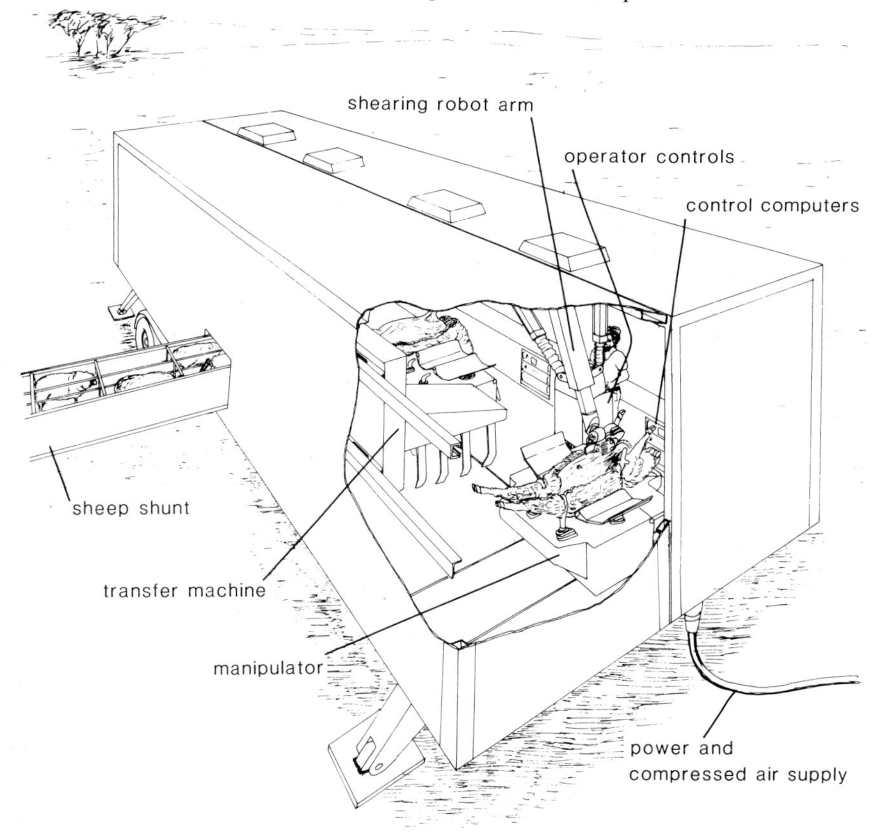

Figure 7.9 Concept for a mobile automated shearing rig—mid 1983.

Shearing positions

The manipulator presented sheep in each of the shearing positions shown in figure 7.4—these were the ones developed for the feasibility trials in 1980 (with the exception of the crutch position).

A shearer uses his free hand and legs to condition the loose skin, stretching either the body of the sheep or the skin to remove folds and presenting a stable smooth surface for each blow, thus avoiding skin cuts.

The manipulator relied on sheep positioning alone; the surface presented for each shearing position was reasonably firm, mostly convex, with the skin stretched by positioning the legs and head. The wool had to be free to peel away from the skin as it was shorn, falling to the floor in chunks. The shearing sequence had to provide a 'clear' initial landing point for the cutter on each part of the sheep (the matting of typical Merino fleeces makes entry through the fleece impractical). The first entry point was either at the crutch, shearing up the belly, or at the armpit, shearing down the belly.

The four basic shearing positions are shown in figure 7.10.

Sheep restraint

Sheep restraint was one of the more intriguing problems to solve. We knew that sheep would lie still provided they were comfortably supported and could *feel* no means of escape. The gentlest restraint was sufficient to keep the sheep still. When provoked by noise or shearing, however, sheep would often struggle for a moment or two before relaxing again. For this a firm grip was needed to prevent the sheep from escaping. We needed gentle, but firm, self-adjusting restraints which would not obscure significant amounts of wool.

Each clamp or restraint acting on the sheep was driven by an hydraulic actuator with a non-return action. The force was preset by adjusting the hydraulic supply pressure for the actuator. The closing speed could also be adjusted—rapidly closing clamps could provoke struggling. The non-return action provided the firmness—the clamp closed gently but could not be reversed, no matter how hard the sheep struggled.

Sheep, like humans, have several points at which moderate pressure will immobilize part of the body. We used these points for the early stages of the manipulation sequence. The front leg centralizer bars (figure 7.8) applied pressure to the front legs, straightening and immobilizing them; the stifle bars worked on equivalent pressure points for the rear legs.

Rapid inversion of the sheep also helped by producing confusion and disorientation; the sheep lay still for a minute or so providing time, during sheep loading, for applying the initial restraints (Ewbank 1968).

Sheep manipulation

The manipulation sequence was as complex as the design itself. It is difficult to describe coherently without copious diagrams for which there is insufficient space here.

1. Two operators lift sheep on to the cradle, holding the legs, belly uppermost. Sheep is registered by pulling forwards until its shoulders rest against the neck rests. Front leg centralizer bars close together, centralizing the brisket, and move forwards onto the front leg pressure points. The rear hocks (ends of rear legs) are placed by hand into clamps which close, securing the rear legs (the sheep is now fully restrained). Stifle bars lift, come together and down, pressing on rear leg pressure point. Length gauge moves against

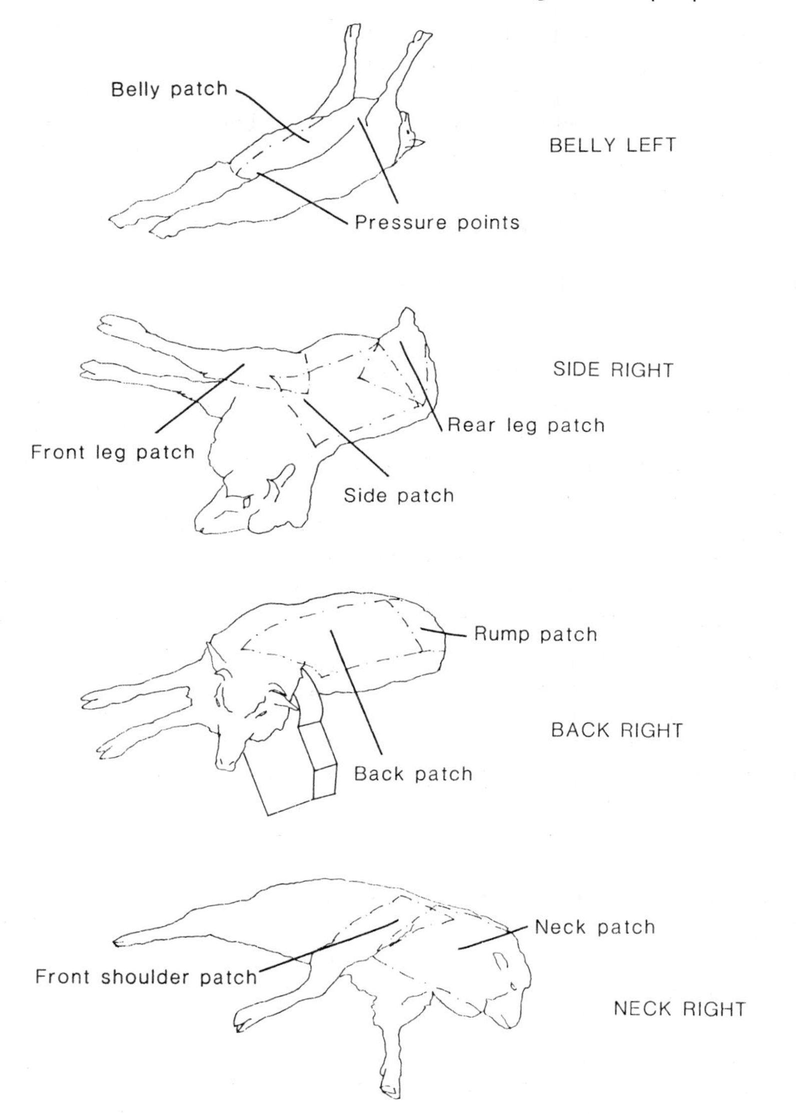

Figure 7.10 Manipulation and shearing positions—mid 1981.

crutch, measuring length of the body. An operator fits and secures the nose clamp.

2. Shear left side of belly.
3. Turn cradle 180°.
4. Shear right side of belly.
5. Raise side panels together; leg manipulator beam is now over sheep. Separate, lift, and then bring rear legs together in rear leg clamp which

turns, locking and stretching the rear legs (figure 7.14, p. 187). The front leg clamps move forward and grasp the front hocks and stretch the front legs (figure 7.15, p. 188). The hock clamps and stifle bars release the rear legs and are folded away. The front leg centralizer bars move back and away from the brisket, and are folded down. The end panels fold in. The side panels close on the belly. Clutches lock the neck rests to the side panel shafts so they will roll with the sheep.

6. The sheep is lifted and rolled out 45° on right side. Head is pulled to right neck rest. Head rest is folded down (figure 7.16, p. 188).
7. Shear right side of sheep.
8. Roll sheep to 120° position (figure 7.8, p. 178). Turn cradle –180°.
9. Shear back of sheep.
10. Turn cradle –45°.
11. Shear rump of sheep.
12. Turn cradle 135° to bring head closest to robot.
13. Shear neck of sheep.
14. Turn cradle –90°.
15. Lift right back rest. Roll sheep back to belly position, raising both side panels and head rest. Transfer leg manipulator bar to left side panel.

16–24. Repeat steps 6 to 14, but on opposite (left) side.

25. Release nose clamp by hand. Leg clamps release. Sheep is lifted off the cradle by two operators.
26. Raise both side panels, folding away end panels. Transfer leg manipulation beam to right side. Lock neck rests to frame so they remain raised when the side rests are lowered. Lower side panels. Raise and register front leg centralizer bars, and separate them to the loading position.

The manipulator is now ready to receive another sheep.

Manipulator control system

The manipulator was controlled by software resident in the computer used to control ORACLE. Each of the hydraulic actuators could be moved in either direction by activating the appropriate solenoid of a four-way valve (figure 7.11). Microswitches were used to sense when the actuator reached the limits of its motion, and sometimes points in between as well. Not all of the actuators had microswitches to sense position—there was provision for 63 limit switches. A few movements were driven by pneumatic actuators with simpler controls.

Many of the moving parts could collide with each other if movements were programmed incorrectly. To simplify programming, we devised a scheme to represent the state of each actuator as one of the following:

(0) unknown
(1) retracting
(2) powered and retracted
(3) retracted
(4) extending

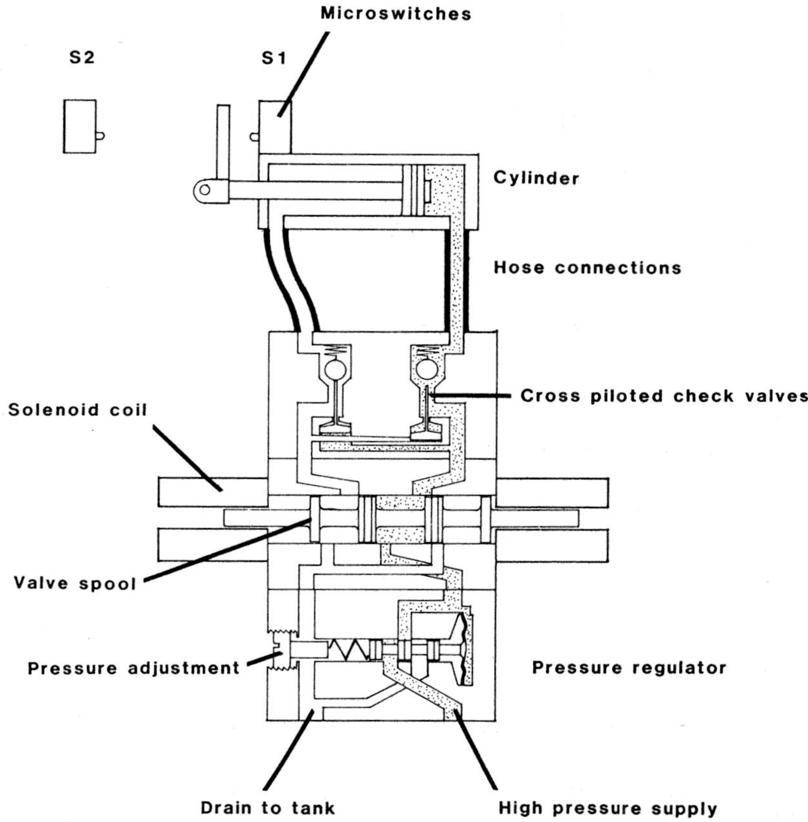

Figure 7.11 Typical hydraulic circuit used on ARAMP manipulator.

(5) powered and extended
(6) extended.

Transitions from states (1) to (2) and from (4) to (5) were determined either by signals from the limit switches, or by elapsed time in the absence of switches. Other transitions occurred as a direct result of switching the valve solenoids. Actuators with intermediate position switches required more states, but the principles were similar (Key and Elford 1983; Key 1983).

The sequence of actuator movements was specified by a series of commands, each of which specified the required end-state for a given actuator. The software inferred the required sequence of valve and switch operations by interpreting a network of permissible state transitions. Each actuator was operated independently—several commands could execute simultaneously. In cases where collisions could occur, the command sequence included a list of contingent states for other actuators—states which had to be achieved before the actuator could be moved. This way, we were able to write reliable programs for sheep manipulation sequences with reasonably simple and concise coding.

There were also commands for branching, and there was provision for procedures defining frequently used groups of commands.

Any sequence could be paused by pressing one of the robot teach pendant buttons. Following a pause, a sequence could be resumed or aborted.

Hydraulic and Pneumatic Controls

A typical hydraulic control circuit consisted of conventional four-way control valves with cross piloted check valves and individually adjustable supply pressure (figure 7.11). Where identical mechanisms were controlled separately (e.g. right and left leg clamps), common pressure control was used.

For pulling the head clamp cable, a single acting cylinder was used; the opposite side circuit simply unloaded the check valve.

Pneumatic components were used on some actuators to save space—the connection hoses were much thinner and more flexible and could be accommodated more easily in confined spaces (figures 7.17, 7.19). We found that pneumatic components rated at 7 Bar could often be used with hydraulic fluid at up to 35 Bar.

Conventional servo hydraulic circuits were used for the platform translation and turn axes, using potentiometers for feedback.

Shearing trials

Construction took much longer than we first expected. The cradle was ready for belly shearing in December 1982 and a further 9 months were needed to complete the leg clamp beam with its intricate internal mechanisms (figure 7.19). The project was reviewed in March 1983, and there were frantic preparations. We used a manually operated leg clamp beam and allowed ourselves plenty of time for demonstration practice. The timing was particularly difficult for me as my wife was expecting the birth of our third child on the demonstration day—fortunately Clare was born ten days early!

Both ARAMP and ORACLE performed well and the WHRAC committee were reasonably pleased. Naturally, one of the committee members asked 'Well, how are you going to catch the sheep and load it automatically?' A valid question of course, but one which we deferred answering. For the time being, we had achieved our aim of demonstrating (almost) automatic manipulation.

Once ARAMP was completed the following September, we refined our manipulation sequence programs and demonstrated reliable manipulation, time after time. A pleasing result and a major step towards shearing with robots.

Catching sheep

At the time, our research was confined to manipulation of the sheep—other research groups were investigating sheep catching and loading. With this disparate approach, it was not surprising that the equipment developed for sheep loading turned out to be incompatible with our manipulator. This caused us some frustration at the time, but was of much less concern to the Corporation.

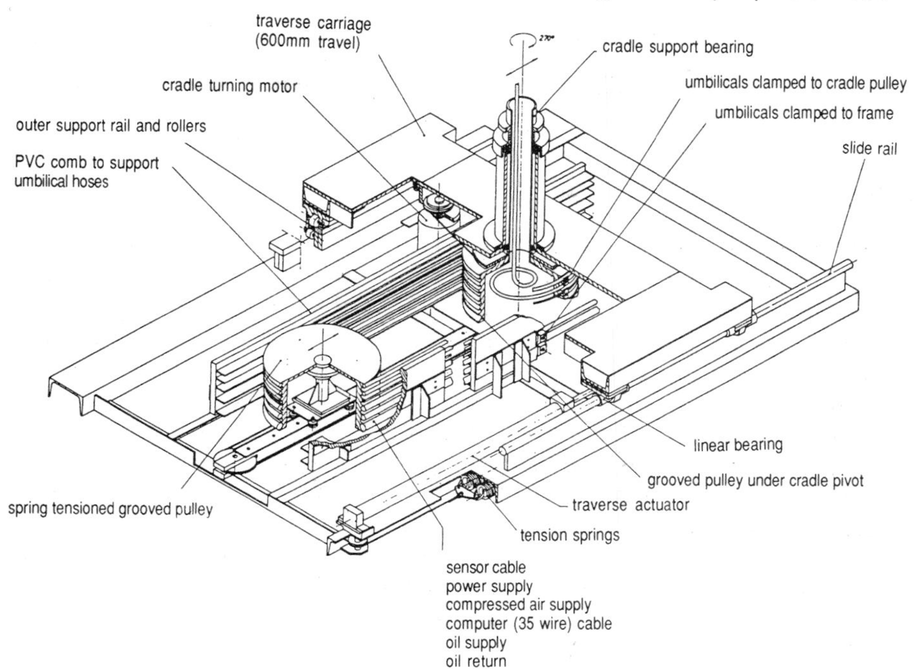

Figure 7.12 ARAMP platform providing 600 mm traverse and 270° turning for cradle.

At shearing time, sheep are mustered into yards near the shearing shed. From there they are driven in smaller mobs up into the back of the shed, and in turn into smaller catching pens where shearers grab them. For a long time, farmers have tried to devise a means of persuading sheep to enter a race (a narrow passage) without having to be there with a dog to force them. A dog could do the job for a while, but they tire after a while and could not keep sheep moving up all day.

There have been many attempts, but so far, no one has claimed absolute success.

One of the most successful arrangements was devised by Leslie Syme of CSIRO (Syme and Durham 1981). She devised a twin race arrangement where sheep move forward in two lines, side by side. Automatic computer controlled gates open in a sequence which ensures that sheep only move up *alongside* another sheep. They are reasonably willing to do this, but computer synthesized dog barks, whoopees, whistles and other sounds are needed to help. Finally, tipping floor and gate mechanisms give the sheep a shove to persuade the most reluctant animals.

The CSIRO race was offered for commercial development but was never taken up.

One related idea which has emerged in several commercial machines is a V-conveyor. Two flat conveyor belts move along the sides of a narrow V-shaped

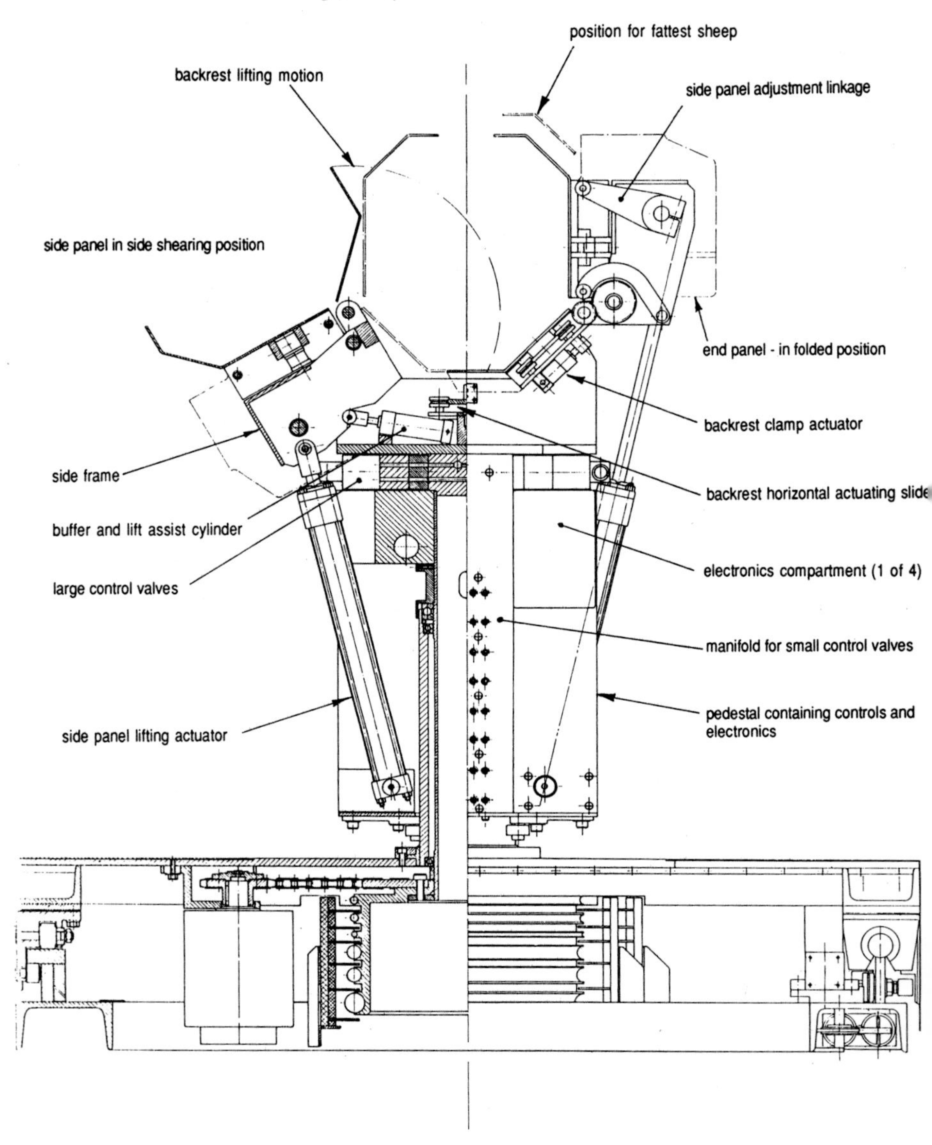

Figure 7.13 Cross section assembly drawing—ARAMP manipulator.

race. The sheep become wedged between the belts and can be moved forward whenever the belt moves. Figure 7.21 (p. 192) shows the delivery end of one such conveyor. The diagram shows a concept for a sheep catcher developed by Burrow (1981) of MUSHEEP at Melbourne University. The sheep are ejected by the V-conveyor into a catching pen where they are grasped by a large multi-finger gripper (the transfer cradle) along the sides. From there they can be

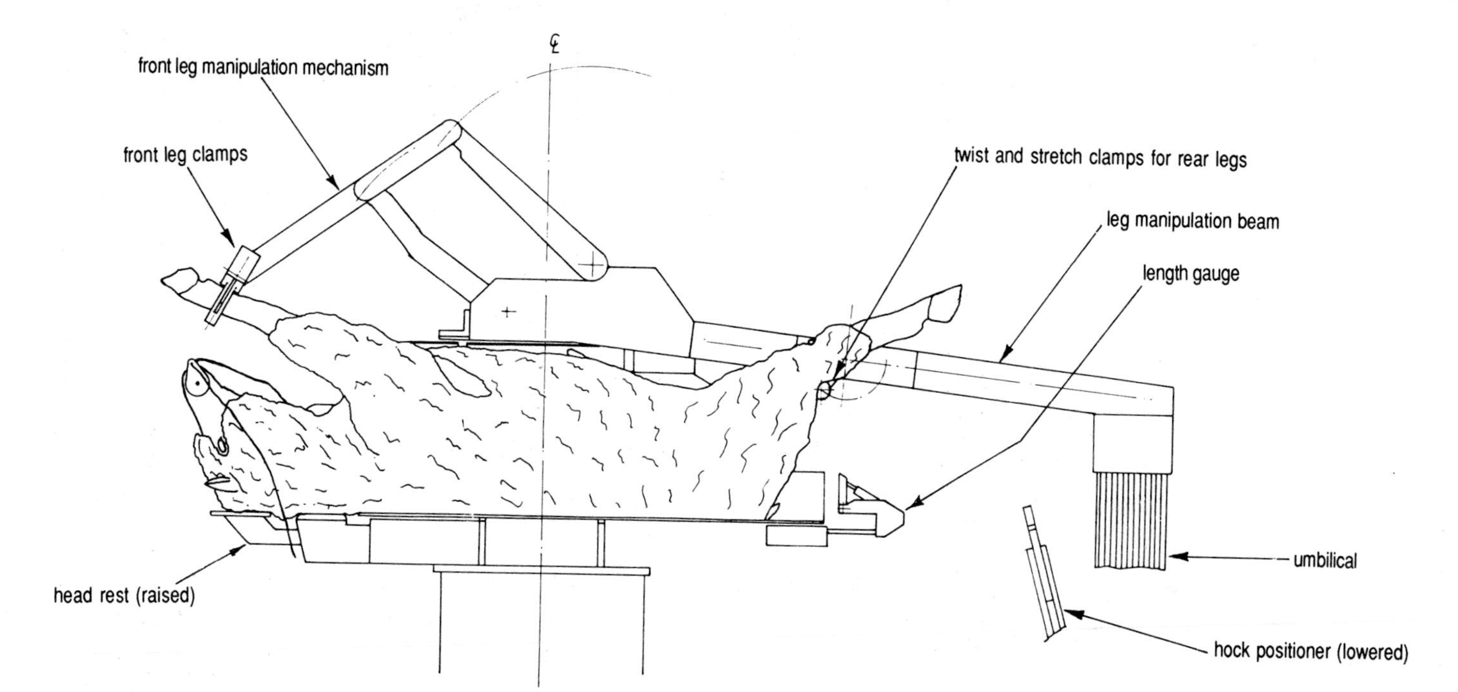

Figure 7.14 Side view of sheep just after transfer of the legs from the initial restraint bars to the front and rear leg clamps. The leg clamps are part of the lag manipulation beam which rolls with the sheep. It is key locked to either the left or right side panel.

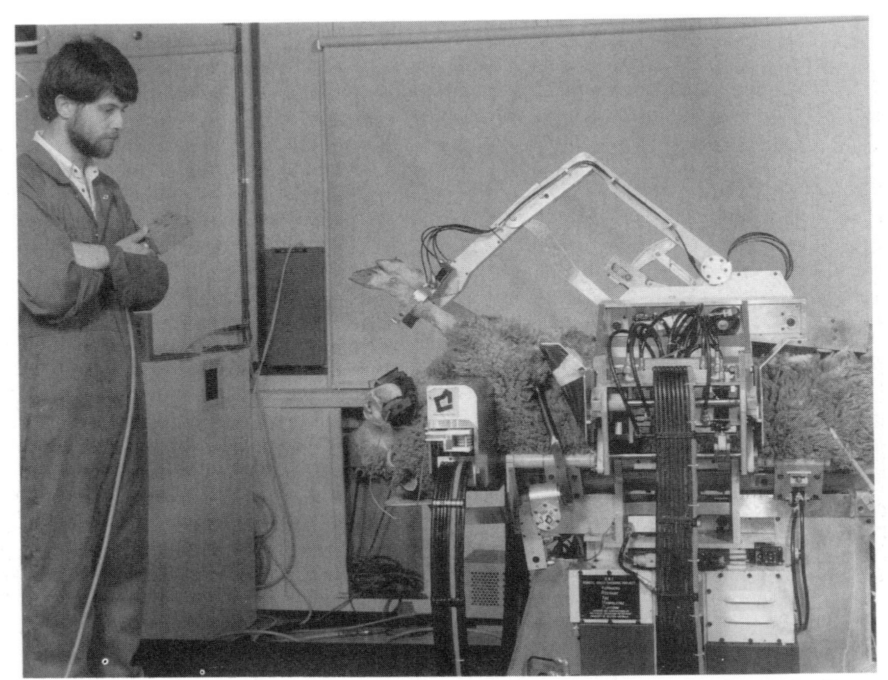

Figure 7.15 Photograph of ARAMP immediately preceding stage shown in figure 7.14. The front leg centralizer bars and stifle bars are still in position, and the front legs are about to be captured by the clamps.

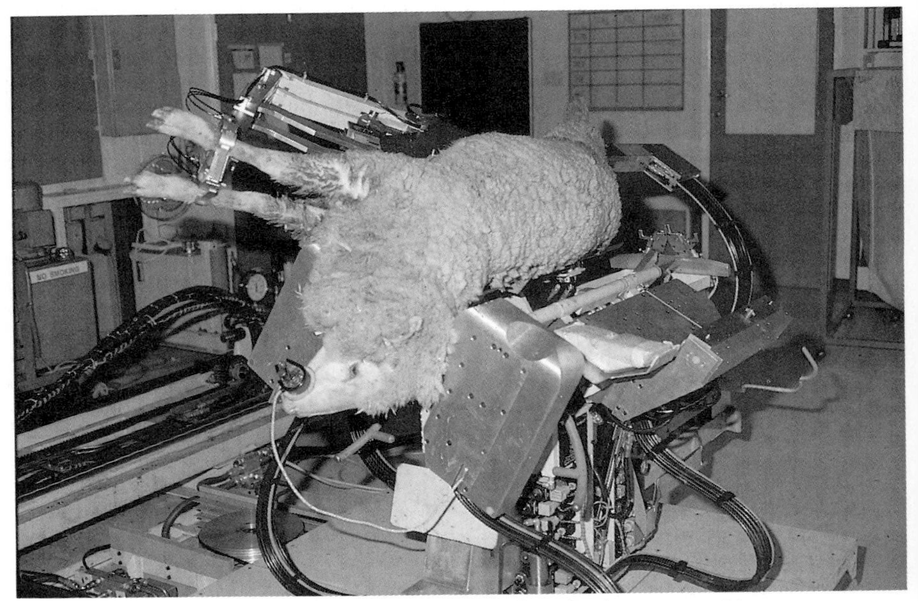

Figure 7.16 Sheep rolled 45° to right side shearing position.

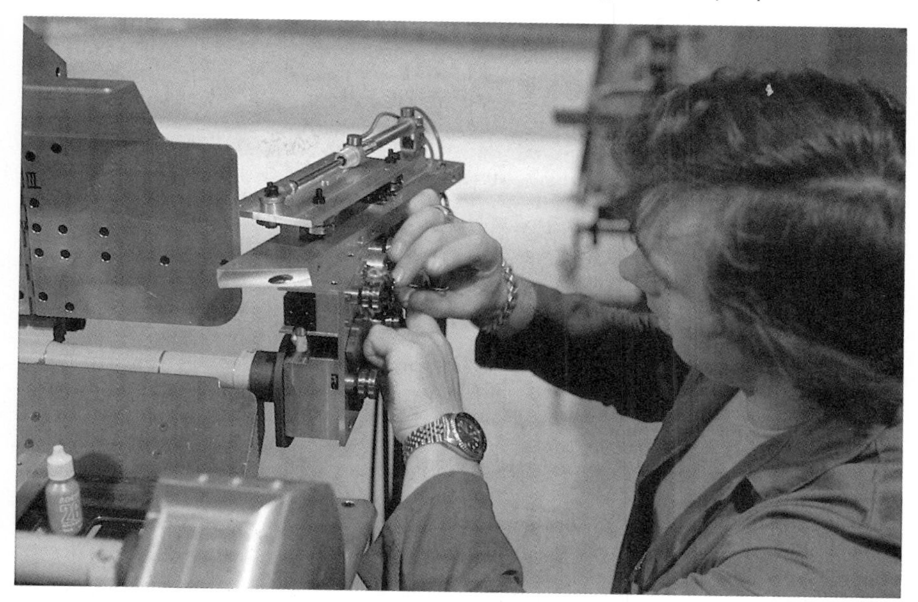

Figure 7.17 Intricate internal neck rest mechanisms being adjusted by technician Stewart Howard.

Figure 7.18 Details of wiring behind space used by electronic circuit boards. The volume of wiring needed for 40 actuators and 32 limit switches proved to be a major problem.

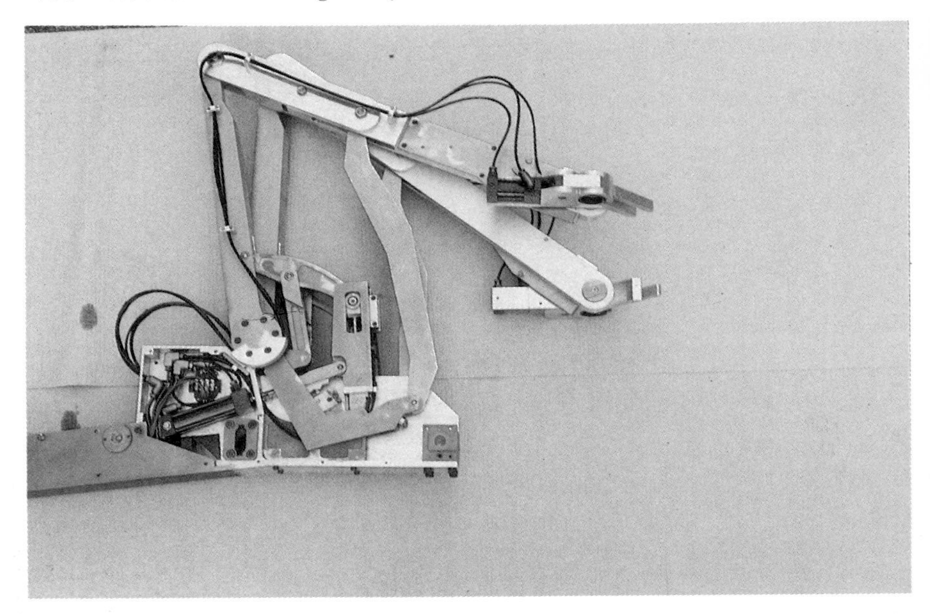

Figure 7.19 Photograph of front leg manipulator mechanism. Two identical mechanisms are accommodated within a total width of 60 mm.

carried forwards to a cradle where the legs and head are clamped automatically. An experimental version of the machine was constructed, but has not been further developed.

A third approach was being developed in Adelaide by a private company. Their aims went well beyond sheep catching; they were intending to build a simple automated shearing plant which farmers could use to shear their flocks. For the first time, we had competition . . .

Competition from Adelaide

Norman Lewis was not alone in dreaming of sheep-shearing robots. Jim Shepherd was an engineer turned farmer, not far from Kojonup in Western Australia. Like many other farmers across Australia, and particularly in Western Australia, Jim was thinking of new ways to shear sheep. An engineer friend, Denton Roberts started tinkering with machinery. Jim was also involved with a new group of farmers who had realized that the classic breeding lines of Merino sheep could do with some improvement to make woolgrowing easier. Called the Australian Merino Society, they were also experimenting with artificial insemination for breeding purposes.

Jim interested some others in his ideas for new shearing machinery and together they formed a wool harvesting research and development subcommittee

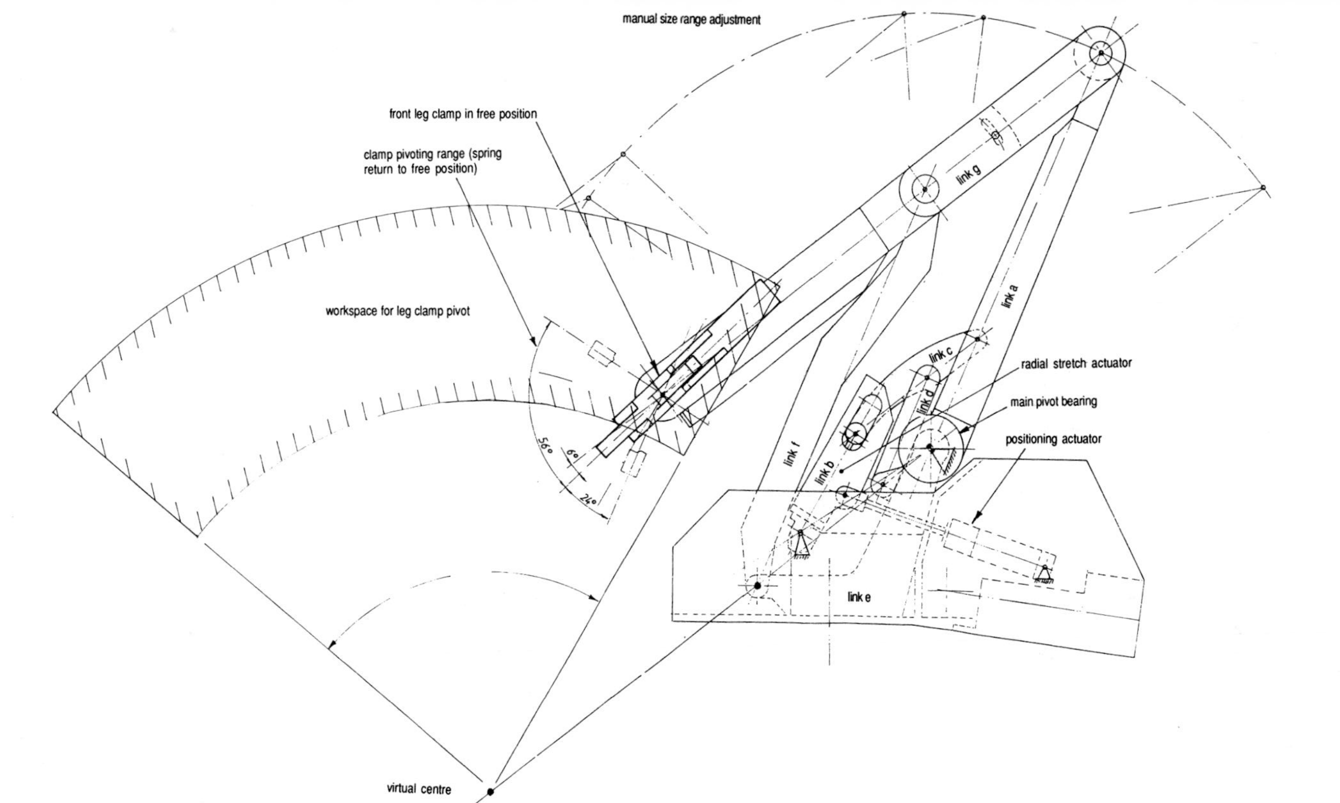

Figure 7.20 Diagram of front leg manipulator mechanism. The pantograph arrangement enables rotation of the ends of the front legs about a virtual centre point located inside the sheep. The legs can be stretched radially at any position. Links a,e,f, and g form a parallelogram, the base of which (link e) is varied by the action of the stretch cylinder in link b. Link g can be loosened and bent through a limited range at the joint with link f to accommodate different ranges of leg length.

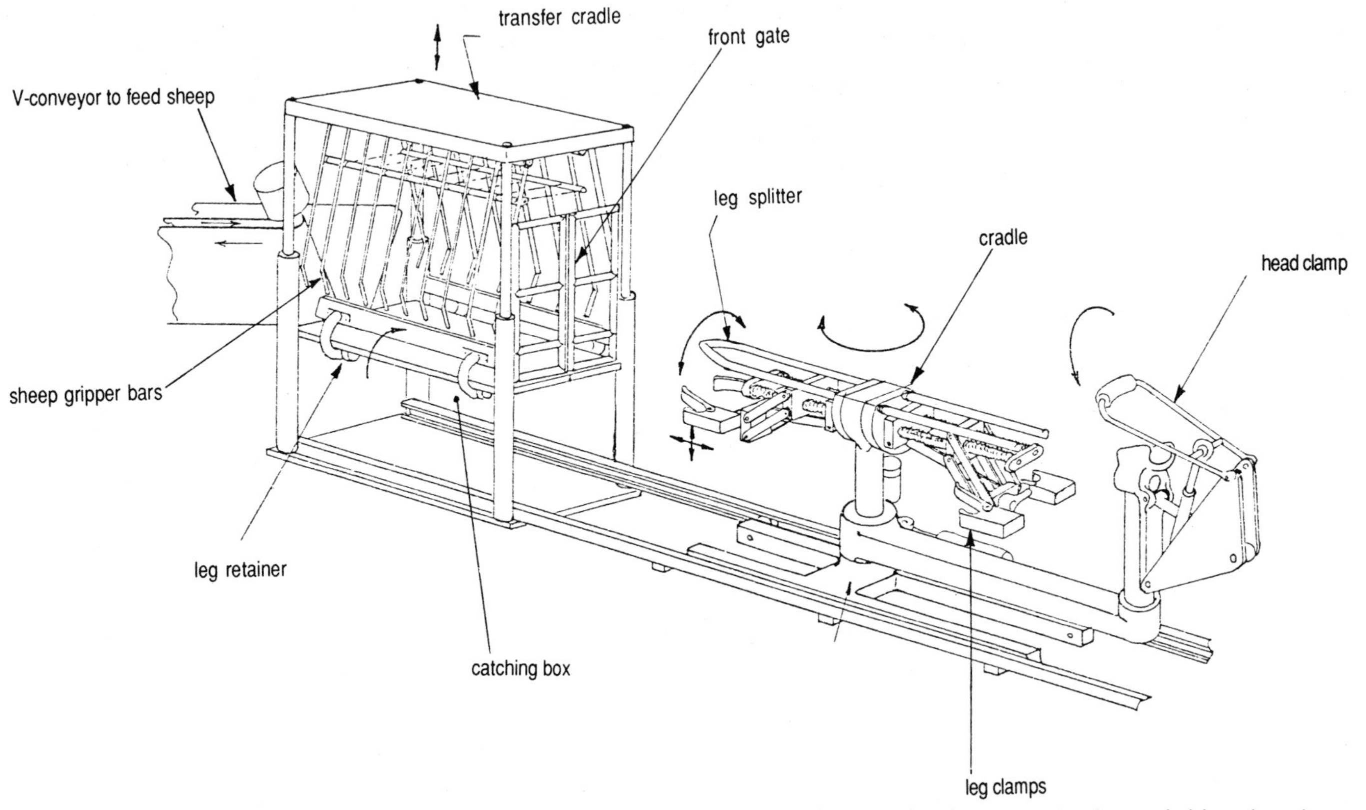

Figure 7.21 Melbourne University sheep catching and loading concept—mid 1981. The sheep remained remarkably calm when picked up by the multi-fingered gripper in laboratory tests. From Burrow (1981) by permission.

of the Australian Merino Society[1] under the chairmanship of a young and influential woolgrower, Bevan Taylor.

These developments soon caught the attention of another engineer turned grazier, Lance Lines, in South Australia. Lance had worked for the Weapons Research Establishment near Adelaide during World War II, and he appreciated the advances in technology which could be applied to farming. A sheep breader like his Western Australian colleagues, he was running the family Gum Hill stud in the mid-north of South Australia. He interested two of his Weapons Research associates in the idea—John Baxter and Gordon Halliwell—and by 1972 they were experimenting with lasers for cutting wool (p. 135).

Like Norman Lewis, they realized that a lot of money was needed to develop sheep-shearing robots, and decided to form a public company in association with the West Australians. The company Australian Merino Wool Harvesting (AMWH) was born. Lance Lines took a big proportion of the stock and the rest was subscribed by woolgrowers throughout Australia.

Lance realized that one essential requirement was to keep the sheep still. With his South Australian colleagues, he hit on the idea of electro-narcosis, passing a small pulsating electric current through the body to freeze the muscles. Originally invented in the USSR for pain control, it had been tried on human beings and animals. They developed this idea into a small electronic device, resembling a torch, with two clips to attach the wires to the skin of the animal. Once attached, the current could be increased until the animal simply stood still, muscles firm and taut, still breathing, though somewhat stressfully.

They realized it would be ideal for a host of animal operations, such as castrating cattle. Normally, the young bullocks were held spreadeagled by ropes, lying on the ground in a blaze of fear. With the 'Stockstill' immobilizer, as it became known, the animal remained standing for the minute or so needed for the operation. Once the clips were removed the animal bounded off, seemingly none the worse.

The device was controversial. The Australian Society of Veterinarians firmly opposed it from the beginning on the grounds that it was no substitute for anaesthesia, and that it was a stressful experience for the animal in any event. But AMWH was undeterred. They obtained worldwide patent cover, and the units started selling, albeit slowly. The aim of selling the device was to generate cash flow to support research and development on shearing robots.

Lance Lines' dream was now emerging. A forthright man, he was disdainful of any government involvement. He had immersed himself in the ethos of the stud-breeders, that they alone were the only source of *really useful* innovation in the wool industry. Government research was seen to be ponderous, slow, mostly irrelevant, and lacking in practical application. Government funding of any kind was to be shunned, as it brought with it restrictions, control, bureaucracy and interference.

[1] A cooperative sheep breeding society based in Western Australia. So called AMS sheep are sought after for their wool production and easy care characteristics.

Our first introduction to AMWH came at a Corporation technical coordination meeting in March 1980. John Baxter represented AMWH, and listened quietly for most of the meeting. Then at the end, in his marvellously relaxed voice, he described their preparations for a prototype robot shearer which would be in action by July 1981 (figure 7.22). The sheep would be supported upright, electrically immobilized, and shorn by two robots, one on each side. The unit would be simple, small, and easily transportable. The immobilizer, he claimed, would hold the sheep still and taut so no manipulation of the sheep would be necessary.

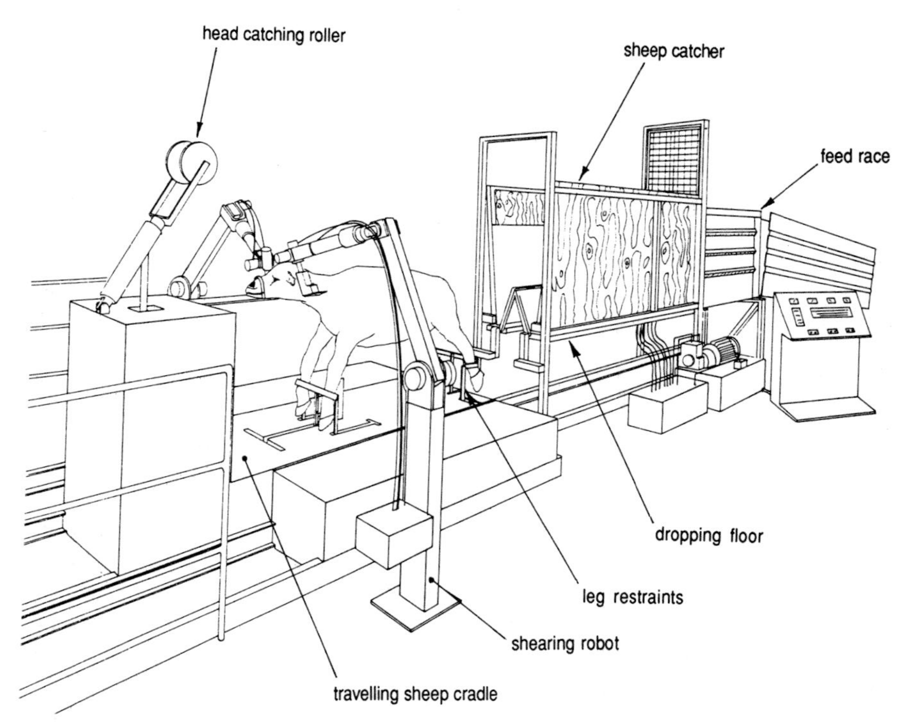

Figure 7.22 Merino Wool Harvesting proposal for a small transportable shearing plant—mid 1981. From Baxter (1981) by permission.

The next brush with AMWH was not quite so relaxing.

The 'National Farmer', a weekly newspaper, published a major article on AMWH; the article[2] featured a photograph of ORACLE in action. Yet in the text just above the photograph the article claimed that AMWH were 'years ahead' of the UWA effort! Before they even had a robot!

This caused a good deal of consternation, particularly as the Corporation had placed major restraints on publicity for our project. On taking the matter up

[2] *National Farmer*, October 2nd 1980, pp. 4–5.

with the editors, we were told that they had simply printed the information supplied to them by AMWH since they had been unaware of our own work. Not surprising, since the Corporation's restrictions on publicity prevented us from providing information.

By now it was clear to us that AMWH were in competition with us. They had not yet applied for research funds, but we thought that was just a matter of time.

In late 1980, AMWH began to step up their robotics work. They obtained a Puma robot for shearing experiments. A television news report showed the Puma performing quite conventional robot-like actions while the voice-over described the shearing cutter which would soon be fitted. Yet the Puma is a small robot for light work—not a good choice for hard physical labour. After John Baxter described their progress at the next meeting in July, Jim Blair showed how difficult it would be for a Puma robot to shear a sheep in a reasonable time. John Baxter was quite unconcerned. He side-stepped the most cogent technical criticism: 'Yes, I can see that might be a problem, but we don't believe it will be. If it really is a problem, then we will deal with it when we come to it'.[3] They soon had to.

Like us, AMWH faced a major shift in direction during 1981. In an optimistic atmosphere, Lance Lines and John Baxter made a strong push for more funding, and Lance Lines wanted to put more capital into the company. The Western Australian board members of AMWH were more cautious and could see that if Lance was allowed to proceed, the balance of power on the board would permanently favour the South Australians. So they decided to pull out, and the focus shifted to Adelaide with a new name—Merino Wool Harvesting Pty Ltd. (MWH)

John Baxter formed an agreement with an American company, Cadillac Gage, to design an hydraulic robot arm for them. Jim Blair's calculations had some significance. Cadillac Gage were to contribute much of the development effort on the understanding that MWH were gearing up for a major production effort. They were manufacturers of rotary hydraulic actuators, and their main market was tanks for the US Army. They were keen to diversify into robotics, but had little control expertise of their own. The arrangement with MWH was to exchange control technology in return for hydraulic arm technology. MWH had enlisted the help of Chris Abell, a brilliant and ambitious engineer who graduated from fluid mechanics research at Adelaide University to form his own consultancy. Chris and his colleague Allan Wallace designed the control system for the new robot. They drew on their defence consulting experience which supplied them with a ready-made digital actuator controller, built for a much more demanding application.

A managing director was appointed, allowing John Baxter to focus on technical direction of the research effort. A detailed development plan was drawn up

[3] Quoted comments are not verbatim but are based on the author's recollections.

to help them obtain a government R & D grant. Elders, the great Australian pastoral company, was also persuaded to inject some much needed capital.

John Baxter was sufficiently confident to discuss the details of the plan with Roy Leslie and myself. We were surprised. For it seemed that they expected to be 'given' the results of our research and receive training from us. Their costings for development work suggested that they would use our research results to design their robot; we were to supply them with free consulting to allow this to happen successfully. Further, the time-scales seemed quite unrealistic. They proposed building a laboratory robot followed by a full field trial version in 18 months.

The Corporation were not at all keen about the idea and promptly agreed with our suggestion that no detailed information be supplied to MWH for the time being. Clearly, the free exchanges and technical coordination meetings we had found so useful could not continue.

In late 1981, MWH applied for Corporation R&D funding. They had called up subscriptions from their woolgrower shareholders. If they sought more capital from Elders, already a major shareholder, they would lose control on the board. Their position was precarious; woolgrowers who had subscribed had little hope of recouping their investment and would have claimed their tax write-offs for research. So, Elders would have had little trouble in purchasing their shares, should they have wished to do so.

The Wool Corporation, on the other hand, seemed reluctant to support MWH. They would not have wanted a repetition of the CSIRO problems, nor duplicated research efforts. They would have been aware of Lance Lines' attitude, yet MWH claimed to be a woolgrower-based company which could not be ignored for political reasons.

MWH made a convincing plea. John Baxter portrayed our work as magnificent but 'esoteric' research work far removed from practical application. The MWH approach was different—it would appeal to the 'electorate', the small farmer, the battler who needed help which new technology could provide. They won their case, twice over, because MWH used the Corporation funding as a lever to attract further government R & D grants.

This development confirmed our worst fears. In February, we made our concern known through our best supporters among local woolgrowers. MWH were geographically closer to centres of power and funding, and used publicity to their advantage. We were concerned for two main reasons. First, we thought that MWH would need some of our research results if they were to keep to their proposed timetable. The confidentiality required to protect aspects of our work could enable others to claim credit for doing it. The memories of their earlier publicity was still fresh in our minds. Second, if MWH failed, they might bring discredit to the idea of robotic shearing. This could have harmed us if not appreciated and anticipated.

While MWH saw us as 'esoteric' and impractical, we saw their approach as equally impractical for very practical reasons. First, we had found *by experi-*

ment very early on that manipulation of the sheep was essential for complete shearing. It was just not possible to shear an upright sheep restrained in just one position. MWH had developed a neat way of supporting a sheep with two stands—one under the crutch and another under the brisket between the front legs. Almost no wool was obscured. Even so, there is more to shearing than just being able to get at the wool. The skin must be reasonably flat and taut, and you can only get that by stretching parts of the animal or moving the skin with the left hand. The immobilizer tightened the muscles but not the skin.

For this reason, they retreated from complete shearing by resorting to a human shearer to clean up the wool left behind by the robot (Baxter 1981). From our own experience, we found that it took a surprisingly long time to clean up. This happens because the robot shears the easy parts, leaving the hard bits behind.

When you think carefully about the consequences, you realize that by using the robot to do the easy shearing, the shearer cleaning up gets a *harder* job, not an easier one. He is always cleaning up the bits and pieces. So the shearer will naturally want to be paid more, and after allowing for the cost of the robot it is difficult to see the economic gain.

MWH were wise in aiming for partial automation of shearing as a first step. Experience has shown that attempts to fully automate any manual process are often unsuccessful. The point which we believe they missed was that partial automation has to make the operator's job more rewarding to obtain a worthwhile improvement.

The Corporation seemed to ignore our concerns and continued funding MWH. We were instructed to collaborate on one or two limited areas of 'common interest' such as the cutter mechanism. At this time, we were running into problems with the cutter on ORACLE, so this made sense.

In 1983, PATS were engaged to perform a shearing system design study for the Corporation and, when we received a draft report, we found that our *experimental* results were being compared with MWH performance *predictions*. The MWH predictions were to justify their project financing, so they had to be optimistic. Particularly in the Australian financial climate of the mid eighties—there was no place for realism. In the subsequent economic study by McGowan International, our concepts for automated shearing were not well received. While the report did not rule out robot shearing, the prospects were not that bright. We responded by criticizing the report for lack of depth, presenting doubtful data of unknown origin, and once again for comparing predictions made by MWH with our experimental results, without any qualifications on MWH's credibility. By then it seemed to us that their confident predictions for field trial tests by July 1983 had been forgotten—their first laboratory robot was still in pieces awaiting components from overseas.

Gently but firmly, we moved quietly towards our own concept for a new shearing robot . . .

8
Shear magic

Our sidetrack into holding sheep may well have delayed the birth of our proper robot, but it brought a finely tuned, creative design team to the job at just the right moment. Six years later, when the robot design was reviewed in depth by Victor Scheinman,[1] he was to pose the question 'If you walked in tomorrow to find a *perfect* robot, would it be any better?' His answer was 'Not much'. And that is tribute enough. Of all the developments in the project, the robot which we christened *Shear Magic* was just that. When we selected the name, and hastily abbreviated it to SM until we were able to unveil the robot, I remember a prescience, a knowing, that this was to be a breakthrough, a pinnacle of the project.

This was to be our *real* robot. ORACLE always had been a test rig—a tool for shearing experiments. Since the beginning we had always assumed that a real shearing robot would look quite different, even though it would incorporate similar ideas.

Few robotics researchers have the chance to design a robot—most have to work within limits imposed by commercial manufacturers. But it is not easy to design a high-performance robot. We had the advantage of a single application (shearing), and almost unlimited freedom in our choice of design.

Concept design

We knew that our new robot would be a radical departure from ORACLE. ORACLE had some excellent design features, but its usable workspace was small in comparison to its apparent size. Figure 8.1 shows the workspace, seen from the side. The small circular arcs show how the cutter can be oriented at each of 100 or so points within the workspace. The only zone usable for shearing is the lower left-hand corner. The main reason for this is the length of the surface follower mechanism—in particular the distance from the cutter tip to the tilt axis (figure 8.2). The workspace must be large enough to accomodate this length when repositioning the cutter outside a full fleece which can be up to 150mm deep.

It is worth recalling the good design features before enumerating the points of departure.

The upper link is driven by a closed chain, pantograph mechanism. This removes most of the dynamic coupling between the upper and lower link. Put

[1] Victor Scheinman—inventor of PUMA robot and ROBOTWORLD assembly machine.

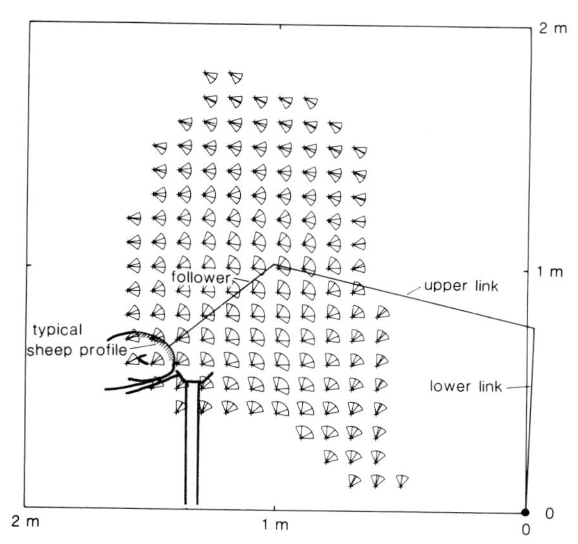

Figure 8.1 Workspace of ORACLE robot. Each point in the workspace is shown with an arc illustrating the orientation range available—the middle section represents the tilt range and the outer sections represent the additional angle from the wrist range.

another way, the reaction forces which result from accelerations of the upper link affect the lower link less than if the upper link actuator were placed across the joint between them—as it would be with an open chain mechanism. The reduction in coupling allowed us to use a simpler control system which ignored coupling effects.

The wrist design separated cutter rotation and translation effects, simplifying the kinematic solution of the arm geometry, substantially reducing the computation loads for a given bandwidth.

The first point of departure from the ORACLE design was to reduce the stroke of the surface follower. Recollect from p. 31 how the stroke was calculated. Our shearing experiments showed that the maximum practical speed for the cutter was about half what we had originally assumed—about 350 mm/sec. A shearer runs the cutter faster for just one or two of the 70 or 80 blows on each sheep. The skin stretches ahead of the cutter, and a slower blow can be more efficient. Top shearers run each blow quite slowly, but make sure they use the full cutting width of the comb. We found that ORACLE needed only about 40 mm stroke out of the 150 mm provided. (With a speed of 350 mm/sec and a bandwidth (achieved) of 3 Hz, the rule of thumb predicts a requirement for ±20 mm stroke.) Assuming that a new robot would have a better bandwidth, the follower stroke could be as little as 20 mm; we aimed to provide 50 mm.

With such a short stroke, the surface follower movement could be relocated to the end of the wrist mechanism. This way, there would be no need to mount

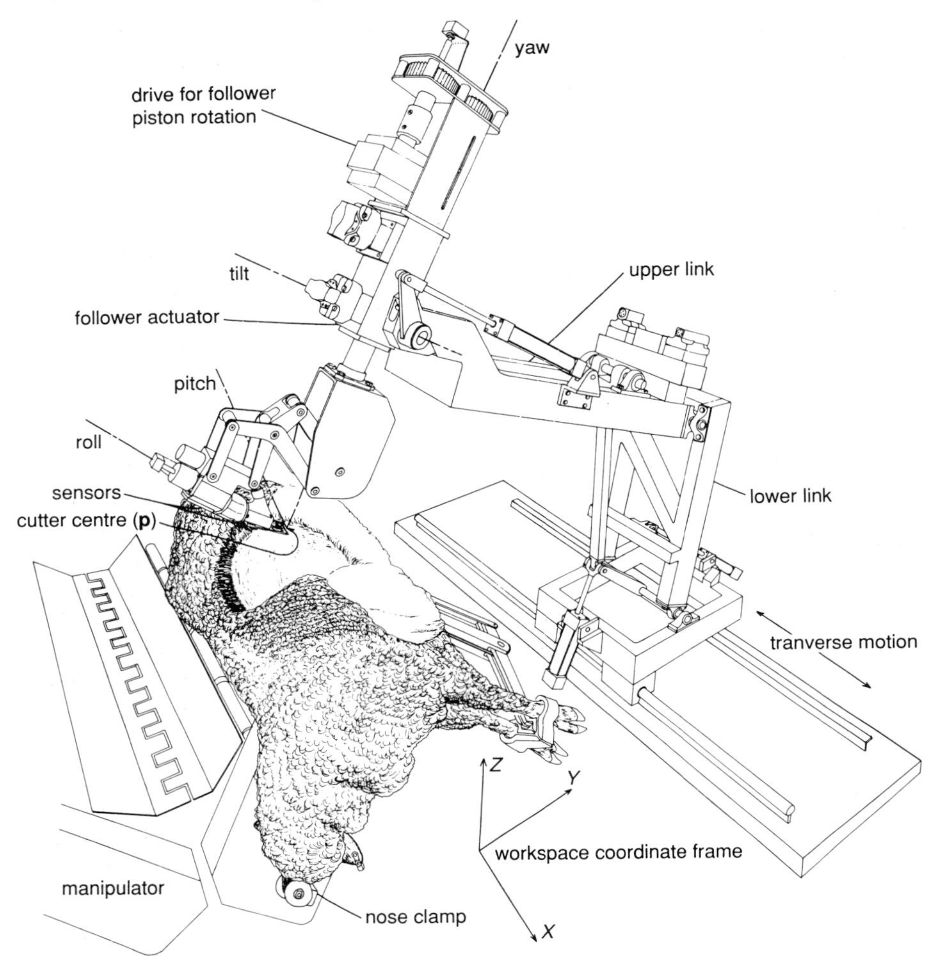

Figure 8.2 ORACLE robot (from chapter 2)

the wrist on the surface follower actuator and subject it to the damaging vibration levels which the ORACLE wrist suffered. (Cable and connector failures due to vibration were a regular part of life with ORACLE.) Given a reasonably light cutter design, the moving mass could be reduced from over 25 kg to about 1 kg.

The next point of departure was equally significant. We needed less bandwidth for wrist orientation than we had expected. ORACLE failed to meet our original expectations, but it was still able to shear with a wrist bandwidth of about 5 Hz. This factor, as well as advances in computer power, meant that we could consider conventional wrist mechanisms. The arm could respond quickly enough to keep up with wrist orientation changes.

Next we decided to mount the new robot over the sheep. A sheep has to be supported on the underside, so the upper side is exposed for shearing. An overhead robot could access all the upper, exposed surface. Unlike ORACLE which had always been constrained to work on one side. And it would leave the floor clear for the sheep and wool handling equipment. Wool, dirt, faeces and urine would run down, away from the robot mechanism. The shorter reach would allow a lighter and faster machine to be designed.

Finally, and above all else, the robot had to look simple. We knew how simple ORACLE was, from an engineering standpoint. Yet to convince any visitor of this, gazing at its mass of exposed cables and piping, was an uphill battle. 'That's too complicated. It'll never work in a shearing shed!' Time after time we heard the same comment, particularly from farmers. SM was to be a turning point—and appearance was a crucial part of this. We knew that the design would have to be slightly more complicated to achieve apparent simplicity. But like ORACLE, SM would always be a laboratory creature. There would be a time for simple engineering, later.

Our economic analysis of a commercial robot shearing system had shown that it was desirable to keep the shearing robots operating as much of the time as possible. Loading sheep into an ARAMP type manipulator was clearly going to take some time, so we developed the idea of moving the robots from one workstation to another. Robots would share work between them. While one manipulator was being unloaded and a woolly sheep was mounted, the robot normally assigned to it would move to another shearing station and assist the robot there to do its job faster. Imagine the complexities of programming! In this way, limited mobility and shared workspace between nearby robots became a feature of the SM design.

Analysis and model building

The first stage of the conceptual process involved diligent observation, and exhaustive analysis of the design requirements. Michael Ong worked with us first as a graduate student, and then as research engineer. For his first two years, he exhaustively analysed every aspect of the performance of the ORACLE robot. Its kinematics, dynamics, control, accuracy, repeatability, hydraulic efficiency, its ungainly structure, and vibrations. Every aspect was documented in meticulous research notes. Then, as ARAMP neared completion, Michael started to look for new ideas. He would arrive about midday after working till the small hours the night before, a flimsy cardboard model of a wrist mechanism cradled in his hands. By the end of 1983, he had analysed every conceivable wrist joint and robot structure.

He built a scale model sheep for his desk, with articulated legs and all the preferred shearing blows painted on. Mechanisms built from a FAC construction kit littered the shelves. He shared an office with fellow graduate student Peter Kovesi, and their room soon became known as the 'dream factory'!

Once ARAMP was completed in September 1983, the attention of the whole team focused on the designs emerging from the dream factory. Performance parameters were quickly confirmed, and a full-scale model of a telescoping pendulum arm was bolted to the ceiling of the laboratory (figures 8.3 and 8.4).

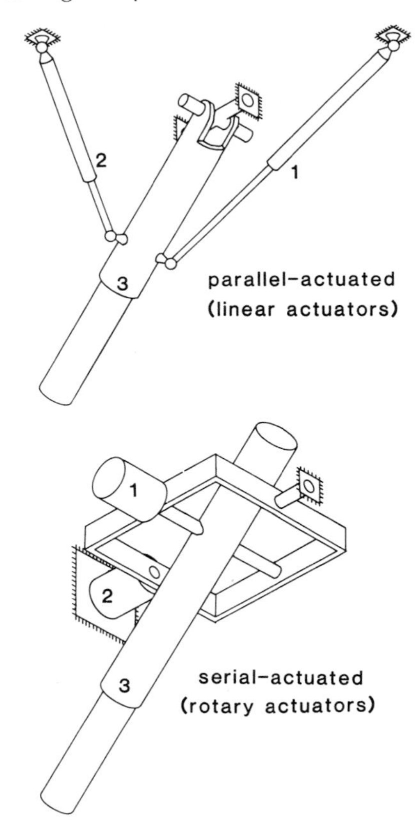

Figure 8.3 Double swing, telescopic pendulum arm configurations. The parallel actuated arm results in dynamic coupling between the actuators. The coupling can be almost completely eliminated by adopting the serial actuated arrangement.

In outline, the performance parameters were

(1) compatibility with ARAMP manipulator
(2) shearing speed—500 mm/sec
(3) reposition speed—2500 mm/sec, wrist rotation 315° /sec.
(4) repeatability—3 mm
(5) accuracy (relative to markers)—5 mm
(6) absolute accuracy—10 mm
(7) arm acceleration—5 g
(8) follower acceleration—50 g
(9) forwards shearing force—120 N
(10) arm actuator bandwidth—10 Hz—mechanically decoupled arm actuators.
(11) arm structure natural frequency—20 Hz

Yet by December we were facing unacceptable complexities.

Michael and Peter had tested every wrist arrangement they had uncovered.

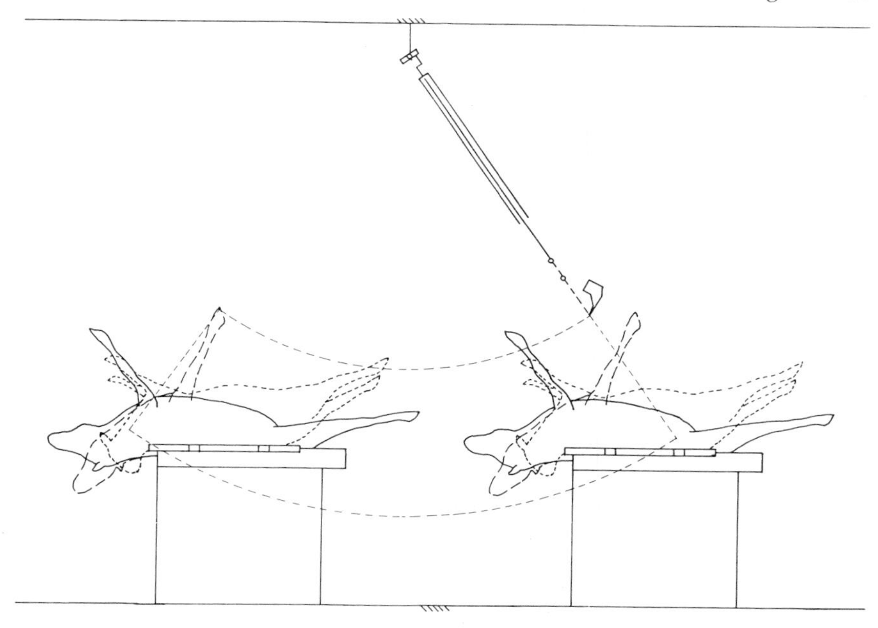

Figure 8.4 Sheep location in relation to pendulum arm. Sheep are supported in ARAMP type manipulators which can be turned about vertical axes. One manipulator can be reloaded while other manipulator is used for shearing.

They ran into the same problem each time—singularities. No matter which way the wrist was arranged, a wrist singularity would interfere with some shearing blows, with all the consequential problems.

Singularities had never been a problem for ORACLE. The range of wrist orientation had to be restricted to about ±30°; at greater angles, the direction of the surface follower was too far from the surface normal. Thus, singularities were avoided by restricting the range of wrist orientation. Additional orientation freedom was provided by the tilt actuator. With SM, the aim was to provide a larger range of usable wrist orientation, and singularities could no longer be avoided.

We wanted to keep at least 30° away from singular wrist configurations during shearing blows—any closer and there was a risk that adapting to a rogue sheep would someday take the robot too close.

The only way out seemed to be an extra joint in the arm. One alternative was a four-axis wrist; the other was a joint towards the bottom of the arm to bend it. Both solutions still required large rotation ranges. Michael thought he needed at least two complete turns either way. But when David Elford heard of this, he asked Michael to think of the consequences for hydraulic hoses and wiring by imagining himself being wrapped twice round a set of bevel gears!

The extra joint was a worry. We had learned a simple rule of thumb—robots double in weight at each joint. With a 1.5 kg payload (the cutter), SM was

Singularities

Imagine watching an aircraft flying straight towards you. You lift your head as it nears the zenith above. Then, you flip around smartly to avoid a nasty fall backwards or lose sight of the plane.

That's the aircraft-watcher's singularity.

If the aircraft is not heading straight for you, but passes off to one side, you turn gently, keeping it in view all the time. You only need to flip round if it passes more or less overhead.

This illustrates the two problems which stem from singularities. First, sudden flips are needed—fast joint motions are demanded: both computational methods and drive motors are strained to their limits.

Second, large ranges of rotary joint movement are used up quickly. For example, the aircraft watcher needs 360° of body rotation to watch an aircraft approaching from any direction. But if the watcher needs to follow the aircraft as it passes, 720° is needed, just to cope with the effects of the singularity! And that assumes that the watcher is correctly positioned within the central 360° of that total movement for the initial approach.

Robots suffer from the same problems, but because they have more joints, they often have more singularities and the consequences are more complex. Sometimes almost impossibly so. Most research papers do not account for the effects of joint rotation limits which practical concerns impose; the few that do show that the singularities interact with these limits to create invisible barriers in the six-dimensional movement space of a robot. Even if you try to avoid the singularities by adding extra joints, the barriers remain, but they are all the more complex and hard to locate.

already weighing about 150 kg; adding an extra joint would add another 150 kg at least. The extra weight would make control more difficult, and the extra joint would complicate the kinematics calculations for the computer.

Michael, Peter and I discussed the problem one warm December afternoon after a long session with models and a very patient sheep. In one of those rare moments, facing what seemed like defeat, or just what some people euphemistically call 'reality', we gained a sudden insight. What if the cutter were mounted on an elephant's trunk, free to pivot at the end? It would not have any singularities! Rotation limits would be no problem! But building an elephant's trunk was a problem.

Michael Ong and I both devised the same solution at about the same time. Two Hookes joints could be used, with their rotations symmetrically coordinated. Two actuators would be sufficient provided mechanical coordination could be devised. We called it the ET (elephant trunk) wrist.

We were all excited by the idea—it could be a breakthrough, but would it work? Michael started searching for suitable mechanisms to coordinate the joints. In the meantime, David Elford pressed on with a three-roll wrist (figure 8.10) mounted on the end of an arm-bending axis as we could ill afford to risk everything on an untried wrist design. The three-roll wrist was chosen because its 'clean' exterior would not damage hoses and cables when they became wrapped around it.

The ET wrist remained in the dream factory. Michael devised a variation of the Clemens joint (Hunt 1978, p. 427) but found that it would be too compliant, and would be susceptible to backlash.

By March, David and I had growing concerns about the robot design. First, with the weight at about 310 kg confirming our rough predictions made earlier, the robot seemed to be too large and powerful for the job. It would need 20 kW of hydraulic power to operate. Second, while the double pendulum design was efficient in reducing the moving mass, much of the advantage was lost when the additional arm-bend joint was introduced. While the pendulum arm could access two sheep from the one position, an arm with a single swing axis on a translating cradle seemed to offer some important practical advantages (figure 8.4, 8.5). One was the ability to run long shearing blows along the body and legs of the sheep. Zbigniew Lambert added a further advantage when he devised the shuffling arrangement for the shunt and carriage—thus allowing the robot to move itself along several metres of track but retaining excellent dynamic response within a limited range. We decided to switch to this latter arrangement using the ET wrist, with the three-roll wrist as a back-up. There would be no arm-bend actuator in either case because the extra weight was unacceptable.

Two weeks later, we reviewed the situation. David had discovered several design complications in the ET wrist. He needed another two weeks to find how to fit in the actuators, evaluate different gearing arrangements and check the design for strength.

On the positive side, Peter Kovesi had discovered a kinematic solution for the wrist mechanism which overcame another difficulty. The axes of conventional wrists meet at a single point; because of this, the kinematic solution of a six-joint robot can be separated into two three-joint problems which are simpler. The ET wrist has no fixed centre of rotation and cannot be treated in the same way. Peter's solution involved much less additional computation than we had expected.

We could have proceeded with a three-roll wrist design immediately, but we decided to wait a little longer.

Even at this stage, we still kept the nascent ET design to ourselves. The Corporation were keenly monitoring our progress and we had briefed them thoroughly on all aspects of the design based on the three-roll wrist. The ET wrist would provide a big performance improvement and simplify programming, but we wanted to be very confident before telling anyone about it—even the Corporation.

Two weeks later, David had completed enough of the design to know that ET was the answer we had all hoped for. Now we had a simple robot with a large workspace uncluttered by singularities. We announced the design to the Corporation, but at the same time asking them to keep it confidential. We

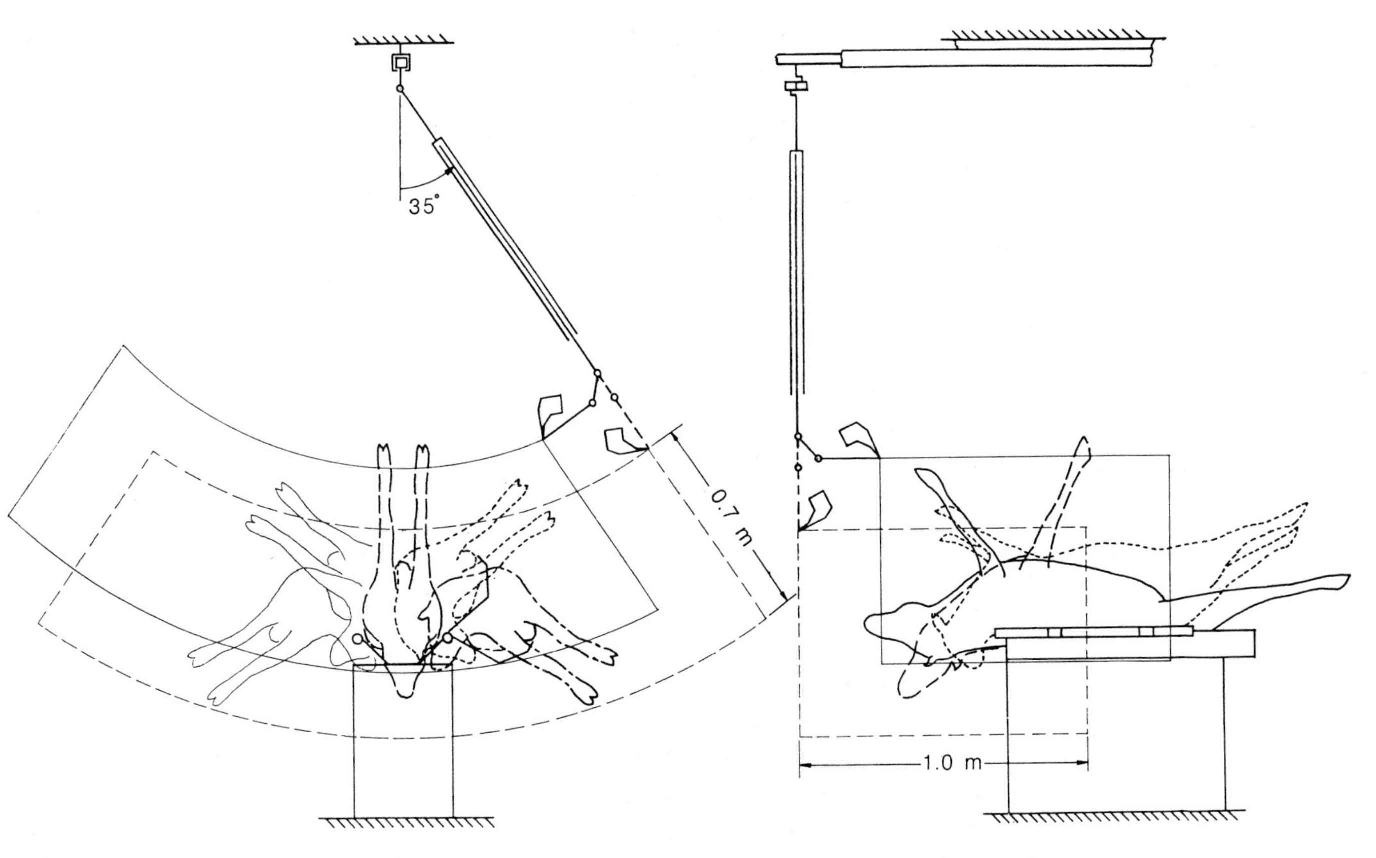

Figure 8.5 Workspace of traverse—swing arm adopted for SM. In 1988 the traverse range was extended to 1.75 m.

caught a wave of excitement as we began to realize the implications of the wrist design. Watching videotapes of factory robots, we realized how much simpler the movements of robots could be without singularities. The twin track design for a robot system (figure 8.6) allowed two robots to share almost all of their workspace. By alternately locking the carriage and shunt to the track, each robot could shuffle along its track to access an almost unlimited workspace, yet within a one metre working range could still display an impressive dynamic response and maximum speed (figures 8.6, 8.31 on p. 245).

In May 1984, we moved into the detailed design phase while the workshop technicians built a full-scale model of SM. The model was used to confirm design details such as actuator ranges, and later to show off the elegant simplicity of the design (figure 8.15, p. 216). The 'Shear Magic' logo on the title page was designed and has appeared on all our work since. The Corporation board met in Perth the following August, and we staged an official 'unveiling' ceremony when board members visited the laboratory after their meeting. John Silcock, a woolgrower and vice-chairman of the Corporation who sat on all the important research committees, was heard to say 'Yes, that looks like it might just work in a shearing shed.'

The details at the heart of the SM robot concept are most succinctly described in the paper 'ET—a wrist mechanism without singular positions', an edited version of which appears in the following pages. The paper also presents mathematical details of the wrist kinematics. I have eliminated introductory material which duplicates the preceding explanations. If the transition in style offends, I apologize—the content is rather more specialized and the reader will be forgiven for skimming ahead to p. 227.

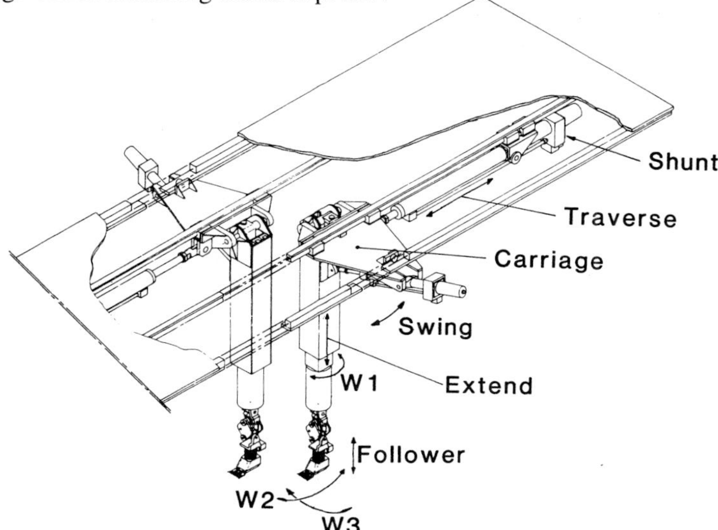

Figure 8.6 Dual SM robot configuration. The arm centrelines are separated by only 250 mm so they share a large common workspace.

ET—A WRIST MECHANISM WITHOUT SINGULAR POSITIONS

James Trevelyan, Peter Kovesi, Michael Ong, David Elford.

From the International Journal of Robotics Research, **4**, (4), p. 71–85, by permission.

1. Importance of wrist design

Experimental shearing of sheep with the ORACLE robot has demonstrated that the design of the wrist is a critical factor in determining the usable workspace of a shearing robot. This factor is common to many other applications for robots (Ong 1981; Trevelyan 1985; Milenkovic 1984).

2. Review of alternative wrist arrangements for SM robot

In order to refer unambiguously to motions of wrist mechanisms in the following discussion, we designate the three actuated motions of each wrist as *W1*, *W2* and *W3* (figure 8.7). Respectively, these are the proximal, mid, and distal motions of the wrist relative to the robot arm. The motions of the robot end-effector generally consist of non-linear combinations of these three motions. Although wrist mechanisms involving four or more actuated motions were considered, the elegance of using no more actuators than are really needed for the task and the presence of some very awkward problems of mechanical design led us to search for a satisfactory design using three motions.

The desired kinematic range of the wrist can be visualized in terms of a workspace cone with an included angle of approximately 210°, centered on the *W1* axis. This cone (figure 8.8) defines the range of directions required of the *W3* axis. Within this cone, the cutter needs to be steered in any direction. However, the proposed wrist mechanism has a singularity cone (Paul and Stephenson 1983) in the centre of the workspace cone. When the axis of the *W3* actuator lies inside this cone, significant additional velocity and torque demands are placed on the wrist actuators, compromising dynamic performance. This type of wrist mechanism and the alternative arrangement shown in figure 8.9 have been widely and successfully used with factory robots because the motions of the robot can be arranged to avoid the singularity cone by either trial and error or more systematic techniques (Yoshikawa 1984). A factory robot merely repeats the originally programmed operations, and so it can always avoid the singularity cone once it is programmed to do so. However, the motion required of a robot that reacts to changes in its task or its environment cannot be predicted in advance other than by concepts such as probability functions. Therefore, the usable workspace of such a robot needs to lie inside the region of the workspace that is well conditioned kinematically. It can be seen that the presence of a singularity cone in the middle of the wrist workspace imposes a severe limitation on the usefulness of the robot in such applications.

Mechanical studies of the wrist arrangement suggested the use of differential gears to drive the *W3* and *W2* motions so that actuators could be placed in the arm of the robot to reduce the dynamic loads on the wrist. However, hydraulic

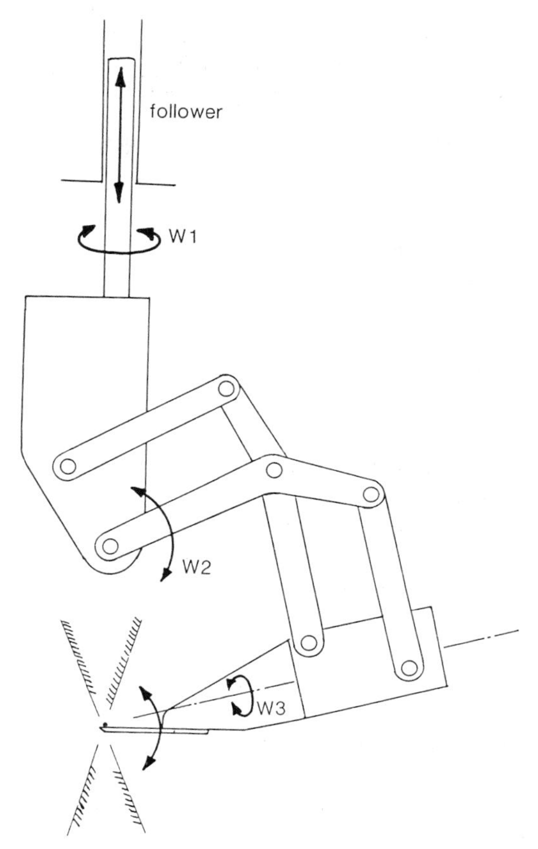

Figure 8.7 ORACLE wrist. The singularity cone for the W3 axis is shown hashed. Motion limits prevent the W3 axis from entering the cone.

services to the follower actuator and the cutter drive motor would have to be carried in an external flexible umbilical cable assembly between the arm and the cutter assembly, outside the wrist mechanism. Further, to overcome the effect of the central singularity cone, an additional actuated motion was required to 'bend' the arm so that the singularity cone could be reoriented. Practical problems with the umbilical cable were explored with the help of models. It was difficult to conceive of an arrangement that would provide the full extent of motion required without exposing the cable to damage from moving parts of the mechanism.

Attention was directed to a 'three-roll wrist' (Stackhouse 1979) which could be arranged with a 'clean' spherical exterior to accommodate an umbilical cable more easily without risk of damage. The mechanism of the wrist could be readily extended to provide a mechanical cutter drive motion through the centre (figure 8.10). The arm-bend motion was still required, because the three-roll

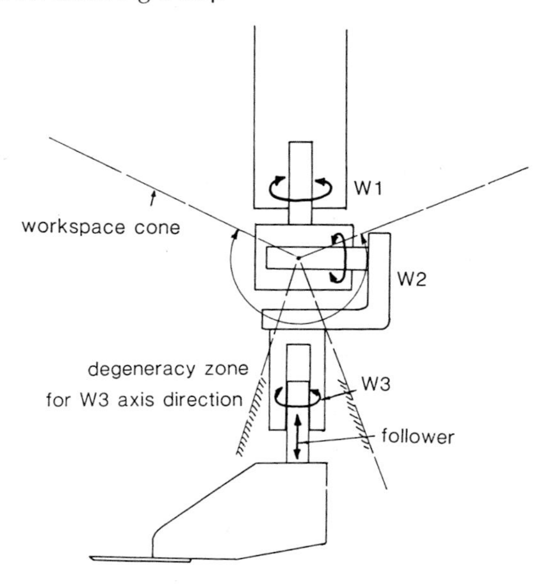

Figure 8.8 Typical orthogonal wrist.

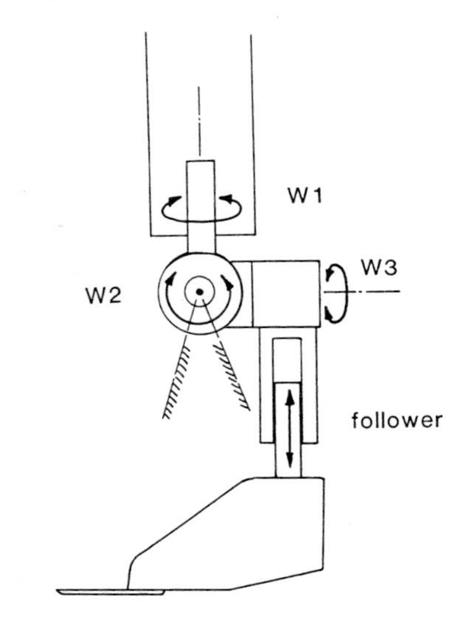

Figure 8.9 Alternative orthogonal wrist.

wrist also has a central singularity cone; however, the mechanical arrangement seemed to be more practical than that for the original proposal. The concentric drives for the wrist originated in a gearbox containing three rotary hydraulic actuators, servo valves, and a hydraulic drive motor for the cutter. The gearbox was pivoted to the end of the arm on the 'bend' axis. The penalty for incorporat-

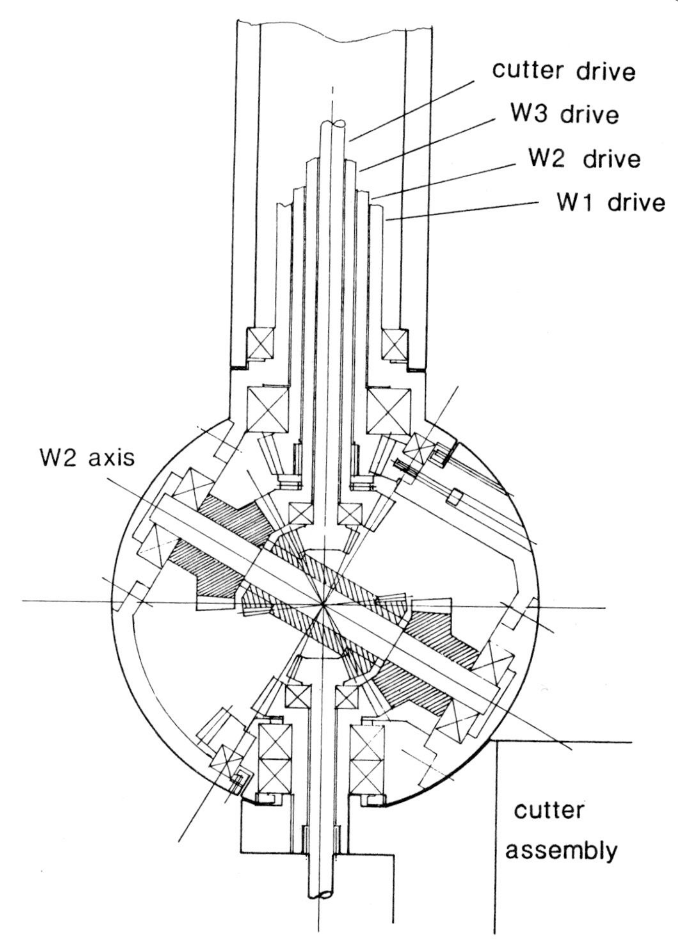

Figure 8.10 Three-roll wrist adapted for additional internal drive for cutter.

ing the bend axis was an increase in the power requirements of the proximal arm actuators (swing, extend, and traverse), which in turn required more rigid supporting members with greater masses.

A full-scale model of the SM robot was constructed for kinematic tests with a sheep in the manipulator. While all the motions were provided, none was actuated except for the arm extension, which was counterbalanced for convenience.

It was soon apparent that a major problem had been overlooked in previous analyses. The limits placed on actuator motions, which were required to prevent excessive wrapping of the flexible umbilical cable round the wrist, were often too narrow even though more than 360° of free movement was available for both the *W1* and *W3* axes. It was evident that certain shearing paths resulted in complete rotation of wrist axes even though the shearing direction had not changed appreciably. Figure 8.11a illustrates this effect; a shearing path is shown traversing a surface, keeping clear of zones in which the wrist would be required to be within a singularity cone. The final shearing direction is almost the same as the initial shearing direction, and the steering angle varies by only about 60° in each direction. However, at the end of the path, the *W1* and *W3* axes have rotated through a complete revolution in opposite directions. Figure 8.11b illustrates the locus of the required *W3* axis direction plotted on a unit

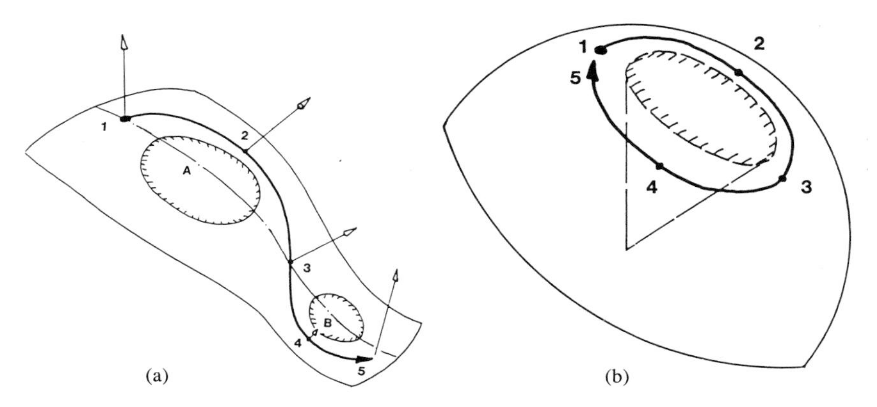

Figure 8.11a Shearing path on a surface avoiding zones A and B where degenerate wrist configurations would be required.

Figure 8.11b Locus of surface normal of shearing path shown in figure 8.11a traced on a unit sphere. The singularity (degeneracy) zone for a typical orthogonal wrist is also shown (figure 8.8). Because the locus encloses the zone, a full 360° rotation will be needed in *W1* and *W3* even though there is no net orientation change.

sphere. The locus encloses the singularity cone. If the locus resulting from a different path does not enclose the singularity cone, then a full rotation of the *W1* and *W3* actuators is no longer required. To avoid encountering limits during the shearing movement, the wrist axes would need to be carefully aligned beforehand in a surface-dependent and path-dependent manner. Depending on surface geometry, some paths would require more rotation than could practically be provided and would require an interruption during the planned path to 'unwind' the actuators (sometimes known as a flip). However, in the case of shearing, resumption of an interrupted blow is neither easy nor desirable and the interruption is a major performance penalty.

A rearrangement of the wrist as shown in figure 8.9 results in a different relation between surface orientation and singularity cones. Tests with the kinematic model of SM showed that the singularity problems were simply relocated from one part of the workspace to another. Shearing is possible on any part of the exposed sheep surface with this arrangement, but it is not possible to shear in all directions without encountering the effects of the singularity cones described above. The design of ORACLE neatly bypassed these problems because the singularity cones were placed (deliberately) outside the range of wrist-actuator motions.

A consensus view was reached: there had to be a better solution. Although the combination of the three-roll wrist and the arm-bend actuator satisfied the workspace requirements, it involved an awkward mechanical arrangement. When dynamic requirements were included, the addition of an extra actuator in the arm led to substantial increases in the size of proximal actuators for the three arm motions and to corresponding increases in stiffness and weight.

3. ET wrist

It was noted that placing a cutter on the end of an 'elephant trunk' as shown in figure 8.12 would provide the necessary workspace without any singularities. At all positions, the *W1*, *W2*, and *W3* motions shown there result in independent motions of the cutter.

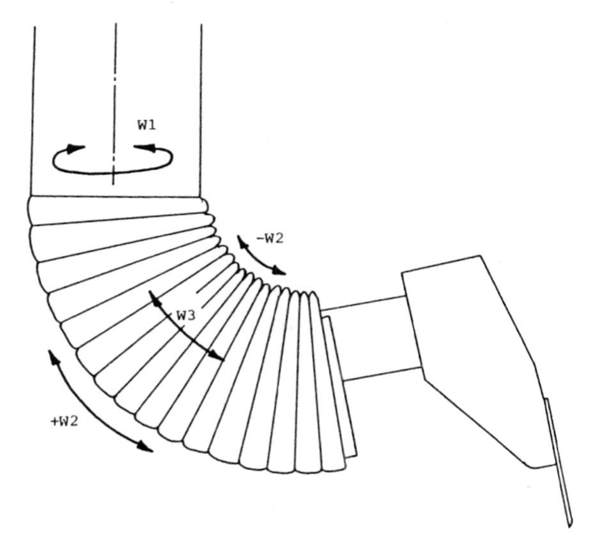

Figure 8.12 Elephant trunk wrist concept

However, disregarding the practical problems of making such a device with the necessary dynamic properties and strength to mount the follower actuator on the end, the kinematics could pose some significant computational problems. After investigating how one could construct such a device with a number of joints similar to the Clemens joint (Hunt 1978, p. 427), we discovered that two Hooke joints could be connected to make a similar device with the required workspace, without singularities. The result, referred to as the Elephant Trunk (ET) wrist, requires only three actuated motions, because the rotations of the upper Hooke joint are mechanically coupled to those of the lower joint. When the ratio R between the rotations of the upper joint and the rotations of the lower joint is arranged to be unity, the effective axes of rotation become parallel only when the end-effector has been rotated through 180°, which is well outside the workspace requirement. Figure 8.13 shows the kinematic arrangement. Figure 8.14 shows a schematic arrangement of gearing that can be used to link the motions of the two Hooke joints in the required manner. Figures 8.15 and 8.16, which make up a photographic study of the prototype wrist model constructed for testing, show the arrangement of two hydraulic actuators to drive the *W2* and *W3* motions. The actuators shown are spatially representative but are not powered. Similar mechanisms have been devised for spray-painting robots (Susnjara and Fleck, patent, 1982). However, the angular range of these devices is inadequate for shearing. Also, the drag forces encountered in shearing, and the movement of the follower mechanism, generate large dynamic and steady loads on the wrist mechanism, as discussed in section 6. Therefore, a much more robust mechanical arrangement was required.

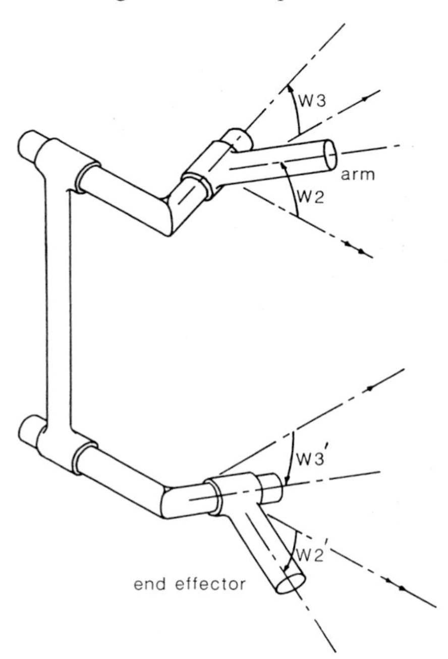

Figure 8.13 Kinematics of ET wrist—two Hooke joints in a mirror image arrangement.

The mechanical arrangement was not easy to design, particularly for the range of motion required. Originally ±110° was required for both the *W2* and *W3* movements, but by mounting the cutter at an offset angle of approximately 25° to the lower yoke axis (figure 8.15) it was possible to reduce the *W2* motion to ±90° . Protection for the gears is yet to be designed; it will take the form of a flexible boot with rigid reinforcing members.

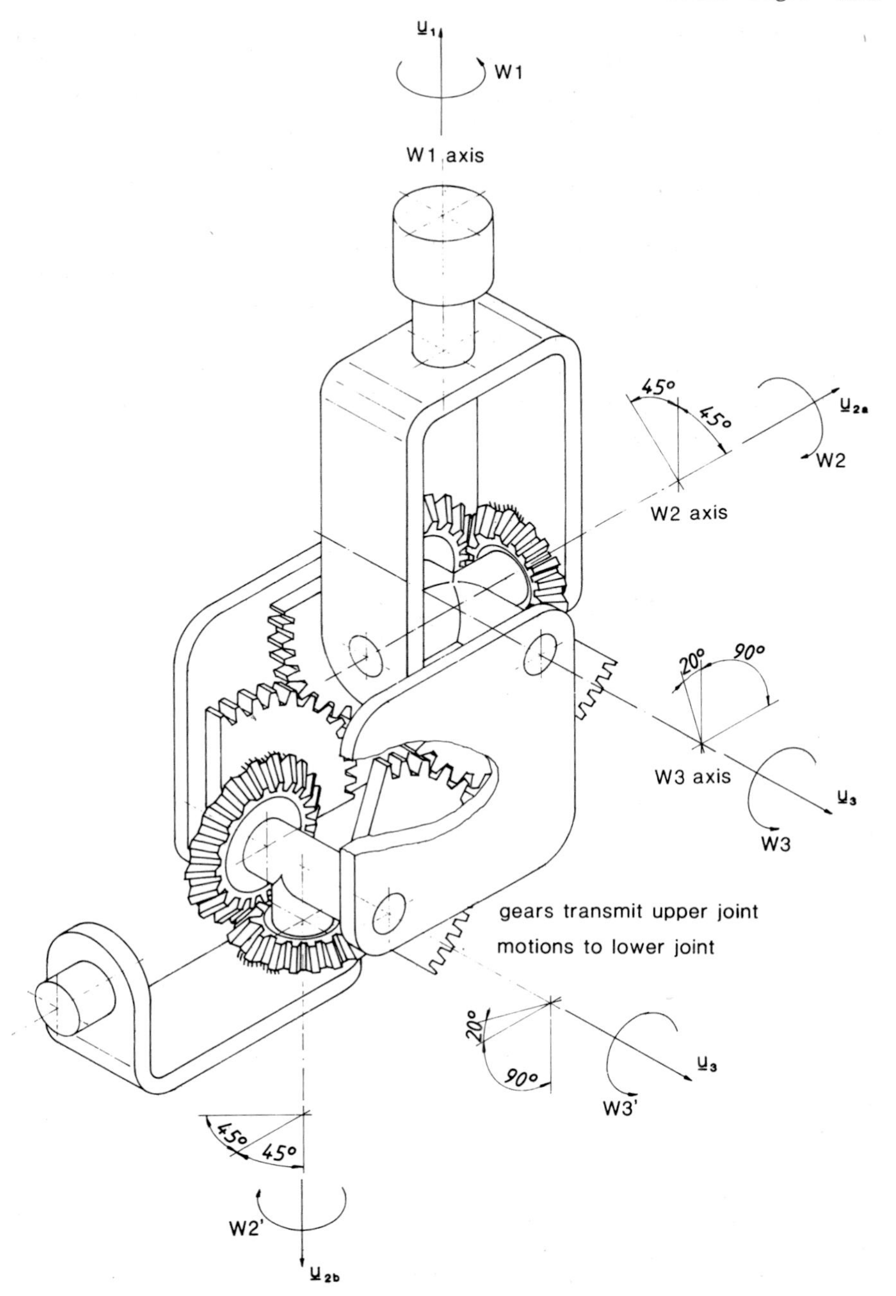

Figure 8.14 ET wrist schematic. Bevel gears on axes $\mathbf{u}_{2a}$ and $\mathbf{u}_{2b}$ are bonded to yoke arms. The other bevels are free to rotate on the two cross shafts, but are bonded to the far side segment gears.

Figure 8.15 Peter Kovesi testing ET wrist movements with full-scale model of SM robot. Model wrist actuators can be seen.

4. Kinematics

Once the ET concept had been established, the singularities of the mechanism had to be studied. The effect of changing the gear ratio was of particular interest. In addition, the maximum actuator velocities that would be required had to be determined for use in the dynamic analysis so that the actuators, valves, and hosing could be specified.

Singularities of ET wrist mechanism

The schematic kinematic arrangement of the wrist is shown in figure 8.14. The rotation axes are defined by the unit vectors $\hat{\mathbf{u}}_1$, $\hat{\mathbf{u}}_{2a}$, $\hat{\mathbf{u}}_{2b}$, and $\hat{\mathbf{u}}_3$. The ratio between the angles $W2$ and $W2'$ and between $W3$ and $W3'$ is fixed by the gear ratio of the mechanism, R:

$$\frac{W2'}{W2} = \frac{W3'}{W3} = R \tag{8.1}$$

At any instant the mechanism can be considered as an equivalent three-axis arrangement. The first axis is $\hat{u}_1$, and the magnitude of rotation about it is $W1$. The second axis is the resultant obtained from a rotation of $W2$ about $\hat{\mathbf{u}}_{2a}$ and a rotation of $W2'$ about $\hat{\mathbf{u}}_{2b}$, giving a total rotation of

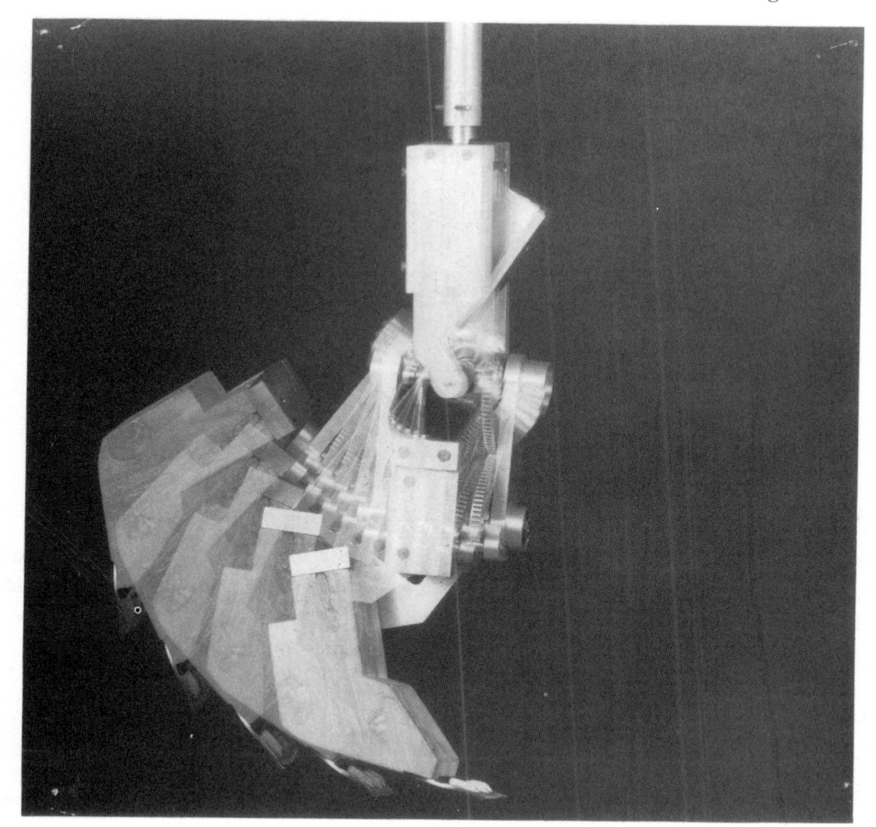

Figure 8.16 Multiple exposure photograph of ET wrist model showing *W2* movement.

$$W2(\hat{\mathbf{u}}_{2a} + R\hat{\mathbf{u}}_{2b}). \tag{8.2}$$

The third rotation is

$$W3(1 + R)\hat{\mathbf{u}}_3. \tag{8.3}$$

Hence, the motion rate control equations are (Whitney 1969):

$$\begin{vmatrix} \delta\text{roll} \\ \delta\text{pitch} \\ \delta\text{yaw} \end{vmatrix} = \begin{vmatrix} & \cdots & \\ \hat{\mathbf{u}}_1 & (\hat{\mathbf{u}}_{2a} + R\hat{\mathbf{u}}_{2b}) & (1 + R)\hat{\mathbf{u}}_3 \\ & \cdots & \end{vmatrix} \begin{vmatrix} \delta W1 \\ \delta W2 \\ \delta W3 \end{vmatrix} \tag{8.4}$$

The wrist will become singular when these three axes become coplanar. This results in two situations that produce singularity:

(1) $\hat{\mathbf{u}}_1$ becoming aligned with $\hat{\mathbf{u}}_3$, and

(2) $(\hat{\mathbf{u}}_{2a} + R\hat{\mathbf{u}}_{2b})$ becoming coplanar with $\hat{\mathbf{u}}_1$ and $\hat{\mathbf{u}}_3$.

It is readily seen that case (1) will occur when $W2 = \pm 90°$. This corresponds to $\pm 180°$ at the end-effector. Case (2) is more complex and will only occur if the gear ratio R is greater than or equal to unity. Figure 8.17 shows the

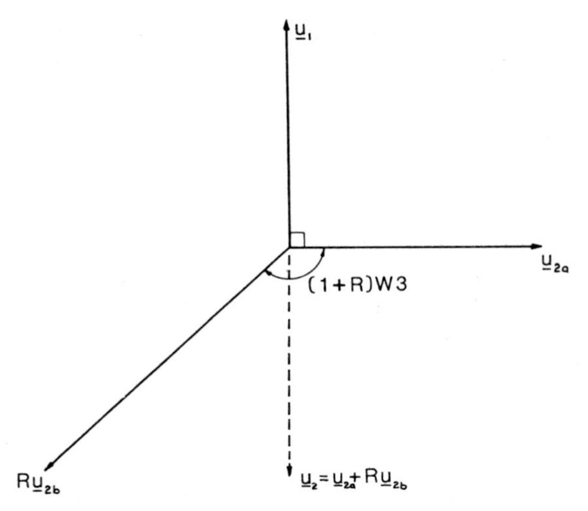

Figure 8.17 Condition for $\mathbf{u}_1$ becoming aligned with $\mathbf{u}_2$ in ET wrist—refer to text.

condition where $\hat{\mathbf{u}}_{2a}$ is coplanar with $\hat{\mathbf{u}}_1$ and $\hat{\mathbf{u}}_3$. Owing to the mechanical arrangement of the system, $\hat{\mathbf{u}}_1$ and $\hat{\mathbf{u}}_{2a}$ are always perpendicular and the angle between $\hat{\mathbf{u}}_{2a}$ and $\hat{\mathbf{u}}_{2b}$ is $(1 + R)W3$. The values of $W3$ resulting in $\hat{\mathbf{u}}_{2a} + R\hat{\mathbf{u}}_{2b}$ having no component perpendicular to the plane defined by $\hat{\mathbf{u}}_1$ and $\hat{\mathbf{u}}_3$ are

$$W3 = \frac{\pm\cos^{-1} - \dfrac{1}{R}}{1+R}. \qquad (8.5)$$

A few examples of singularity angles for various gear ratios are tabulated below:

R	$W3$
1	±90°
2	40° or 80°
3	27.35° or 62.63°
10	8.7° or 24.03°

Note that for $R = 1$, $\hat{\mathbf{u}}_{2a} + R\hat{\mathbf{u}}_{2b}$ disappears to zero for $W3 = \pm 90°$. Another point to note is that this condition cannot occur if $R < 1$, as $\cos^{-1}x$ is undefined for $|x| > 1$.

Overall it would appear that it is best to use a gear ratio of unity, as this results in only one position of singularity: that of the end-effector pointing backward along the $W1$ axis.

The useful working range of a wrist

Analysis of the useful working range of a wrist (as distinct from its points of singularity) is more complex. Paul and Stephenson (1983) define the useful operating range of a wrist as the region where the joint rates are always less than or equal to twice their minimum values. It is generally accepted that, for any series, the joint rates for a three-joint wrist are inversely proportional to the determinant of the Jacobian. However, it was found that this did not apply to the ET wrist.

Least favourable reorientation direction

For each joint, at any given configuration of the wrist, there is a least favourable direction to reorient the wrist that will cause maximum velocity in the joint. These least favourable reorientation directions are readily determined from the inverse of the Jacobian. Inspection of

$$\begin{vmatrix} \delta W1 \\ \delta W2 \\ \delta W3 \end{vmatrix} = \begin{vmatrix} \cdots \\ \mathbf{J}^{-1} \\ \cdots \end{vmatrix} \begin{vmatrix} \delta\phi_1 \\ \delta\phi_2 \\ \delta\phi_3 \end{vmatrix} \quad (8.6)$$

shows that the value of $\delta W1$ is equal to the dot product of the first row of the inverse Jacobian and the reorientation vector, $\delta\phi$. This will be a maximum when the reorientation vector is in the same direction as the first row of the inverse Jacobian. Thus, for a unit orientation change of the wrist, the maximum possible movement of any joint is equal to the magnitude of the corresponding row of the inverse Jacobian. Using a base coordinate frame attached to the centre frame of the ET wrist and assuming a gear ratio of unity, one can write the motion rate control equation as

$$\begin{vmatrix} \delta\phi 1 \\ \delta\phi 2 \\ \delta\phi 3 \end{vmatrix} = \begin{vmatrix} -\sin W2 & 0 & 2 \\ \sin W3 \cos W2 & 2\cos W3 & 0 \\ \cos W3 \cos W2 & 0 & 0 \end{vmatrix} \begin{vmatrix} \delta W1 \\ \delta W2 \\ \delta W3 \end{vmatrix} . \quad (8.7)$$

The Jacobian can be inverted to produce the equation

$$\begin{vmatrix} \delta W1 \\ \delta W2 \\ \delta W3 \end{vmatrix} = \begin{vmatrix} 0 & 0 & \dfrac{1}{\cos W3 \cos W2} \\ 0 & \dfrac{1}{2\cos W3} & \dfrac{-\sin W3}{2\cos^2 W3} \\ 0.500 & 0 & \dfrac{\sin W2}{2\cos W2 \cos W3} \end{vmatrix} \begin{vmatrix} \delta\phi 1 \\ \delta\phi 2 \\ \delta\phi 3 \end{vmatrix} \quad (8.8)$$

where $\det J = -4\cos^2 W3 \cos W2$.

The maximum possible joint movements for a unit reorientation are

$$\delta W1_{max} = \frac{1}{|\cos W3 \cos W2|} \quad (8.9)$$

$$\delta W2_{max} = \frac{1}{|2\cos^2 W3\,|} \quad (8.10)$$

$$\delta W3_{max} = \frac{\sqrt{(\cos^2 W3 \cos^2 W2 + \sin^2 W2)}}{|2\cos W3 \ \cos W2|} \tag{8.11}$$

Figure 8.18 shows contour lines where the maximum possible joint velocities are twice their minimum. (The minimum occurs at $W2 = W3 = 0$.)

The working zone is roughly defined by $-45° < W3 < 45°$ and $-55° < W2 < 55°$. These ranges correspond to 90° and 110° at the end effector, respectively.

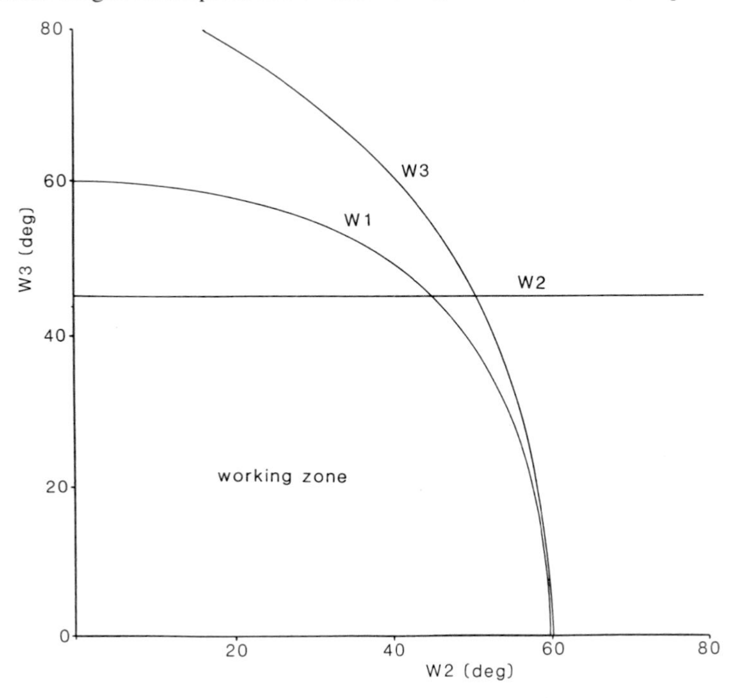

Figure 8.18 Working zone of ET wrist—refer to text.

Comparison of workspaces

Figure 8.20 illustrates the workspaces of the three-roll wrist and of the proposed ET wrist mechanism. Shaded areas show zones of degeneracy. The diagrams were obtained by plotting the range of available directions for a vector normal to the sheep surface relative to the *W1* axis. The unusual geometry of the ET wrist requires two views for adequate representation.

5. Implications for arm control and inverse kinematics

The overall kinematic solution for the arm and wrist is awkward because the ET wrist has no fixed orientation centre where all the rotation axes intersect at a single point.

Note: If a six-degree-of-freedom arm has a wrist with an orientation centre that is fixed relative to both the end-effector and the forearm, the kinematic problem can be

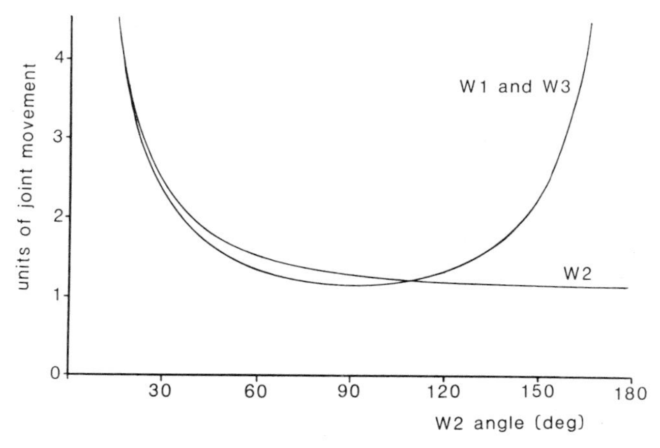

Figure 8.19 Maximum possible joint movements for a unit reorientation of the three-roll wrist.

decomposed into two three-degree-of-freedom systems. Specifying the position and orientation of the end-effector defines the necessary position of the orientation centre, which enables the arm kinematics to be solved. Once the orientation of the forearm is known, the wrist kinematics can then be solved.

Rather than attempt to solve the whole system simultaneously using motion rate control, we use the iterative technique proposed by Milenkovic and Huang (1983). Basically the technique is to move the three arm actuators so as to position the end-effector in the desired position, then move the three wrist actuators to turn the wrist to the desired orientation; the two steps are performed in an alternating sequence until the solution converges to sufficient accuracy. The combination of the SM arm and the ET wrist is suited to this form of solution, as movements of the SM's pendulum-type arm cause only minor changes in the end-effector orientation, resulting in rapid convergence. Two solutions for the system based on this technique have been developed. One uses motion-rate control solutions for the arm and the wrist; the other is based on direct inverse solutions. Terminating the iterative procedure with a position correction step ensures that the end-effector will have error only in orientation. It has been estimated that under the most extreme conditions anticipated for the robot the use of three steps—a position correction, an orientation correction, and a final position correction—will result in a maximum orientation error of the order of one degree for a calculation cycle time of 70 msec. Joint space interpolation is used to generate intermediate joint positions at an update rate of 100 Hz.

6. Dynamics

Dynamic analysis of the ET wrist was limited to consideration of actuator sizes, stiffness, and stresses on gears, bearings, and other highly stressed components.

A complete analysis of the ET wrist would involve three-dimensional

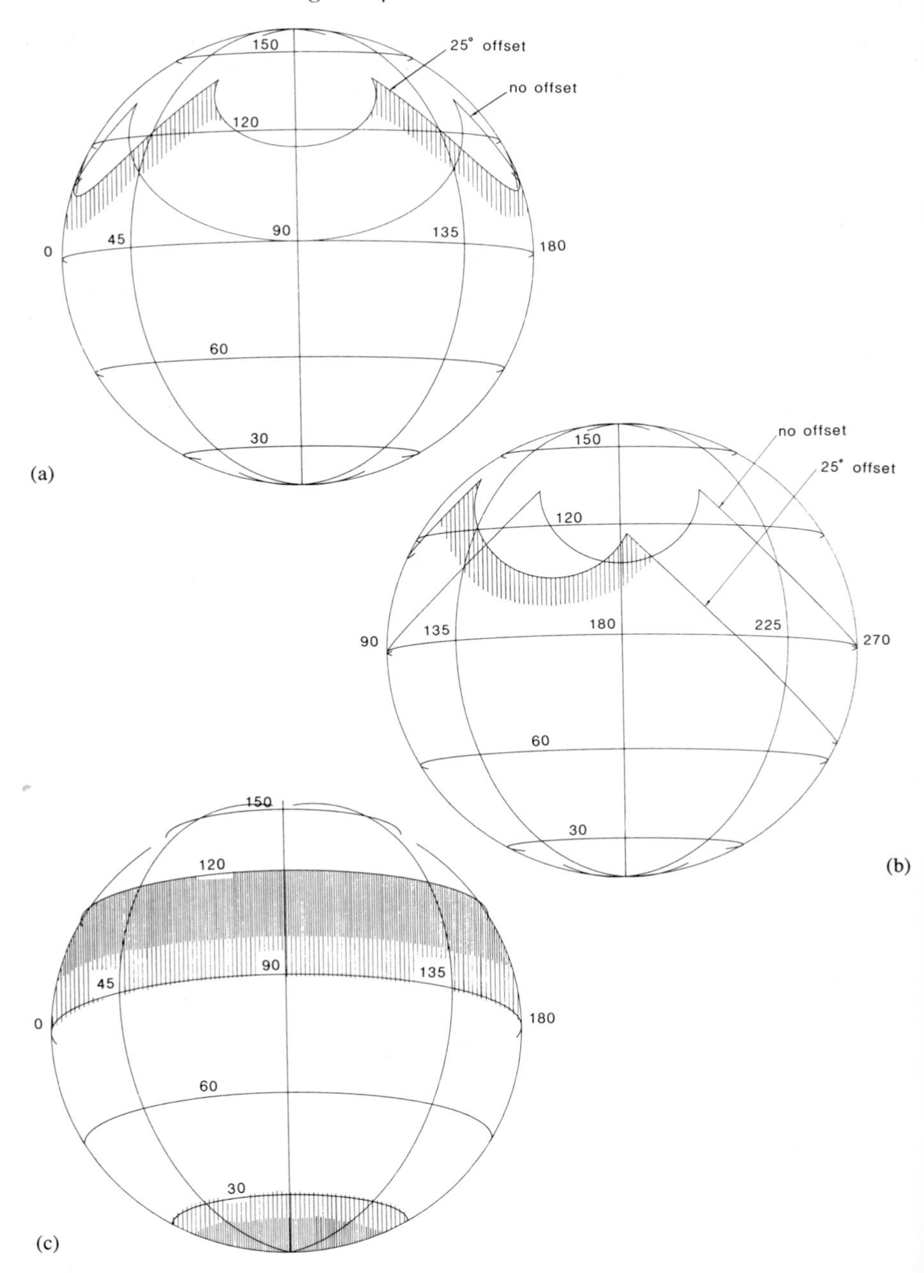

Figure 8.20 Workspace of ET wrist and three-roll wrist projected on to a sphere. (a) ET wrist—shading applies to workspace with 25° offset only. (b) ET wrist—shading applies to workspace with 25° offset only. (c) Three-roll wrist

multibody dynamics of a closed chain mechanism. A much simpler approach is to make the sweeping (yet intuitively reasonable) assumption that the most critical loading situations are as follows.

1. For *W1* actuator: when the wrist is configured such that its inertia about the *W1* axis is a maximum.
2. For *W2* actuator: when the *W3* actuator is at mid-range.
3. For *W3* actuator: when the *W2* actuator is at mid-range.

A pivoted body serves as a model for situation 1. For situations 2 and 3, free-body diagrams of the relevant members are shown in figure 8.21. The diagram is strictly correct only for situation 3, since for situation 2 active forces are actually transmitted through another pair of gears perpendicular to the plane of the diagram. In spite of this, however, energy considerations can be used to show that the actuator force calculated from this diagram is equally applicable to situations 2 and 3.

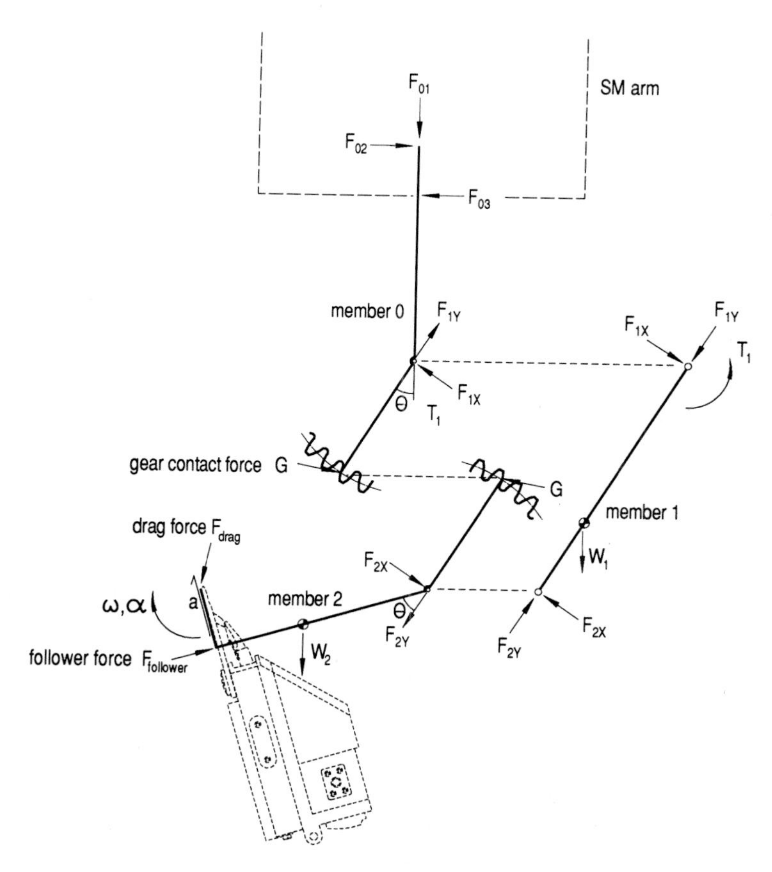

Figure 8.21 Dynamics model of ET wrist—separated into free-body diagrams.

Algebraic expressions for the unknown forces are formulated in the following manner. The geometrical configuration is a function of θ. Starting from member 2, we specify the desired linear acceleration, a, at the leading edge of the cutter; the angular acceleration and velocity, α and ω; the shearing drag force, F_{drag}; and the follower actuator inertial load, $F_{follower}$. Call this the 'performance specification'; typical maximum values are a = 10 m/sec^2, α = 170 rad/sec^2, ω= 5 rad/sec, F_{drag} = 115 N, $F_{follower}$ = 1000 N, although not simultaneously. The follower load affects wrist strength but not actuator sizing, since it is an impulsive load. The three equations for planar dynamic equilibrium (force balance in X and Y directions and moment balance about centroid in Z direction) are applied, and the equations are solved for the gear contact force, G, and components of the bearing force, F_{2x} and F_{2y}. Now for member 1, with the forces at the lower bearing known, the three equilibrium equations can be used to solve for the actuator torque T_1 as well as the two components of force F_{2x} and F_{2y} at the upper bearing. Reaction forces at the arm connection points can be solved in similar fashion on member *0*. With T_1 and the geometrical configuration known, the linear actuator force can be calculated.

The analysis outlined above was used to calculate the required actuator forces when the geometrical configuration and the performance specifications were defined. In a hydraulic servo system the available actuator force is a function of supply pressure, actuator and servo valve dimensions, and actuator speed (which is a function of the angular velocity and the geometrical configuration of the cutter). Given the speed and the piston-rod and bore diameters of the actuator, the required flow rate is calculated. Then, given the rated flow of the servo valve and the supply pressure, the pressure drop across the servo valve when it is fully open can be estimated. The pressure drop across hoses (and other constrictions between the servo valve and the actuator) can also be estimated from the flow rate and the relevant dimensions, using well-known empirical hydraulics formulae. The remaining pressure difference after deducting servo valve and hose pressure drops appears across the actuator piston. Finally, an allowance is made for actuator friction, so that an 'available actuator force' results.

An actuator and a servo valve sized according to these calculations should be capable of providing the required forces. A computer program was written to enable us to impose many sets of 'performance specifications' on a tentative ET wrist design for which the actuator bore and piston rod diameter as well as the servo valve rating and hose dimensions were to be found. For each set of performance specifications, the required and available actuator forces were computed over the full range of θ and for a range of likely servo valve ratings. The numerical results served mainly to confirm or refute the choice of actuator dimensions and to indicate the most suitable servo valve rating.

The basic logic in selecting actuator and servo valve capacity was this: space limitations in the wrist dictated the actuator's geometry, so that only its bore diameter could be tailored to give the required torque across the joint. Ideally, the bore diameter should be just sufficient to meet all the performance specifications; excessively large actuators make for unnecessary mass, space, and power consumption. The servo valve capacity was likewise selected; excessive capacity results in coarser performance. In addition to providing the required force without too much excess, the system should have adequate rigidity to allow fairly high response control of the wrist. Compliance in oil and hoses between actuator and servo valve interact with wrist inertia to resonate at a characteristic 'natural frequency.' By using Lagrange's equations (Meriam 1975), an expression for this natural frequency was derived in terms of wrist dimensions and an effective 'spring constant' to model the compliance. The spring constant is a function of actuator geometry, bulk modulus of the oil, and elastic properties of the hoses. From such calculations it was found that the natural frequency is significantly lower when flexible hoses are used instead of rigid piping, although the former was still acceptable.[2]

Besides sizing the actuators and the servo valves, it was necessary to design the wrist to withstand the dynamic loads it will be subjected to. In particular, the gear contact forces, bearing and shaft loads, and actuator rod compressive loads were calculated. The calculations were based on the simplified dynamics model explained earlier but modified as required; for instance, contact forces on the bevel gears can be deduced from the contact force G simply by multiplying by the gear ratio.

7. Mechanical Arrangement (see also later part of chapter)

The mechanical design of the ET wrist mechanism is much more difficult than that of other wrist mechanisms, such as the three-roll wrist mentioned above. The principal difficulties arise from the need to provide bearings of sufficient size and stiffness for backlash-free gear operation and the need to make the wrist withstand the peak actuator loads. The bevel gears are the most highly stressed components, and they must be as large as possible in order to obtain satisfactory design loading.

Given these requirements, the mechanical arrangement of the components was obtained by considering the central member containing the gears to be fixed and the two yokes to rotate about it. Nevertheless, it is not easy to visualize the spatial requirements for free movement of each of the yokes and of the segment gears throughout the full range of *W2* and *W3* movement.

The actuator for the *W2* motion, clearly visible in the upper yoke shown in figure 8.15, consists of a miniature linear hydraulic actuator with integrated linear position transducer servo valve and cross port relief valve. The *W3* actuator can be seen between the two side plates. Because there is insufficient space for a linear position transducer, a potentiometer has been attached to the cross shaft bearing on one of the side plates. High-performance miniature servo valves are used.

[2] This turned out to be a mistake—see later developments on pp. 231, 240.

At the time of the writing of this paper, the detailed mechanical design is still not completed. The diagrams and photographs refer to full-scale models which have been used to verify kinematic performance and to help visualize the detailed mechanical design problems. A more detailed exposition of the mechanical design and a report of the actual performance of the wrist will be given in a forthcoming paper.[3]

Conclusion

A novel arrangement of a wrist mechanism has been proposed for a robot that requires a well-conditioned kinematic and dynamic performance throughout large workspace angles without enclosing singularity cones. The wrist mechanism is light and strong and could be used, with different actuation, for precision applications. Though the wrist mechanism has been proposed for use where three end-effector rotations are required, it is equally applicable, and exhibits comparable advantages, when used in applications where only two end-effector rotations are required.

In the kinematic analysis, it was shown that the ET wrist exhibits more desirable properties than the other wrist arrangements investigated. It can provide the required workspace for the SM robot in all the necessary shearing directions. The use of linear hydraulic actuators for the *W2* and *W3* motions results in substantial weight savings when compared with rotary actuators but is possible only because of the limited angular range through which the actuated joints work. Weight savings in the wrist result in much greater weight and performance reductions in the proximal actuators of the arm. An umbilical cable is still required, but there is no need for it to be wrapped around the wrist mechanism, since there is adequate space for rotary hydraulic and electrical connections in the *W1* axis joint at the end of the arm. The umbilical can be routed to the cutter assembly along the wrist sideplates.

Surprisingly, the kinematic control equations are only slightly more complicated than for a conventional wrist, because of the symmetry of rotation about the two joints. The arm arrangement permits rapid convergence of an approximate inverse kinematic solution method.

The disadvantages of this wrist mechanism include the relative complexity of the mechanical design and the difficulty of maintaining adequate stiffness and freedom from backlash. A typical mechanical arrangement consists of complex chains of gears and other elements (Susnjara and Fleck, patent, 1982). These difficulties have been referred to by other authors, such as Milenkovic (1984), but we believe that our simpler mechanical arrangement represents a substantial advance in the state of the art and provides adequate stiffness and freedom from backlash.

[3] Such a paper has not appeared. Full details appeared in the patent by Elford *et al* (1988). Reproductions of assembly drawings appear in figures 8.27a and b, p. 234–5.

Mechanical design of SM

The SM robot design deserves an entire book to itself; this chapter presents just a few interesting details. The first part of the chapter describes the concepts which provide the foundation for the design. And it is the mechanical aspects which presented the greatest design challenges. Of all the major components, the wrist was the most difficult but several other elegant solutions were found for particular design problems.

Like ARAMP, the entire robot was designed on drawing boards; only limited use was made of computer techniques. We raised the possibility of buying a computer-aided design system with the Corporation, with little success. Apart from his overall responsibilities, David Elford took on the detailed design of the lower arm, cutter and wrist while Zbigniew Lambert tackled the upper parts of the robot. There was a two-way tussle for a while—we needed an extend actuator stroke of 700mm, but this turned out to be very difficult to fit in the gap between the ARAMP manipulator and the ceiling of the laboratory. The ceiling consisted of 300 mm thick doubly reinforced concrete—there was no question of raising it! The floor was just as thick, and ARAMP could not be lowered without a lot of work. Between the two, we had to squeeze track mounting plates, precision tracks, a carriage, a swing bearing, an upper arm carrying the telescoping lower arm, a rotary actuator, a wrist, a cutter, 120mm of wool and a sheep! David tried his best to shorten the lower arm and wrist, while Zbigniew shaved millimetres off the top end. We were all relieved when they finally announced there was enough space.

Traverse actuator

The traverse actuator presented an intriguing design problem. We needed a stroke of 1 metre, and a bandwidth of 10 Hz. A cylinder diameter of 27 mm would provide enough force but this raised a major difficulty. A 1 metre long piston rod needed to be at least 25 mm in diameter if it were not to buckle under the compressive loads which could be expected. We could have enlarged the cylinder diameter to about 37 mm to compensate for the piston rod area. But unless we adopted a double-ended cylinder we would then have a big difference between the piston areas at each end. The difference would have led to non-linearities making control more difficult. (Furthermore, the increased cylinder diameter would have required a still larger piston rod to be safe from buckling failure.) We could have adopted a double-ended cylinder but the piston rod would have been over 2.5 m long. Instead, Zbigniew devised his own twin cylinder arrangement using a push–pull principle (figure 8.22). Later we were able to lengthen the stroke to 1.75 m with minimal modifications. Part of the elegance of his solution lies in the porting—both oil ports to the cylinders are at the rod end and this eliminates much of the extra piping needed for a conventional design. Only one piston is needed—the thick rod is simply a ram without a piston.

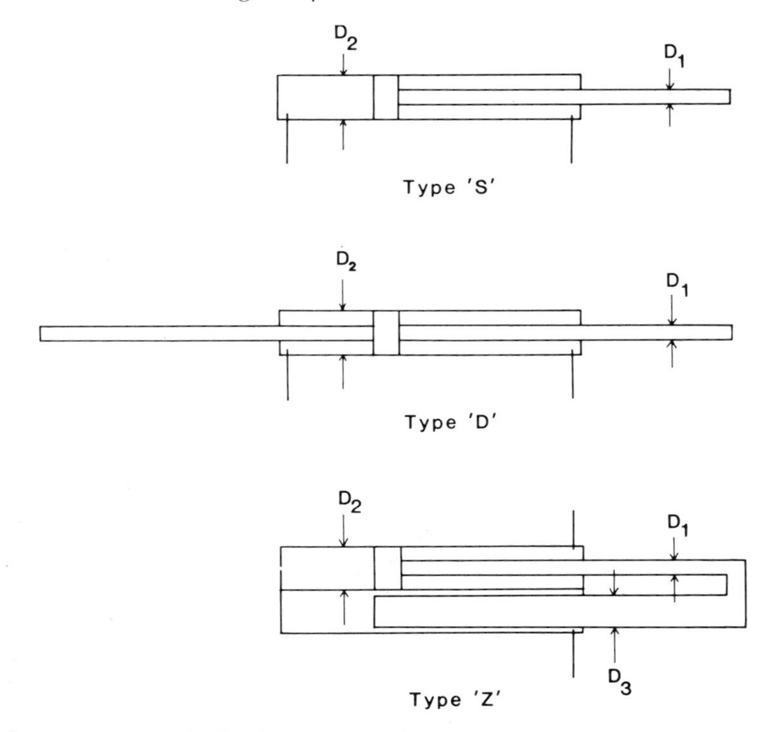

Figure 8.22 Linear hydraulic actuator configurations. Type S—single ended—cross-section areas differ by rod area. Type D—double ended—equal areas. Type Z—single ended—equal areas. In type Z design, $D_3^2=D_2^2-D_1^2$. The thick rod is in compression and the thin rod is in tension. Note that both oil ports are at the same end, simplifying connections to the servo valve.

In contrast, we were able to use an entirely conventional single-ended design for the swing actuator.

Arm

The upper arm provides the sliding mechanism for telescoping the lower arm. The principal design requirement was stiffness—the natural frequency had to be more than 20 Hz with the lower arm extended. Space was at a premium to accommodate the 700 mm stroke within a minimum length.

Figure 8.25 shows the internal arrangement of the upper arm in schematic form, in particular how the service lines pass through the upper arm. The lower drawing also shows the principal structural elements to scale.

The main structure is a thick fabricated steel box section forming the outer cover. The tracks for the THK linear bearings carrying the lower arm require a precise mounting. It was not possible to machine the inside of the steel box with enough accuracy. Instead, the outside was machined on one face. The tracks are mounted on a machined aluminium supporting piece with bearers which project

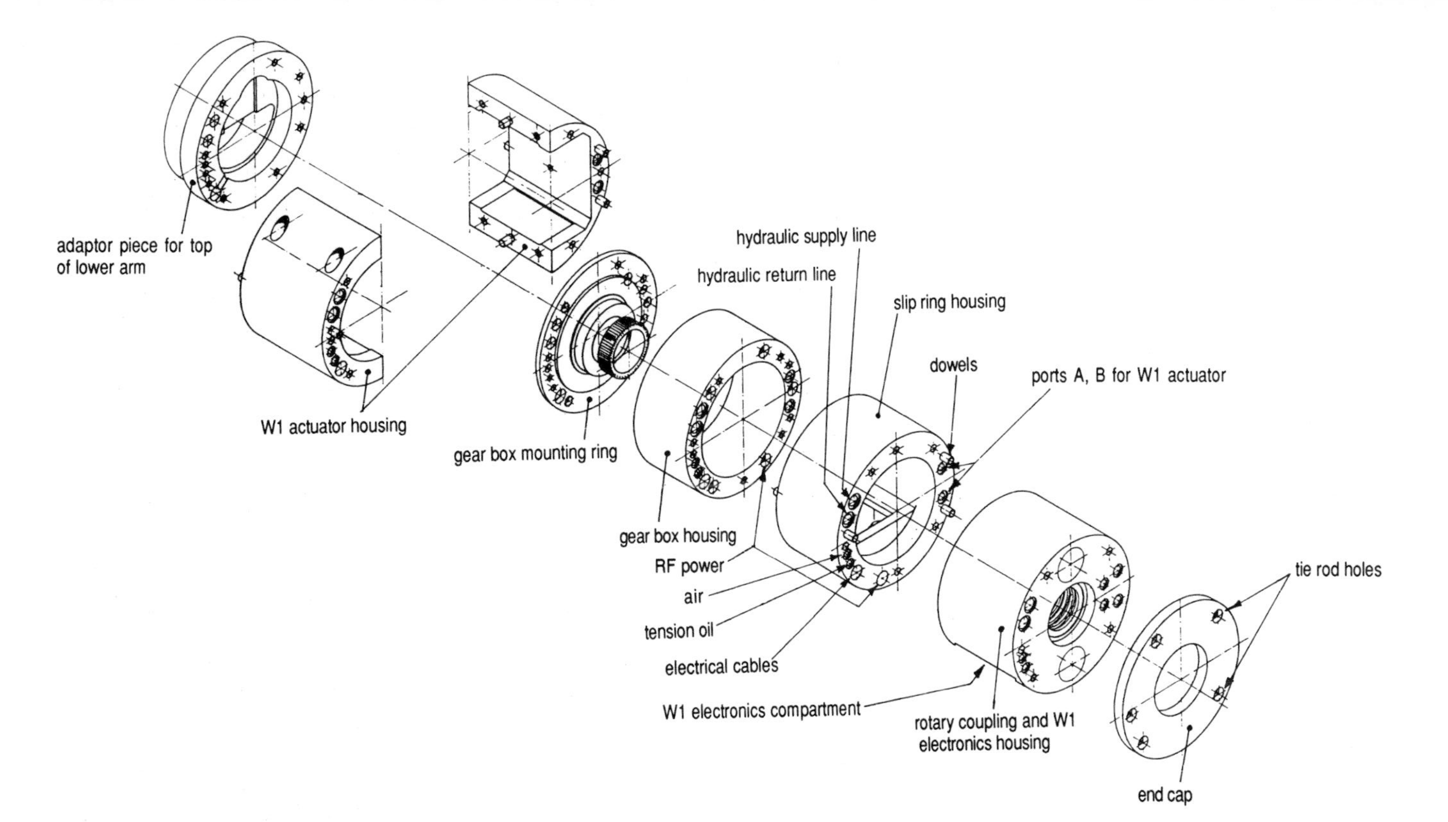

Figure 8.23 Elements of lower arm. Peripheral holes carry electrical cables, tie rods, compressed air, cutter tension oil pressure, supply oil (high pressure), drain (to tank) and two control ports for *W1* actuator and location dowels.

through holes in the steel plate. This way, we achieved the necessary accuracy and structural stiffness.

The lower arm is entirely aluminium alloy (T6063), machined from solid ingots, and finished by clear anodizing. Again, figure 8.25 illustrates how the service lines pass through the top section, and figure 8.23 shows the lower sections of the lower arm.

We wanted to keep all the services inside the arm, as far as possible. Apart from the need for a simple appearance, there were other compelling reasons. We wanted to avoid wrapping cables around the arm otherwise we risked pinching them at the extremes of the arm extend motion. If we provided sufficient length for the expected range of movement, they could have hung around the wrist in certain positions, with the risk of colliding with the sharp teeth of the cutter during rapid repositioning movements.

The six main services were:

(1) oil supply line
(2) oil return line
(3) oil pressure to set cutter tension[4]
(4) compressed air
(5) cables for multiplexed electronics
(6) RF power for inductance sensor.

The multiplexing system reduced the diameter of the cable bundle to 8mm and is described in the next section. Without this arrangement, the volume of electrical connections would have been too great for internal cabling. We needed to make provision for inductance sensing and ultrasonic sensing experiments. For these we needed a separate RF cable suppling 800 MHz power and two additional coaxial cables for the ultrasonic transmitter and receiver. Compressed air was needed to carry an oil mist to lubricate and cool the cutter.

Figure 8.23 shows the elements of the lower arm column. When assembled (figure 8.24) they contain the *W1* actuator, the gearbox to transform its 270° range to 440°, electrical slip rings and a rotary hydraulic coupling to the wrist at the bottom end. The *W1* servo amplifier electronics is also mounted inside the rotary coupling piece. The hydraulic and electrical service lines pass through holes around the periphery of these pieces. Seals at the mating surfaces prevent leaks. Five tie rods keep the slices together, and dowels keep them accurately aligned.

The wrist yoke is mounted on the shaft which passes through the rotary coupling. Five service ducts pass through the 30 mm diameter shaft—oil supply and return, electrical cables, compressed air and tension oil. Provision was made for a further external coupling for the inductance sensing and ultrasonic sensing cables but they have not yet been used.

The elegance of the arrangement reflects years of design experience. No amount of computer equipment could have made up for the experience, but it could have reduced the drawing effort needed.

[4] tension—the term used for the compressive force between the comb and the cutter

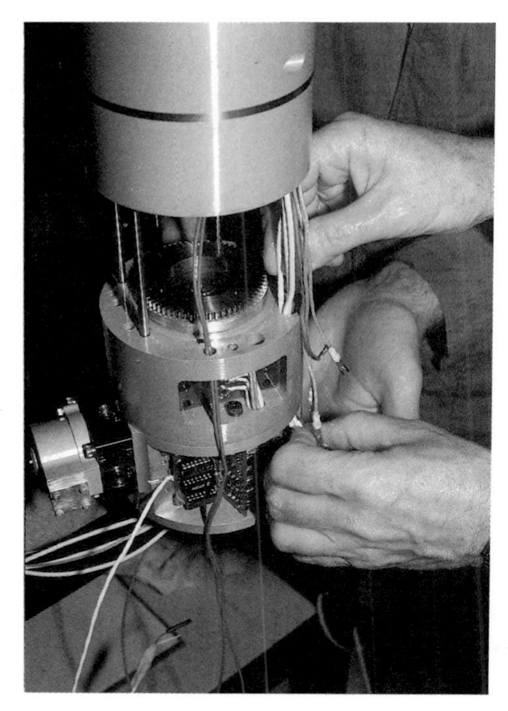

Figure 8.24 Assembly of lower arm. The gear is a spline which engages with the output ring of an epicyclic gearbox on the output of the *W1* actuator. The *W1* servo valve is at the lower left side. Note the *W1* servo electronics in the cavity at the bottom.

Confirmation of design

Analysis of the design by the Rayleigh–Ritz method predicted a natural frequency of 27 Hz for the arm (Ong et al 1989). A more thorough analysis by the receptance method (Howard 1987) predicted frequencies of 23, 42, 59 and 82 Hz, all within 3% of measured values. This later analysis took factors such as bearing stiffness into account.

ET wrist

The ET wrist paper which appears in this chapter was written before detailed mechanical design commenced. It was also written before we discovered that some major design changes would be needed; first the gearing needed to be 4 times stronger than planned and then the structure required greater rigidity. Further, the servo valve for the *W3* actuator inside the wrist had to be mounted on the actuator; the flexible hoses we planned to use introduced too much compliance in the oil. At the same time, experiments with the wrist model (figure 8.15) had shown that the full range of *W2* and *W3* motions would be required *independently* of each other. Several other non-singular wrist designs have

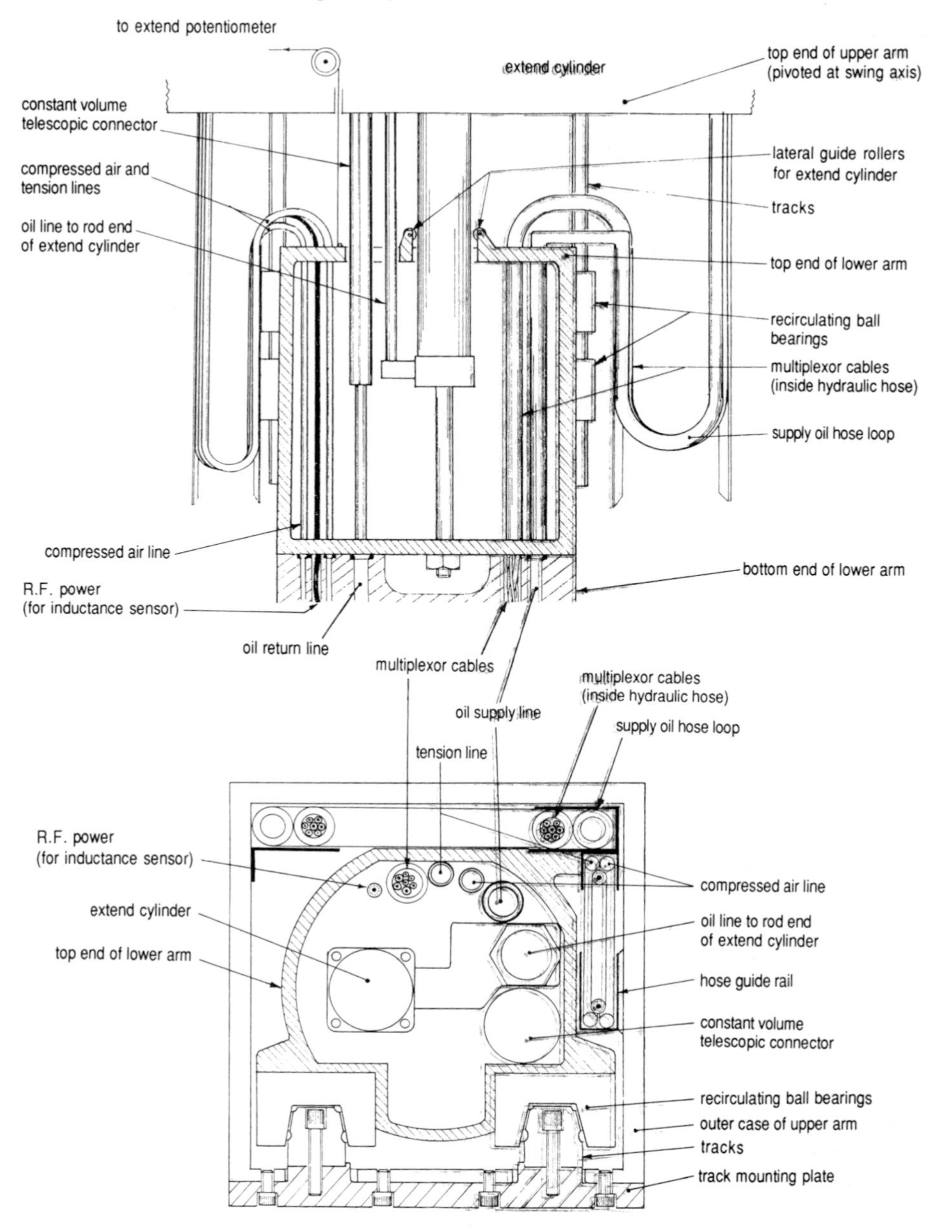

Figure 8.25 Arrangement of service lines in upper arm. Top diagram is schematic, lower diagram is a simplified cross-section view of the SM arrangement. The extension of the lower arm is measured by a potentiometer driven by the spring loaded cable.

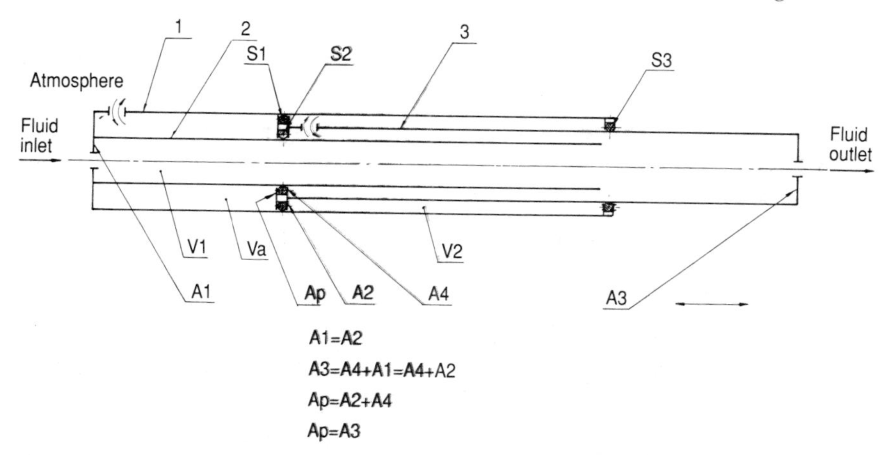

Figure 8.26 Telescopic connector. A mechanical curiosity—the contained volume is independent of length.

appeared in recent years (Milenkovic 1988, 1990, Rosheim 1989) but all of these aim to provide a 180° hemispherical workspace. The cusps of the ET workspace at the extremes of the *W2* and *W3* motions provide extra cutter roll which is important for shearing parts of the legs and neck of the sheep.

The resulting wrist design is shown by assembly drawings in figure 8.27. Figure 8.30 shows some of the components and figure 8.32 shows the final stages of assembly. The distance between the top and bottom cross shaft centres gives the scale—100 mm.

The structure of the wrist is asymmetric so that the strength and rigidity is concentrated into the *W2* gear transmission which is most heavily loaded and most sensitive to backlash with three geared pairs. Figure 8.27b (section AA) shows this most clearly. The front side plate is made from high tensile strength steel; the two hollow stub shafts carry quadruple angular contact bearings for the cross shafts on the inside and needle roller bearings for the moving bevel gears on the outside. The much lighter rear side plate is linked by three structural connecting pieces; one forms the cover for the *W3* actuator, and another at the bottom forms the mounting for the *W3* actuator pivot. This piece is sculpted to accommodate the full range of yoke movement, and to allow the outer sections of the undercut spur gear B to pass around it (figure 8.30).

The wrist asymmetry is carried over to the wrist yoke above, where the left arm carrying the fixed bevel gear is much stronger (high tensile steel) than the right arm (aluminium). The lower yoke, carrying the follower assembly and the cutter, is one sided.

The *W2* actuator is housed in the yoke and acts on the upper cross shaft, causing the upper moving bevel to roll around the fixed bevel on the yoke arm (figure 8.27b). This motion is transmitted through the spur gears A and B to the bottom bevels; the spur gears are bonded to the moving bevel gears. The bottom

(a)

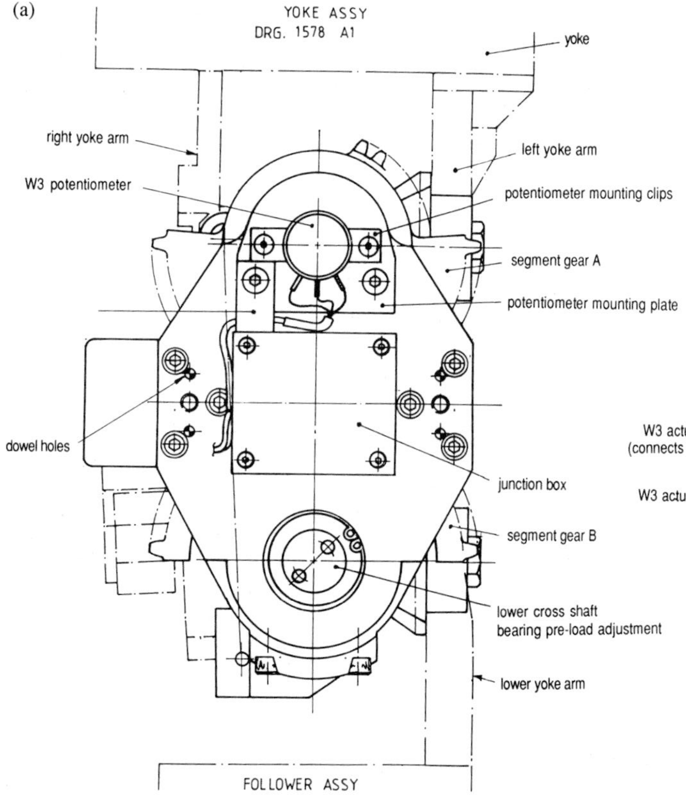

YOKE ASSY
DRG. 1578 A1

yoke

right yoke arm

W3 potentiometer

left yoke arm

potentiometer mounting clips

segment gear A

potentiometer mounting plate

dowel holes

junction box

segment gear B

lower cross shaft
bearing pre-load adjustment

lower yoke arm

FOLLOWER ASSY

View from front

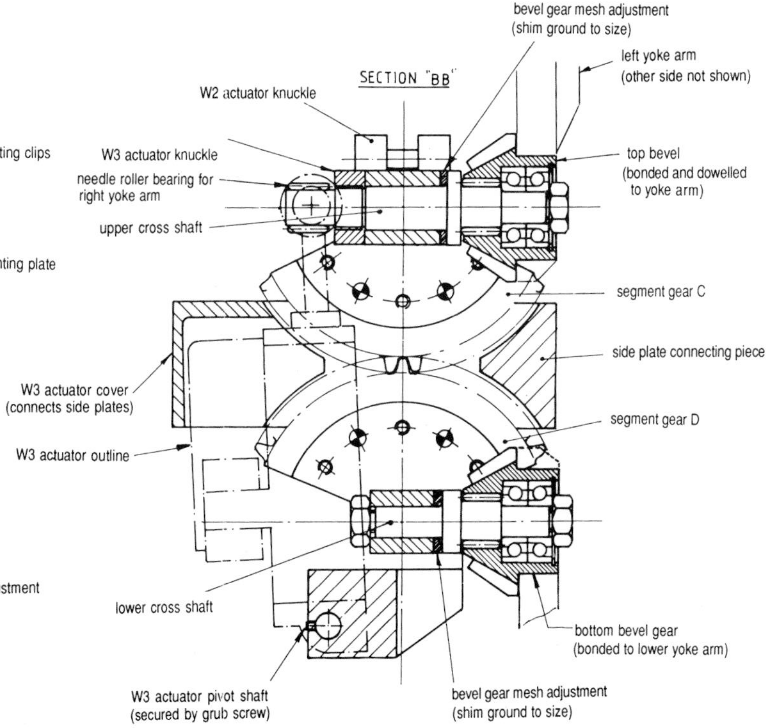

SECTION "BB"

bevel gear mesh adjustment
(shim ground to size)

left yoke arm
(other side not shown)

W2 actuator knuckle

W3 actuator knuckle

needle roller bearing for
right yoke arm

upper cross shaft

top bevel
(bonded and dowelled
to yoke arm)

segment gear C

side plate connecting piece

W3 actuator cover
(connects side plates)

segment gear D

W3 actuator outline

lower cross shaft

bottom bevel gear
(bonded to lower yoke arm)

W3 actuator pivot shaft
(secured by grub screw)

bevel gear mesh adjustment
(shim ground to size)

Section BB (viewed from front)

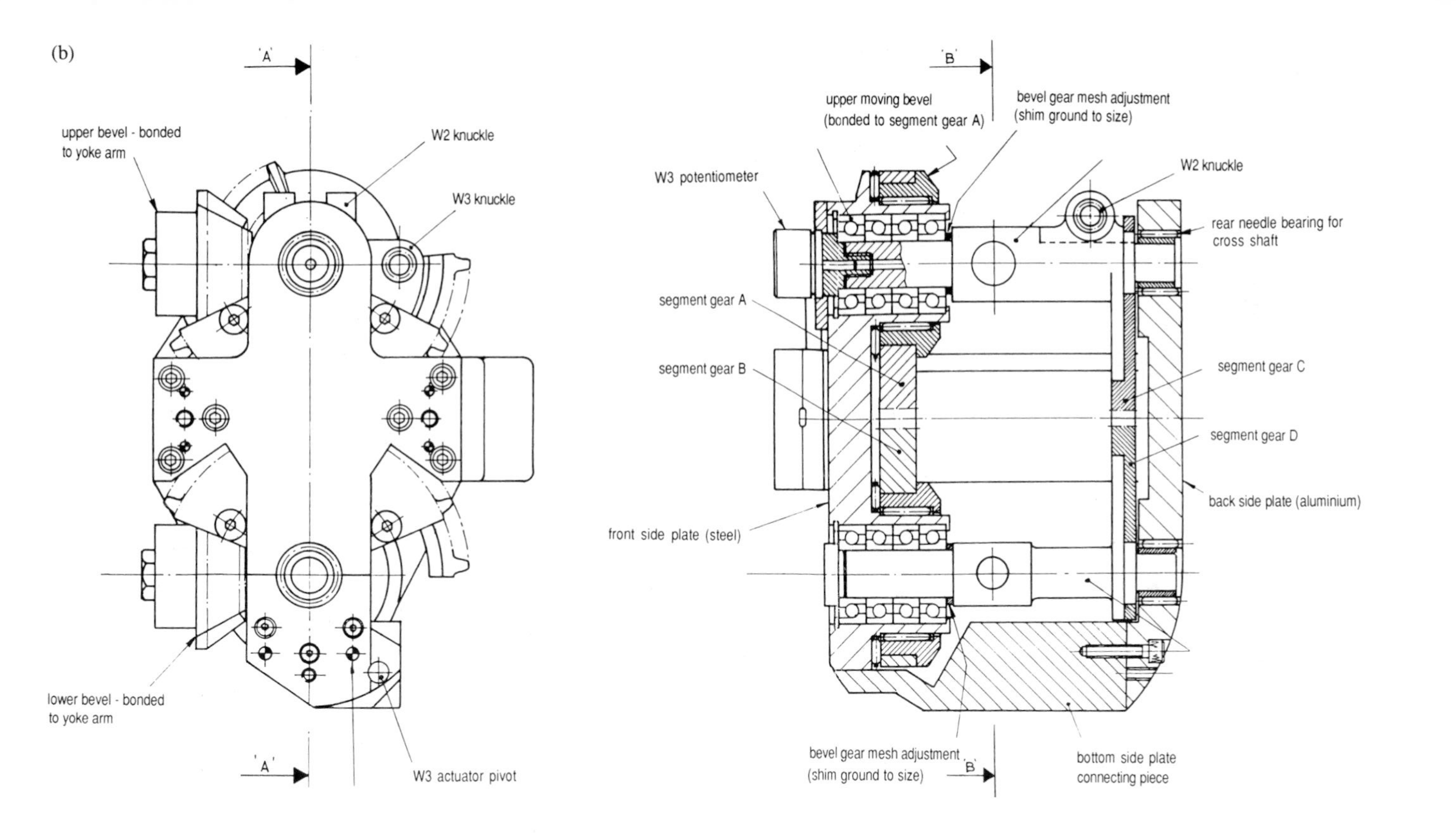

Figure 8.27a and b. E.T. Wrist assembly drawings.

bevel is bonded to the lower yoke arm; the top bevel is bonded and pinned. Thus the bonding of the lower bevel serves as a weakness and fails in the event of a collision.

The meshing distance of the bevel gears needs to be carefully set to minimize backlash. Shim washers on the cross shafts were ground to the appropriate thickness for this adjustment (sections AA, BB). The gears were originally meant to be hardened and ground for better precision, but they have given 5 years of satisfactory service without that extra treatment so far.

The *W3* actuator outline is shown in section BB. A single high tensile steel housing forms the cylinder and the manifold block. A Moog 30 series servo valve is mounted on the top of the manifold. The manifold block contains 3 miniature custom designed cross port relief valves, two custom pressure transducers, and two custom hose connectors bonded into the housing. The cylinder has an internal diameter of 20 mm; the rod diameter is 8 mm and the stroke is approximately 70 mm. The aluminium cylinder end contains twin Shamban step seals. This in itself is an amazing degree of miniaturization by design.

The *W2* actuator is similar in specification, but is slightly less specialized since more space is available in the yoke. The actuator position is measured by a linear potentiometer alongside the cylinder (figure 8.32). The *W3* position is measured by a rotary potentiometer bonded to the upper cross shaft (section AA). A hole through the cross shaft was provided to enable the potentiometer shaft to be knocked out if it needs replacing.

A special feature of the design is the access for maintenance. Both the *W2* and *W3* actuators can be removed easily without having to dismantle any of the wrist. The lower yoke can be taken off by undoing a single retaining nut when the follower assembly needs to be removed.

Several needle roller bearings are used to save space. The bearing surfaces had to be heat-treated steel, ground carefully to size.

This description reveals the relative simplicity of the design. Some of the components are expensive and demanding to manufacture. Yet of all the features of the robot design, the wrist was one of the most significant contributions to its dexterity and dynamic performance. The freedom from singularities enormously simplifies robot control and programming, both of which reduce the cost and increase the performance of the robot.

Cables

One of the threads running just below the surface through the conceptual design phase was the issue of electrical cables. SM was to have more sensor capacity than ORACLE, and that meant more wiring. We planned on 35 sensor inputs and about 25 outputs. Yet at the same time, we needed quite a different cable arrangement; ORACLE had suffered innumerable breakdowns caused by cable or connector failures.

First, we decided to place the servo amplifier circuits for controlling the actuators on the robot structure, near each servo valve. Each circuit consisted of a more or less standard arrangement, using analogue amplifier circuits. Like ORACLE, the actuators were largely decoupled by the kinematic design of the arm; each actuator could be controlled independently by its own feedback loop. Our experience with ORACLE taught us valuable lessons about the design of the amplifiers, and about using the computer to compensate for drift and some dynamic effects by preprocessing the set point signals.

Apart from cable failures, ORACLE had been surprisingly reliable. In view of this, we could afford fewer in-line connectors in the wiring. They had been included to permit disassembly of the major components for maintenance, and yet they were the source of many failures. Often, these connectors, or the joints between them and the cables had failed due to vibration, flexure or oil contamination.

Jan Baranski designed an analogue time-division multiplexed communication system to eliminate the heavy multi-cored cables which had caused so much trouble on ORACLE. Each of the actuator control circuits was connected by sample-and-hold circuits. Standardized timing and logic circuits were a part of each of the control circuits.

Cables with large numbers of individual wires are commonplace in the electronics industry, of course. And they can be quite small—a 10 mm cable can contain 50 or more pairs of wires. The factor which causes so much complication for robots is flexure. Cables in robots need to flex repeatedly, and must endure millions of cycles without risk of failure. To do this, the individual wires inside the cables need to be able to slide past each other. Normal computer cables are quite unsuitable.

The cables which we used in ORACLE were heavy industrial grade communications cables, each with 12 twisted pairs and an overall screen. Each pair carried one signal; one wire carried an analogue voltage and the other wire was connected to a common grounding point inside the junction box next to the analogue interface circuits and the computer. Each cable was about 20 mm in diameter, and the minimum specified bending radius was approximately 300 mm. They were subjected to a bending radius of about 200 mm and, predictably, failed after some 10,000 or so cycles. Replacing them was not easy.

Jan selected high strength teflon sheath coaxial cables for SM. With oversize finely stranded cores they are specifically designed for applications requiring flexure, and have smooth sheaths so they can easily slide past each other when constrained in a bundle. The large cores also helped to reduce resistance losses to a minimum which was important for the power supplies. The screens protected all wires from electromagnetic interference. The total length of cable from the computer to the tip of the arm was about 20 m and interference could have been a major problem.

The cables were expensive. But so was the cost of a cable failure deep inside the robot structure. David wanted to integrate the wiring and hydraulic connections into the structure of the robot to maintain an elegantly simple outer appearance. Without Jan's multiplexing arrangement, allowing all the cables to

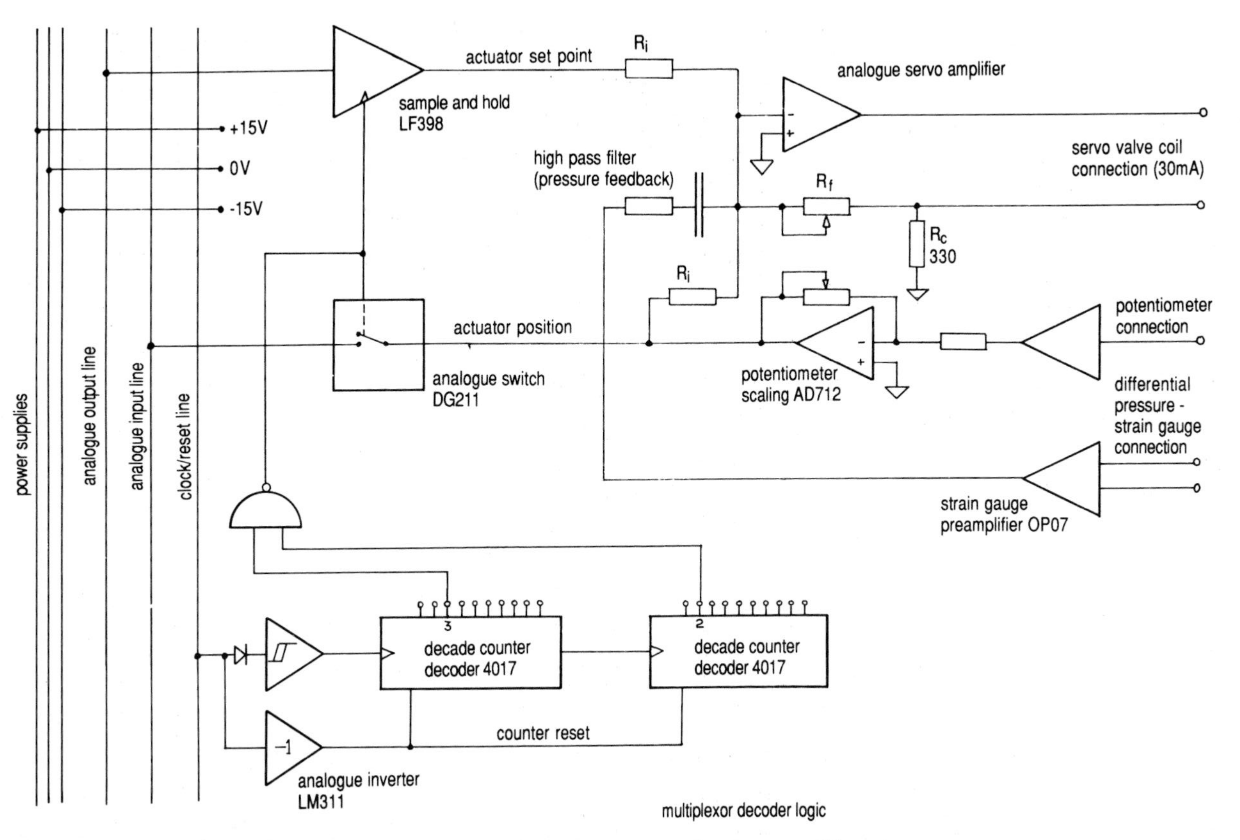

Figure 8.28 Typical multiplexor circuit with servo amplifier electronics.

be passed through ducts as small as 8 mm diameter, this would have been impossible. Cable strength and durability was equally important of course. Repairing a cable failure would require complete disassembly of the arm.

Project control issues

The story of SM would not be complete without some mention of project management. Project management techniques are not easy to apply in a research environment, particularly a university. Fellow robotics researchers have expressed amazement when I have told them about using critical path and other planning methods.

ARAMP had been a frustrating experience for me—interminable delays and postponements, finally missing our original demonstration deadline by some six months.

I was keen to keep the SM robot construction on schedule. In May 1984, we listed all the jobs needed to finish the robot, with the time and resources needed. The main task was the construction of the mechanical components, but there were countless others too. I wrote a critical path and resource allocation program to suit our particular constraints—a fixed complement of staff with commitments to maintenance, holidays, and a certain time component reserved for meetings and other team interactions. My analysis predicted that component manufacture would be completed by March 1985, and assembly would be completed by the beginning of May. I allowed a month for final commissioning.

Once manufacture of the component parts began, I soon realized that quite a different method was going to be needed. With hundreds of parts being made in our own workshops and three sub-contractors, we soon lost track of progress. It was impossible to estimate how much work had been completed.

Michael Ong and I wrote a new program to keep track of each component, and to compare the actual manufacture time with the original estimates. The actual times for each part were not known, and impossible to measure since several parts were usually being worked on at the same time. While a technician was waiting for a lathe, he was marking out other parts, or drilling, welding or finishing.

But each week, we could compare the total working time for all the parts completed with the sum of the estimates. It soon became apparent that we had underestimated the throughput of our workshop technicians, and overestimated that of the commercial sub-contractors. Within a few weeks of starting to make parts, I knew that they would not be finished until May or possibly June 1985.

By March 1985 we had accepted a deadline—the Wool Corporation expected a 'convincing' demonstration of SM shearing by mid November. But now some real problems were becoming apparent.

Michael was near to panic. He had analysed the control and actuator requirements in minute detail. He had specified a dozen or more 'worst case' situations to derive performance specifications and to select actuator and servo valves. He now realized that he had mis-interpreted servo valve specifications and he was appalled to find that the cutter positioning error could be more than 25 mm! This was far beyond what we could tolerate; he was so ashamed of his error that he offered to resign immediately.

I had a feel for the control system design, and thought we could not be that far out. At the same time there was some concern because Michael was always so meticulous and careful. I repeated the analysis using different methods, and found that he had made an interpretation error, but not as bad as he feared. But the results showed clearly that we could not tolerate the compliance arising from flexible hoses between the wrist actuators and the servo valves.

David reconsidered the wrist design to mount the servo valves on the actuators (see wrist paper again). This was not easy, but we had recently received data on a miniature Moog servo valve which could fit inside the wrist. Meanwhile, we accelerated the manufacture of one wrist actuator so Michael could test it as soon as possible.

April came, and brought more delays and problems. Some parts were wrecked when they were anodized incorrectly. One sub-contractor simply kept our work at the bottom of his list—unfortunately his was the only workshop capable of producing certain parts. And still David had not been able to finish the design—the time needed to chase sub-contractors kept him away from his drawing board.

With SM half completed, we had taken the annual renewal of our funding for granted. For no apparent reason, the Corporation decided to limit support for Peter Kovesi and Stewart Key to six months from July; their positions would be reconsidered after the demonstration in November! We were shocked by this, but too busy to do much more than renew our determination to meet the deadline.

We gave ourselves some light relief by celebrating the retirement of ORACLE after 400 trials. It had served us well, and was soon sitting proudly on public exhibition at the Royal Agricultural Show. Team spirit picked up as Stewart Key, Peter Kovesi and myself switched from shearing trials to help with the more mundane assembly tasks. The entire team (and some family members) helped to install the massive track mountings on the ceiling of the laboratory—an operation involving the precise positioning of two tons of steel frame carrying the two steel plates to be glued and bolted to the ceiling.

Then in May, Michael found a new problem. He had analysed the strength and stiffness of the wrist gearing and produced a detailed recommendation to double the size of the bevel gears and the bearings. He neatly shaded cross-section drawings to show where he thought there would be space. A difficult design job became nearly impossible. David looked quizzically at Michael, over the top of his new reading glasses, and simply said 'I'll see what I can do'.

There was now a large new hole in my plans. David had originally laid out the wrist design a year before, but the new layout proposed by Michael would need another four weeks. There was little anyone could do, and David was still confident about finishing the robot by August.

We very nearly did.

There had been setbacks in the upper part of the robot too. The first time oil pressure was applied to the upper arm, Zbigniew discovered that one of the myriad of oil passages drilled through the manifold blocks had accidentally been left unplugged. On the second try, there was an oil leak. By the time these had been overcome, the wrist was complete and mounted on the bottom of the arm but without its actuators. The cutter was still unfinished, so the mockup from the SM display model was attached instead.

Figure 8.29 Michael Ong and Zbigniew Lambert assembling shunt to overhead tracks—May 1985.

Just before the first scheduled commissioning tests, we found a new problem. Each of the two wrist actuators had been fitted with three miniature relief valves built into the valve manifolds. Each valve was a mere 6 mm in diameter and 9 mm long. They were a late addition to the design. Michael had calculated that they were crucial—the inertia of the wrist was sufficient to blow the actuator end caps off if the servo valve closed during a rapid movement—we could not afford such a failure at any time.

The valves were very simple—small steel balls held against hardened seats by retaining springs. There was no room for adjustment—the springs were to be gradually shortened by grinding the ends until the valves opened at the right pressure.

Unfortunately they failed to reseat after reaching their opening pressure. Examination of the balls and the seats revealed no problems. Both surfaces were polished and true to form. Because of the shortage of space, the valves could

Figure 8.30 Machined wrist components—a tribute to the skills of our mechanical technicians Ian Hamilton, Derek Goad, Dennis Brown and Stewart Howard.

only be accessed by dismantling the manifold block. Time after time, they would be patiently cleaned and reassembled, and found to seal perfectly. But as soon as the pressure broke the seal, they would not close again. Days went by. At first I left David and Michael to sort out the problem. But as weeks passed with no solution I became more anxious—there were now only 10 weeks left. Finally, the problem became clear—the slightest misalignment of the spring would be sufficient to prevent the ball from seating properly. The solution was to add guide bushes to keep the balls aligned. On 29th August, the problem was solved and commissioning could commence.

The delay was not wasted. If the relief valves had worked first time, there would have been a host of other problems to solve—it was just that they took longest to fix. Jan Baranski worked long hours on the multiplexor electronics, chasing unexpected noise on sensor signals. I spent three weekends designing and sewing the complex bellows for the wrist mechanism cover. Peter Kovesi and Stewart worked on measuring and predicting sheep surface models. Zbigniew and the workshop technicians still had many design details to complete—the main oil supply umbilicals and emergency cut-off valve were needed to replace temporary ones being used for tests. And disconcerting amounts of metal fragments were still being flushed out from the arm after exercising the actuators—the cause would remain hidden until four years later.

Four weeks later, and three months late, SM took on its rightful name 'Shear Magic' when the first sheep was mounted for a trial. It started well enough, but ended in disgrace—a specially designed hose connection popped out of a wrist actuator, showering the rear end of the unfortunate sheep with warm foaming oil.

In theory, we should have been able to run the ORACLE shearing programs with SM without any changes. After ORACLE was retired, we completely overhauled our methods for predicting sheep surface models so the new models for SM were slightly different. We also wanted to exploit the dramatic increase in workspace so we made many changes to the shearing sequences. Constraints on repositioning were different, and blows could be extended past limits imposed by the ORACLE workspace. There were also countless minor software and hardware problems. It seems remarkable looking back, but just four weeks after the first shearing test, with three weeks left to go, we had our first trouble-free shearing run. The sheep was cleanly shorn on all except its neck and face, and a large mohawk of wool which was left over the rump.

The remaining weeks were used for practice sessions. There were still one or two hardware problems, and an interruption when the main oil supplies were properly connected for the first time. During the week before the demonstration we performed to invited 'friendly' audiences to help overcome stage fright. By the end of that week, we had fourteen cleanly shorn sheep, including two ewes with healthy lambs born just after shearing.

For the first time, we had a demonstration ready, well prepared and practised, and we had met our deadline. Not with as much time to spare as we would have liked. In retrospect, I learned to be more cautious with time estimates and work schedules, but I was pleased that the revised estimates of completion had not been too far out. Computer-based project control techniques had played an important part in this.

The demonstration—climax and anticlimax

The Corporation needed a good demonstration from us to reassure themselves about their decision to have funded us. Or at least that is what I now believe in retrospect.

We knew that it was important, and went to a great deal of trouble to ensure it was successful. We selected a group of sheep months in advance, and selected from them to ensure we had the best sheep available on the demonstration day.

There was an element of theatre about the occasion. The committee members were seated in the passageway outside the laboratory—we installed a stepped platform to make sure that they all had a good view. As soon as they were seated, blinds on the passageway windows were lifted and immediately the robot started shearing. The sheep was brightly lit so they would not miss any of the action (also for a good video picture so the team members not involved could watch on a monitor in a nearby laboratory). At the end of the slow, but successful shearing run, the robot smoothly lifted itself away from the sheep and pulled down a rolled banner displaying 'Shear Magic!'

They were impressed. They knew we had only just started shearing with the robot, and remembered earlier disappointing demonstrations. There were the inevitable detractors too—'Well that's OK but how are you going to load the sheep?'

I think we misunderstood their thinking, though. We believed that they wanted a good demonstration to give them the confidence to move on to a more commercially oriented development programme. With hindsight, I think that they wanted reassurance that their decision to allow us to build the new robot had been correct. They still wanted to see 95% of the sheep shorn automatically, even though we had explained many times that both the robot *and manipulator* would have to be redesigned to achieve that.

There were other issues to resolve as well of course. We had just a few days pay left for both Peter Kovesi and Stewart Key and needed a prompt decision.

Prof. Brian Stone, who had taken over as officer in charge on Prof. A-W's retirement, took them to task on the issue of communication and on keeping Stewart Key and Peter Kovesi on six month contracts. He asked for encouragement which had been rarely received, particularly in writing. He queried their arbitrary decisions which had been made without consultation or discussion afterwards. He asked for longer-term contracts.

They immediately extended funding for Peter and Stewart. But we had expected important decisions and announcements to follow our successful showing and were to be disappointed. After weeks of waiting, a letter arrived pointing out that we had still not achieved 95% shearing, and that the status quo would remain. There would be no thought of commercial development.

The result was devastating for the team. Stewart Key devoted his energies to his own consultancy and though he remained with the team for another 18 months his heart lay elsewhere. As Peter Kovesi was to say later, 'You're only part of this team if you think about it while you're under the shower each morning!' Peter decided to visit the USA with Robyn Owens who was taking study leave at Berkeley for six months. Zbigniew Lambert found a government job with more security. Michael Ong, who had put more of his emotional energy into SM than any of us, decided he wanted to think about music and farming for at least a year and went trekking in Nepal.

After the Christmas break, we regrouped and passed yet another milestone. With great patience, we programmed SM to shear about 97% of the wool from two sheep. We knew that ARAMP was not the machine to do this, but we decided to demonstrate that it could be done. The results were ready in time for the WHRAC meeting in March, and once again we thought they would be impressed. But it made no difference—the status quo was maintained—and they decided on yet another review of the project later in the year.

In the years since 1985, the team has been rebuilt. But the level of enthusiasm, energy, drive and creativity which produced the SM robot has never been recaptured.

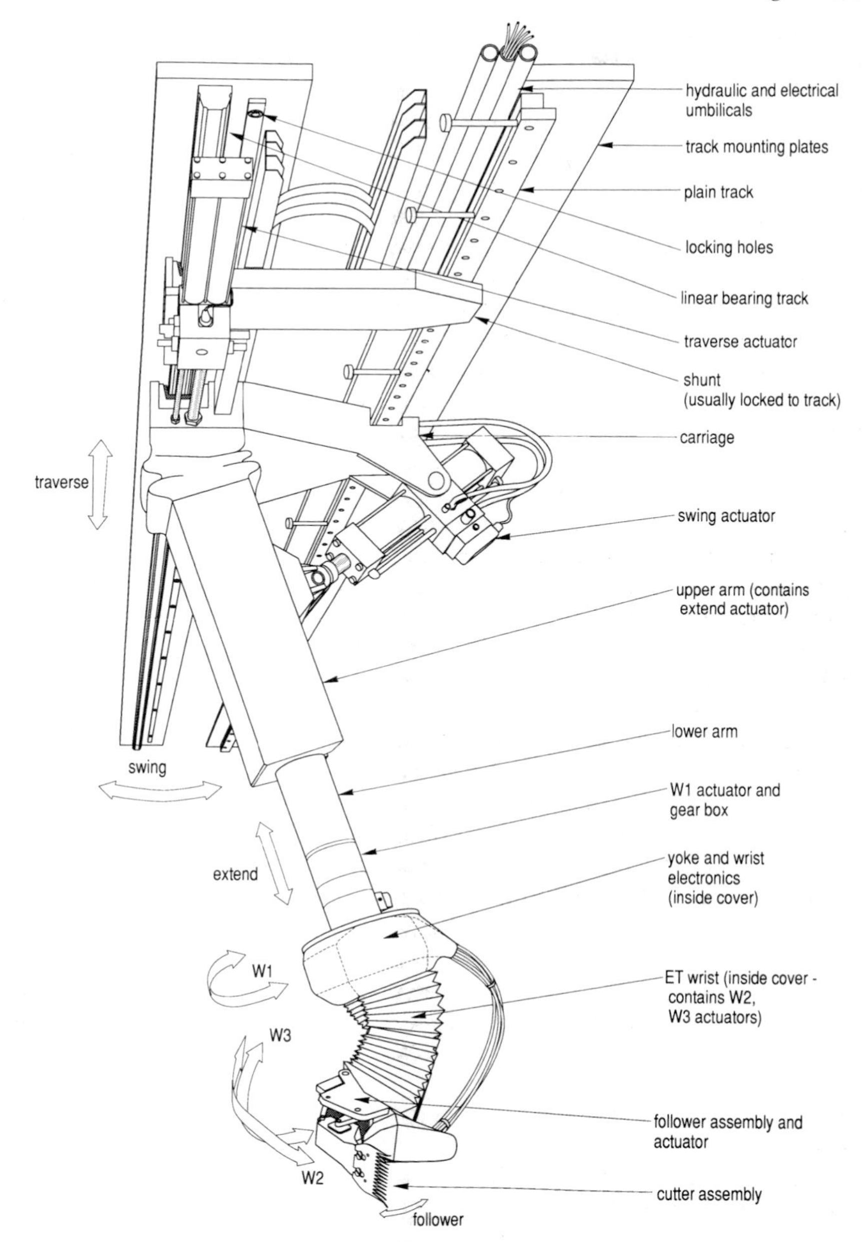

Figure 8.31 SM Robot—major parts. The carriage can be locked to the track and the shunt unlocked. In this state the robot can move the shunt to a new position, and with a sequence of shuffling moves, can reposition its traverse workspace anywhere along the 5.4 m track.

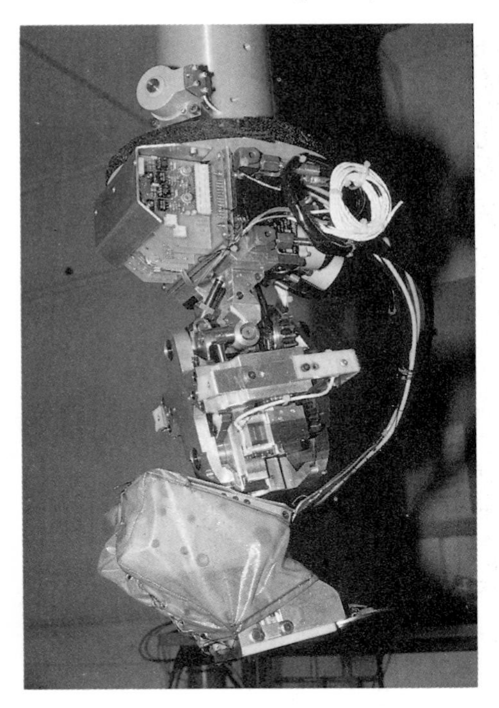

Figure 8.32 Assembled wrist and cutter mechanism—September 1985.

Figure 8.33 Peter Kovesi and David Elford watch during demonstration practice trials.

Figure 8.34 Stewart Key watches clearances between SM and ARAMP during collision avoidance tests.

Figure 8.35 Sheep shorn before November demonstration to WHRAC committee.

Figure 8.36 Team photograph after November demonstration. From left: author, David Elford, Dennis Brown, Roberto diBiaggio, Jan Baranski, Ian Howard, Peter Kovesi, Ian Hamilton, Michael Ong, Stewart Key, Stewart Howard. Absent: Lynette Lynn, Wai Chee Yao, Virginia Skipworth, Brian Stone, Dan Pitic, Zbigniew Lambert.

9

Holding the sheep—part 2

We took two and a half years to conceive, design, build and commission the SM robot. It looked simple and elegant compared with its forerunner. SM gave us a vastly increased workspace and dexterity yet behind the simple exterior it was no simpler conceptually than ORACLE and was a great deal more mechanically complex.

Within a few short weeks of commissioning SM we had shorn 97% of the wool off two sheep in a painfully complicated and tedious operation. The sheep were restrained by the ARAMP manipulator but several new tricky positions had to be devised to expose all of the wool for shearing.

David was now faced with the expectation that we would be able to simplify the manipulation platform and improve its performance in the same way we had for the robot. Yet this time there had to be a real mechanical simplification. With 42 actuators ARAMP was unacceptably complex—too much even to be considered as the basis of a simpler design. The new manipulator also had to be much faster, fit a larger range of sheep, hold the head and fit within the workspace of the robot. At the same time it had to embody practical ideas for loading the sheep and handling the wool, neither of which were considered for ARAMP. We had to make a greater conceptual leap than any we had so far attempted.

System design issues

Wool handling

We had carefully sidestepped wool handling since we started. At the same time we could not ignore it because shorn wool was often lying in the way of the cutter. When this was a problem, we pulled it out of the way by hand, or programmed the robot to push it out of the way.

John Bryson, working at the MUSHEEP laboratory, devised and experimented with a vacuum suction machine to remove wool from the cutter as it was being shorn. He soon found that some mechanical assistance was required. The fleece wool of most sheep hangs together like a fluffy blanket. Shorn wool cannot easily be separated from the rest and when it does separate a large clump of wool comes away at the same time—these clumps immediately blocked the vacuum system. To counter this he devised a wool stripper—a miniature conveyor belt with projecting spikes that pulled the wool away from the unshorn fleece just above the cutter (Bryson and Field 1981). He used a double belt—an outer belt carrying the wool and an inner one carrying the spikes running on a smaller pulley at the top end. The spikes were retracted at the top end allowing the wool to be caught by an air suction stream.

It was a clumsy device which required height adjustment for different lengths of wool. It was not entirely effective and was never seen as a practical device to incorporate on the SM robot. We switched our preference to gravity by arranging the shearing pattern such that large chunks of fleece peeled away from the sheep as it was shorn. We used this as best we could on the ARAMP manipulator.

Fleece lying on top of the sheep would not fall away of course, and this led to the question of second cuts. If wool was left in place over the sheep would it not be cut again when running another shearing blow underneath it? Fortunately the answer was mostly 'no'—the tendency of the fleece to hang together like a woolly jumper protected it from being shorn twice. However, if the stubble was shorn twice then second cuts were a problem—very short pieces of stubble which were no use to anybody and a direct production loss. Long stubble is best left on the sheep for harvesting the following season.

Contrast this with manual shearing. The human shearer manipulates and moves the animal not only to get the sheep into the right position for shearing but also to allow the fleece to peel off the animal in one continuous piece. As shearing proceeds the animal is rolled on the floor away from the growing pile of shorn wool. After shearing, the fleece is gathered up by a rouseabout and tossed over a classing table. A good rouseabout can throw even the flimsiest fleece without it flying into pieces in mid-air.

In the early days we thought that keeping the fleece intact was just too ambitious, too much of a constraint on design. Yet perhaps this was a case of narrow thinking; maybe we should have looked at other possibilities.

Sheep loading

When we started to explore sheep loading we immediately crossed Corporation demarcation lines. Other research groups such as MUSHEEP at Melbourne University had been well funded to investigate sheep loading—why was it necessary for us to reinvent their solutions? It was not easy to explain that the details of sheep loading had to be worked out before we could be sure we had a practical manipulator design. In turn sheep loading depends on the arrangement of the manipulator. It was not easy for the Corporation to accept that the results of earlier research had such limited value. Research work often tells you more about how not to design machinery!

Size range

The ARAMP manipulator was designed around a live weight range of 30 to 45 kg. Initially we thought that this would be a typical medium size range for Australian sheep. Alas, some months into the detailed design phase of ARAMP we discovered that the flock we used for size measurement was in very poor condition at the time. They had been cared for by a dairy farmer on quite unsuitable land. Figure 9.1 shows a distribution of sheep weights and lengths which we obtained from measurements between 1981 and 1985 and reveals how much larger the size range for our new manipulator was to be.

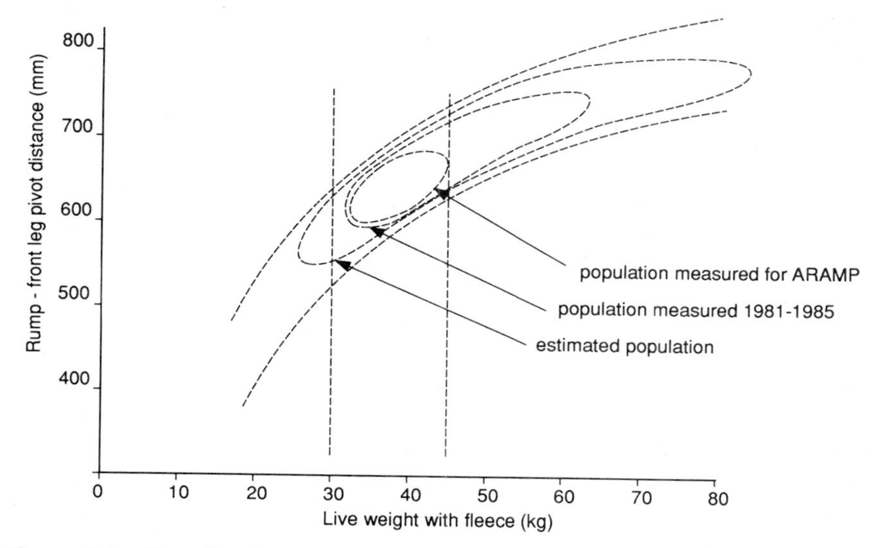

Figure 9.1 Size distribution of sheep. The vertical coordinate roughly corresponds to the length of the sheep's body. Leg lengths do not correlate well with weight or body length (redrawn from computer plots).

Manipulation time

The speed of manipulation was also of concern. ARAMP required 12 minutes of sheep manipulation time during which the animal was inaccessible to the robot and extra time was needed for sheep loading. We wanted to reduce the manipulation time to less than two minutes. We had conceived a multiple robot arrangement in which individual robots could move from one manipulator to another while sheep were being loaded or turned over. Two robots could work on each sheep at a time. This way we could keep all the robots and manipulators busy and improve the productivity of the shearing plant.

Head restraint

The ARAMP head restraint consisted of a jaw clamp attached to the two neck rests by ropes containing the sensing wire. Not only was head positioning erratic; the jaw was not a comfortable means of holding the head. Head shearing would require accurate location of the eyes, ears and horns and we supposed that this would be beyond any practical sensing system. So accurate restraint of the head was going to be essential, preferably without obscuring any of the face wool.

Workspace and repeatability

Not everything was going to be more difficult. Changes in technology relaxed two vital constraints. First, the SM robot had a huge usable workspace and could freely access all the upper exposed half of the sheep. ARAMP on the other hand had to be designed around the narrow constraints of ORACLE's workspace and limited dexterity. The second factor was machine vision which we planned to use to find the position of the sheep. ARAMP had been designed around the idea of repeatably positioning animals relative to reference positions. This was essential for predicting surface models. A machine vision system would allow us to measure the position of the sheep on the cradle—repeatable positioning was no longer a requirement.

SLAMP concept

David began a patient search through all the sheep-handling concepts which had emerged during the previous decade (Beadle 1987). Fortunately luck was on our side.

Ian Campbell was an enthusiastic woolgrower who had long been a keen supporter of the project. Apart from his constant efforts to engender political support for the project among woolgrowers he was always ready to help find sheep for shearing trials and to provide thoughtful advice when called on. He found a simple and apparently elegant sheep-handling device at the Wagin Woolarama—a major rural wool show. He persuaded the New Zealand inventor Ian Moffatt to visit us and show his impressive videotape. David was enthusiastic and immediately ordered a Moffatt cradle.

The cradle seemed a remarkable match to our requirements, at least as a conceptual starting point. The sheep lay horizontal, on rollers, entirely within the robot's workspace. The entire fleece peeled away, intact. The cradle worked well with a variety of sheep. It was well positioned for manual shearing, should the need arise. Sheep loading was simple—the cradle was rotated and tilted to meet a sheep being dragged out of a catching pen. Once leaning against the tilted rollers, the sheep is tilted and lifted to the horizontal position for shearing (figures 9.2, 9.3, 9.4).

Roller cradles were not new of course. In the last century a Mr T. Millear invented a roller cradle and a device similar to the 'baa-baa chair' (figure 7.4, p. 169) for hand shearing and claimed that skilled shearers were no longer needed! (Adam-Smith 1982, p. 154)

Moffatt had designed his cradle for New Zealand sheep with plain necks, no skin wrinkles and little if any wool on the legs and face. Merino sheep were quite different. It was no easy matter to translate Moffatt's flowing, easy shearing style from the promotion videotape on to a wrinkly, greasy, dirty Merino wether. Our first attempts at shearing them were marked by hard work and plenty of perspiration in pursuit of inspiration!

The major difficulty was neck shearing. It was soon clear that the cradle concept had to be extended: there were no easy answers. Along with other team

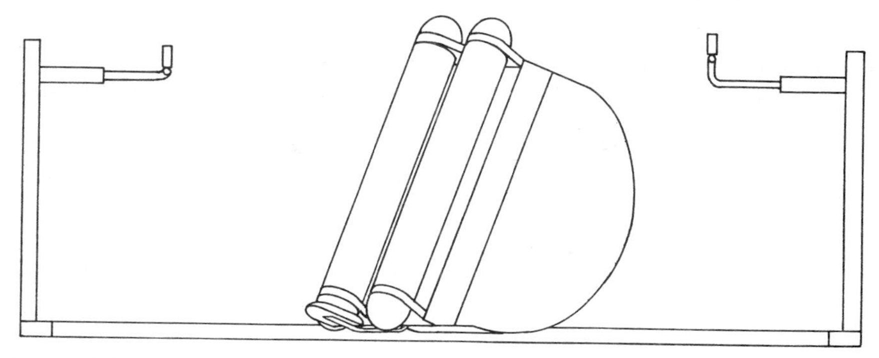

Figure 9.2 SLAMP 1 cradle (1986). The rollers have been tilted into the loading position.

Figure 9.3 SLAMP 1 cradle. Sheep has been dragged to cradle.

members, David and I joined a shearing course for novices to learn how shearers handle Merino necks. This was a valuable experience at the expense of more perspiration and aching joints for a week afterwards. Apart from the modicum of skill gained, it helped to strengthen team morale through shared and unforgettable moments like our electronics technician Roberto Di Biaggio being kissed on the nipple by an amorous ewe.

Within weeks of the Moffatt cradle arriving in the laboratory David had designed the first experimental SLAMP cradle (Simplified Loading and Manipulation Platform). We learned an important lesson from our ARAMP experience. We had to build an experimental cradle to check the design before proceeding with a mechanically elaborate cradle which could not be changed later.

In spite of the apparent simplicity our progress was slower than expected. 1986 was a year overshadowed by deep misunderstandings between ourselves and the Corporation. Changes in the Wool Industry Act governing the Corporation created additional uncertainties and no doubt contributed to a rather different understanding of our problems by the Corporation.

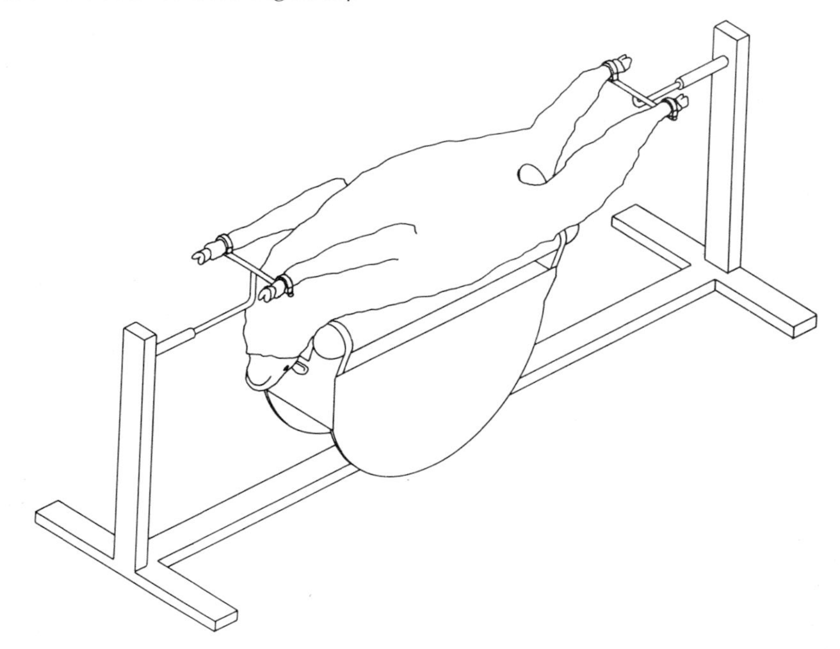

Figure 9.4 SLAMP 1 cradle. Cradle has been returned to the horizontal position for shearing.

After the budget review in March 1986 we were told there was no money for documentation, inductance sensing, vision or sheep loading work and we were not to pursue these lines of work. Instead a major technical and economic review was to be commissioned before expanding the scope of our work. In the meantime, we were to focus on the problem of removing 95% of the wool 'under automatic control'. We were dismayed—we could not design a new manipulator without taking sheep loading into account. Machine vision for locating the sheep was a key part of our concept freeing us from the constraints of repeatable positioning. Without inductance sensing the reactive wool problem would remain unanswered.

To the Corporation, their intention to commission a review by external consultants was an expression of confidence in the project. To us, it was yet another hurdle to cross. More demonstrations and more time spent justifying our concepts and ideas. It appeared more like a loss of confidence.

At least the way was clear to fill the vacancies created by the resignations of Zbigniew and Michael Ong. Here too we were to run into administrative problems with the university. With three months of the financial year remaining we were constrained to advertise in terms of

> 'the (Research Engineer) appointment is for the period to the 30th June 1986 with some prospects for continuation depending on the availability of research funding'.

One applicant was on holiday from New Zealand and he needed some money; he appeared for his interview dressed in a tee shirt, shorts and thongs! After readvertising the positions with slightly improved wording we were able to appoint a mechanical designer but the only promising applicant for the research position switched to a higher paid job at the last moment. We were also constrained to make research appointments at the graduate level so we were unable to recruit experienced staff.

By the time our resources should have been focussed on the construction of our first experimental cradle much of our effort was focussed on the review by Russell Flack of Aptech, a firm of technology consultants, and Alex Holzer, head of the department of Mechanical and Production Engineering at the Royal Melbourne Institute of Technology.

Once again we were placed in a competitive position with Merino Wool Harvesting (MWH). Once again we feared that our experimental results and prototype equipment costs would be compared with the bold commercial forecast which Merino Wool Harvesting were bound to claim to attract their funding. This time our fears were unfounded—the consultants concluded that our approach was a more direct path to robotic shearing.

In January 1987 the project administration changed for the better. The Corporation had appointed a sub-committee of three prominent technologists to oversee our work. The chairman Ken Langley had been president of the Institution of Engineers Australia and operated a major consultancy to the petroleum industry. Jan Kolm, director of research at ICI, was able to contribute his long experience of industrial research and development. Ken Bishop brought manufacturing experience from his direct association with major appliance manufacturers. They soon grasped the essence of our problems and by early March had provided us with the resources we needed to make some real progress. They provided funds for documentation for all aspects of our work, particularly computer software. We were soon joined by a computer scientist, Steve Ridout, who was to make major contributions in his own right within a short time. Money was supplied for a first-class computer-aided design system, and Russell Flack from Aptech was to visit regularly to ensure that we were no longer hampered by lack of material resources.

With renewed vigour and with the help of measurements from the by now completed SLAMP 1 cradle David was able to complete the design details of the SLAMP 2 manipulator (figure 9.5).

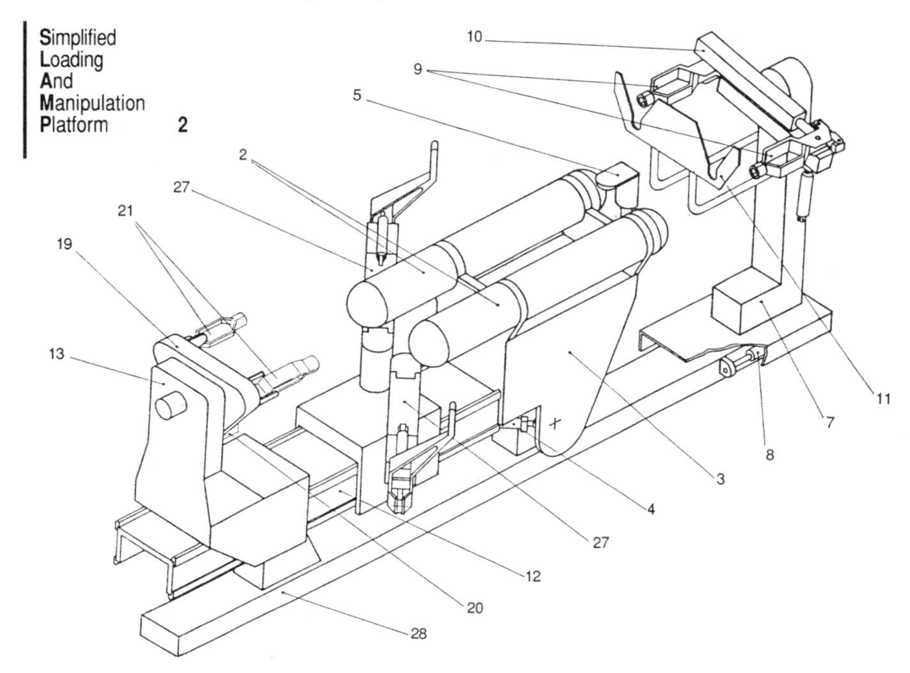

Figure 9.5 SLAMP 2 cradle (1987). The swing arms, 27, have been shown in the neck shearing position. Some details have been omitted for clarity.

List of sub-assemblies shown in figure 9.5.

The diagram was part of the technical report detailing the SLAMP 2 design.

(2) Interchangeable roller extensions. Different length extension rollers were to be available for tailoring the manipulator to different size ranges of sheep—e.g. small, medium and large.

(3) Main cradle frame and roller mounting.

(4) Cradle turning actuator (used for loading).

(5) Crutch gauge—shown in raised position. This engages with the sheep's crutch when the animal is raised from the tilted to the horizontal position.

(7) Rear leg carriage—incorporates a drive to rotate the rear leg clamp assembly when the sheep is rolled.

(8) Rear leg tensioning cylinder.

(9) Rear leg clamps. The design allows the clamps to pivot through about 30° when closed; when open the clamps are constrained so that they are in the right position for the rear legs to be inserted.

(10) Rear leg spread cylinders—spread the rear legs for crutch shearing.

(11) Rear leg locator—raised during loading sequence to guide the rear legs into the open clamps for grasping. Normally lowered out of the way.

(12) Main front end track beam—part of the tilting cradle. The track carries both the swing arm carriage and the front turret carriage.

(13) Front end turret drive. Contains a drive motor to move it on the track (12), a rotation drive, an engagement key lock and release drives for the leg and head clamps.

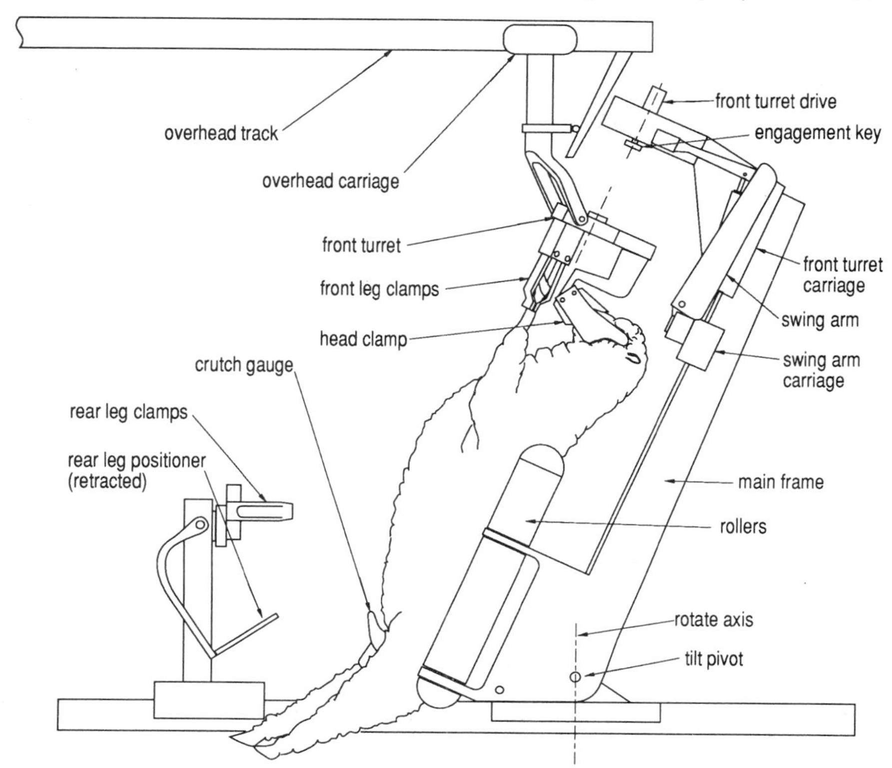

Figure 9.6 SLAMP 2 sheep loading scheme. The sheep is dragged, supported by the head and leg clamps. The clamp fixture is carried by a powered overhead conveyor. When the sheep is dragged to the cradle, the front turret descends to engage the clamp fixture, releasing it from the conveyor. When the cradle lifts the sheep to the horizontal position, the rear legs are trapped and caught by the rear leg clamps and the locator bars. Once this has been done, the crutch gauge is folded down out of the way.

(19) Turret carrying the front leg clamps and head clamp. The clamps are normally locked to the turret but may be released and transferred to the swing arms for neck and front leg shearing. The drives for operating the release/engagement mechanisms were located in (13).

(20) Head clamp—not detailed.

(21) Front leg clamps. The clamps grasped the hocks and so must contain the hooves.

(27) Swing arms and carriage. Two identically symmetric arms can rotate about pivots on the carriage, be raised and lowered, and can extend (stretch) the legs when attached. Arranging the arms to precisely dock with the head and leg clamp fixtures on (19) was a difficult design problem to solve. The carriage contains a drive to move it along the track (12) so that it is positioned under the brisket of the sheep.

(28) Base pedestal.

David was pleased with his design but expressed private reservations about its complexity and some of the gaps which had still to be resolved. SLAMP 2 depended on loading the sheep by dragging it backwards towards the tilted rollers (figure 9.6). Front leg and head clamps would be attached at a central restraining station; the sheep would be partly suspended by the clamps and dragged along the floor on its backside, just as a shearer drags a sheep from a catching pen. Yet we still had no satisfactory head clamp. A device which grasped the nose and the underside of the jawbone had been tested but we found that it slipped off many of the sheep we tried it on. At least 27 actuated move ments would be needed and in contrast to ARAMP several of these required mid-stroke position control.

It would be easy to overlook the design breakthroughs in the SLAMP 2 design by focussing too much on its shortcomings. We had solved the neck shearing problem. Figure 9.14 (p. 267) shows the neck shearing position we were aiming for and to achieve this we needed to roll the sheep on its side and then stretch the head out on one side of the cradle and the upper front leg on the other side. The head and neck turn through about 60° and the leg turns 120°. This way the head, shoulders and front leg form a straight line at approximately 60° to the sheep's longitudinal axis. The lower front leg remains outstretched. If it is swung back with the upper front leg the sheep can pull itself along the rollers by its rear legs which are still restrained. The crutch position is fixed by the crutch gauge used to help lift the sheep. Therefore the turret carrying the front head needs a long adjustment range. However, it also means that the two swing arms which manipulate the head or front legs into the neck shearing position also need to be carried on a moving carriage. The rotation centres of the swing arms need to be positioned under the brisket of the sheep.

As soon as the design was clarified we started work on an improved experimental prototype, SLAMP 1.5. We wanted to test head and neck manipulation before proceeding too far with detailed design. We also had to test sheep loading.

Once again we were blessed with good fortune which arrived with the tall, imposing figure of Allen White. Allen had earned a reputation not only for shearing but also for his ability to teach shearing skills for wide combs. Once the bitterness of the 1984 shearers strike had receded and the wide combs became popular it was apparent that few shearers knew how to use them. Allen had successfully trained several shearing teams in the Geraldton area north of Perth. He had visited our laboratory out of curiosity the previous year and was immediately able to help us with comb and cutter sharpening techniques. We managed to persuade the Corporation to hire him as a consultant for two months full-time to help us with machine shearing techniques.

And so SLAMP 1.5 was transmuted in Allen's hands to SLAMP 1.75 before it had even left the drawing board. Allen showed us how we could turn the rollers in opposite directions to tighten the skin around the sheep (figure 9.11, p. 266). This way the wool was opened out and easier to shear and simply pulled out of places which are hard to get to. Allen also suggested a new pivot to swing the rear legs from side to side, improving the skin stretching.

With David on extended leave taking a much needed rest I decided it was time we resolved the head clamp issue. For some time I had thought about holding the sheep by its eyes. Squatting in front of the animal I found that by forming cups with my hands and placing them over the eyes I could insert my fingers into depressions at the back of the skull behind the eye sockets. With its eyes covered in this way the sheep subsides into a torpor. It relaxes and stops struggling, its legs gradually folding as it settles on the floor. It is a similar idea to a trick farmers have known for centuries. A sheep can be left immobile for an hour or longer by simply lying it down on a slope, its back downhill with a coat over its head.

With the help of Andy Whitehead, David's design draftsman, I soon had a workable prototype (figure 9.14). To refine our fitting we made a plaster cast of a medium-sized sheep's head—it was soon christened Yorick. Most of the sheep didn't fit the head clamp quite as well as Yorick but there was real excitement. For we now had a way of holding the sheep's head rigidly yet comfortably in a way which protected the sheep's eyes, relaxed the sheep and yet didn't obscure any of the wool which had to be shorn from the head (Trevelyan and Whitehead, patent, 1988).

With the head clamp problem solved the way was now open for sheep loading tests but there were further diversions. David was back from leave and coming to terms with the design changes which had taken place in his absence. The changes were not well coordinated and had to be pulled back into line. One of the graduate engineers unfortunately interpreted this redirection as direct criticism of his competence. He resigned two weeks later but his outburst and departure caused further distractions at a time when we could least afford them.

The computer-aided design system arrived and both David and Andy became completely absorbed in learning how to use it. Both had been reared on drawing boards, and it is a great credit to the designers of the system[1] that they both consigned their boards to a storeroom some six weeks later and have never missed them since.

Other team members were also concerned at the delay in testing, among them Peter Kovesi. Peter had greater reservations than most of us about sheep loading and had been advocating tests for some time. He and David finally set up a test early in November. The sheep was clamped by its head and front legs. In spite of the calming effect of the eye covers the sheep kicked and struggled violently as soon as its weight was taken by the clamps. Even when the sheep was calmer it was still difficult to drag it to the rollers and it was soon clear that the loading scheme on which the SLAMP design depended was far from promising.

Peter suggested loading the sheep on a horizontal stretcher instead,

[1] Hewlett Packard ME-10. It was selected as a computer-aided drafting system.

eliminating the need to tilt the cradle. David refined Peter's idea but soon concluded that it would introduce as many complications as it would eliminate. At the same time there was no going back. Sheep loading had to change.

David emerged from his office two weeks later with a broad smile on his face. He decided to split SLAMP and bring the sheep on to the manipulation platform already mounted on a simplified roller cradle. Once on the platform the cradle became part of the manipulator. Drives engaged with the rollers to turn them and mechanical arms manipulated the front and rear legs. Once shearing was completed the sheep would leave the manipulation platform still on its cradle.

It was the final breakthrough we needed. David had entirely removed the complex and uncertain tasks of restraining the sheep in its clamps away from the manipulator to a loading station. Here an operator would roll each sheep out of a feed race on to a roller cradle (with mechanical assistance) and insert its legs and head into restraint clamps. The loading station would serve as an inspection and preparation station at the same time. The manipulator was left with the more certain task of manipulation. Shorn sheep would return to the operator on their cradles for inspection and unloading. With a manipulator change-over time of just a few seconds, moving the robot to another shearing station to keep it working productively was no longer a requirement. Though this was a significant feature of the SM design, it was one we were glad to dismiss.

Within three weeks of the abortive sheep loading test David produced an exciting new concept and infused the team with a wave of enthusiasm. By March 1988 the design was completed and the first components started to emerge from the workshop and 11 months later the first sheep were being shorn completely automatically.

The paper which follows was written in June 1988 and describes the technical details of the SLAMP 3 manipulator. Some introductory material has been eliminated.

SHEEP HANDLING AND MANIPULATION FOR AUTOMATED SHEARING

James Trevelyan, Technical Director,
David Elford, Senior Mechanical Designer

Presented at 19th ISIR Sydney, 1988 and reproduced and edited by permission. Winner of the innaugural Japan Industrial Robot Association Prize for best paper at 18th and 19th ISIR's in application technology.

Principles of sheep restraint and manipulation

We have devised sheep-shearing robots that are capable of shearing on a sheep which is substantially, but not completely, immobilized. Clearly the sheep must continue to breathe during shearing and will move itself slightly from time to time to relieve itself from some discomfort. Therefore, we have designed our robot to be able to shear sheep which are mechanically restrained and which move within certain limits from time to time.

We found during early experiments that a sheep will stay immobilized sufficiently for automated shearing as long as it is comfortably, yet firmly, restrained. Comfort is obtained by providing as large an area of support as possible and by the selected use of cushioning where there is the possibility of hard surfaces coming into direct contact with the bony parts of the animal. Firm restraint is applied by clamping the head and legs of the animal. Ideally, the legs should be immobilized by the application of pressure just above the 'knee' joint of the back legs and just above and behind the elbow joint of the front legs. A human shearer uses this technique, but can compensate for deflection of the outer parts of the legs by a combination of experience and tactile and vision sensing. For a robot, the ends of the legs must be firmly located while they are being shorn, and further knee and elbow restraint makes the upper parts of the legs difficult, if not impossible, to access. Therefore, we have elected to clamp just the ends of the legs even though this provides a means for the sheep to push against the machine and encourages a degree of struggling.

Our first development in automated animal restraint and manipulation was the ARAMP manipulator (automated restraint and manipulation platform) which was described in Key and Elford (1983). This machine was constructed specifically for the ORACLE robot in use for shearing experiments at the time and demonstrated that sheep could be automatically restrained and manipulated and presented to the robot for shearing in different positions. It was successfully used for shearing experiments on over 200 sheep with a consistently high level of reliability.

It was recognized right from the start of the ARAMP development that a much simpler and faster manipulating device would eventually be needed for a commercial shearing system. With the completion of the Shear Magic (SM) robot in 1985 the focus of our research and development was turned once again to animal restraint and manipulation with the object of designing such a simplified machine.

Manual shearing cradles

A survey of patents related to devices for holding and restraining sheep runs to several hundred pages (Beadle 1987). Many inventive farmers and innovative agricultural engineers have turned their attention to the problem of restraining sheep to make the job of shearing by hand a less arduous one. However, only two or three devices have had any commercial success and the major market has been farmers with small flocks which they can shear themselves without having to acquire the skills and level of physical fitness required for traditional shearing. One type of cradle appeared to us to be an obvious starting point for designing a simplified cradle for automated shearing. In its most common form it consists of two parallel horizontal rollers which support the body of the sheep, and leg clamps which rotate about an axis parallel to, and above, the rollers so the sheep can be turned over by rotating the rollers and leg clamps (figure. 9.4 and patents: Anderson 1976; 1987; Brooker 1983; Drew 1976; Moffatt 1981; Tunsey 1974; Van der Heyden 1977; Wytkin 1981).

This type of cradle is an attractive arrangement for automated shearing for several reasons. Firstly, the sheep is horizontal and this suits the workspace of an overhead robot such as the SM robot shown in figure 9.10 (p. 265). Secondly, the sheep is turned about its longitudinal axis on the rollers and therefore does not move sideways when it is turned over. In the case of the earlier SLAMP manipulator the sheep was displaced by approximately 600 mm sideways when it was inverted and this meant that the robot needed a larger workspace.

The third advantage was that such a cradle would provide excellent access for manual shearing to handle sheep which were too difficult to shear by robot.

A fourth advantage is that the fleece peels off the animal during shearing and remains in one piece. It has become apparent to us that the retention of the fleece in one piece is an important wool trade requirement. Wool fibre characteristics vary from one part of the animal to another but wool can only be sorted on this basis after shearing if the fleece is kept in one piece. There is a further benefit of this particular method of handling the wool after shearing—as the fleece peels away from the animal it also stays out of the way of shearing movements which makes the shearing movements easier for both human and machine shearer alike.

Typical roller cradles for hand shearing restrain all four legs of the sheep by a simple arrangement of ropes (figure 9.7). It has been said that this technique has been known to South American Indians for hundreds of years for restraining Llamas. The head of the sheep is unrestrained—the operator of the machine is expected to hold the head by the left hand while shearing the head and neck areas of the sheep.

Roller type cradles have been successfully used for shearing small to medium sized English and New Zealand sheep which have round, plump bodies and relatively open wool. Australian Merino sheep on the other hand have dense wool over a larger part of the body of the sheep and practically all of them have

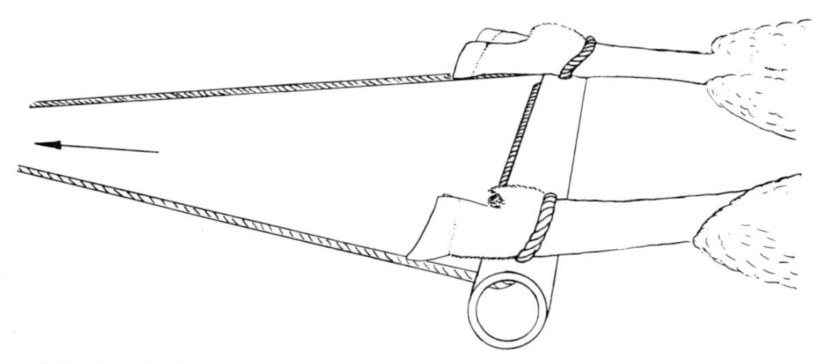

Figure 9.7 Typical arrangement of rope clamps used for hand shearing cradles.

extensive folds of skin around the neck which make shearing much more difficult.

Assessment of roller cradles
Several hand shearing trials were conducted with a commercial roller cradle. Our first attempts at shearing Merino sheep were marked by the considerable expenditure of physical effort and perspiration in the quest for inspiration. However, the positioning of the animal for shearing all the main fleece areas, apart from the head and neck, was better than had been obtained with the earlier ARAMP cradle. The big problem which had to be solved was how to manipulate the head and front legs to provide good access and skin conditioning for shearing the neck.

Skin conditioning for shearing
Before describing how the sheep is positioned for neck shearing it is necessary to provide some background information on the techniques evolved in manual shearing for conditioning the skin of the animal to make the task easier. A human shearer places the animal into as many as 30 or 40 different positions in the course of shearing a sheep. This is done to make shearing easier by adjusting the shape of the surface to be shorn, stretching the skin, and pushing the unshorn wool with the handpiece in one direction or another so that it is shorn more easily. The shearer aims to have a surface which is either flat or slightly convex. Concave surfaces need to be avoided to maintain the consistency of shorn fibre lengths and to leave a short, even stubble. The shearing handpiece also works more efficiently and is easier to push through the wool if all of the comb teeth are running through wool fibres close to the skin.

The skin is stretched to smooth out skin wrinkles and to avoid clogging of the handpiece. If the skin is not stretched sufficiently the drag force caused by the comb points penetrating between wool fibres causes the skin to move forward and the wool fibres to become more densely packed just in front of the comb. This in turn increases the penetration drag and in some situations can stop the comb from penetrating the fleece completely, making shearing seem impossible to the uninitiated. The skin can be stretched either sideways or forwards in the direction of shearing to avoid this problem.

Unlike human skin, the skin of a sheep is relatively mobile on the underlying tissues. Part of the aim of manipulation is to move skin over the underlying tissues to drag it from the very sharply curved parts of the sheep's body to flatter areas where shearing is much easier. Some of these techniques are illustrated in figures 9.8 and 9.9.

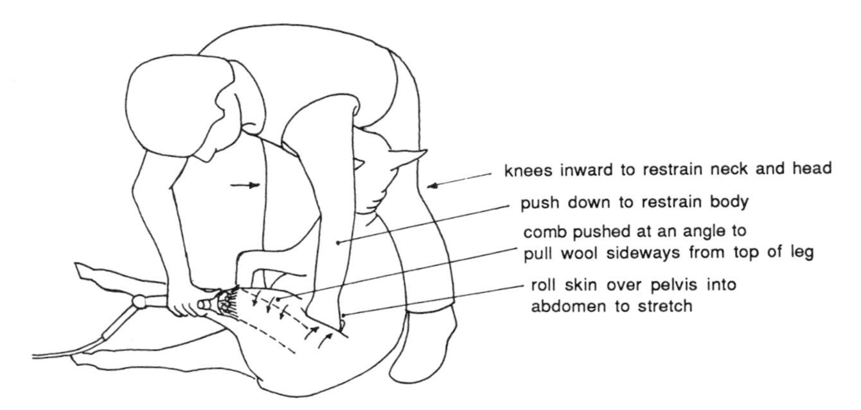

Figure 9.8 Manipulation of the sheep for shearing the rear leg—hand shearing.

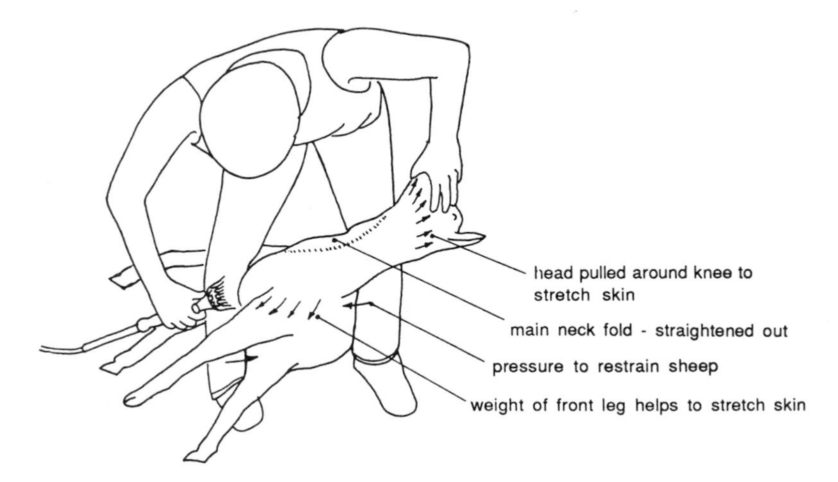

Figure 9.9 Manipulation for shearing the neck—hand shearing.

These techniques, which have been learned through decades of refinement by human shearers, are an essential contribution to automated shearing. They make it possible to shear a sheep cleanly with relatively few, skilfully executed blows, which seem to be executed fairly slowly to the uninitiated observer. This is the key to relaxed and easy shearing by both man and machine.

Positioning the head and neck

The greatest challenge faced in the design of the SLAMP manipulator was achieving a satisfactory shearing position for the neck. The neck of a Merino sheep has a substantial longitudinal skin fold along the underside of the neck and several circumferential folds around the sides of the neck. The extent of the skin folds varies markedly between sheep with large folds being favoured by some breeders in the belief that the greatest skin area produces more wool. More recently breeders have favoured sheep with fewer and smaller skin folds as they are easier to shear and less trouble to look after. Nevertheless, it is still essential to manipulate the sheep into the position shown in figure 9.9. This type of position is simply not feasible in a simple roller cradle because the legs remain stretched between the two end posts. To achieve this type of sheep position, independent manipulation of the front legs and head is required. We have devised a position which provides good access to the neck for shearing with most if not all of the circumferential skin folds stretched out and with

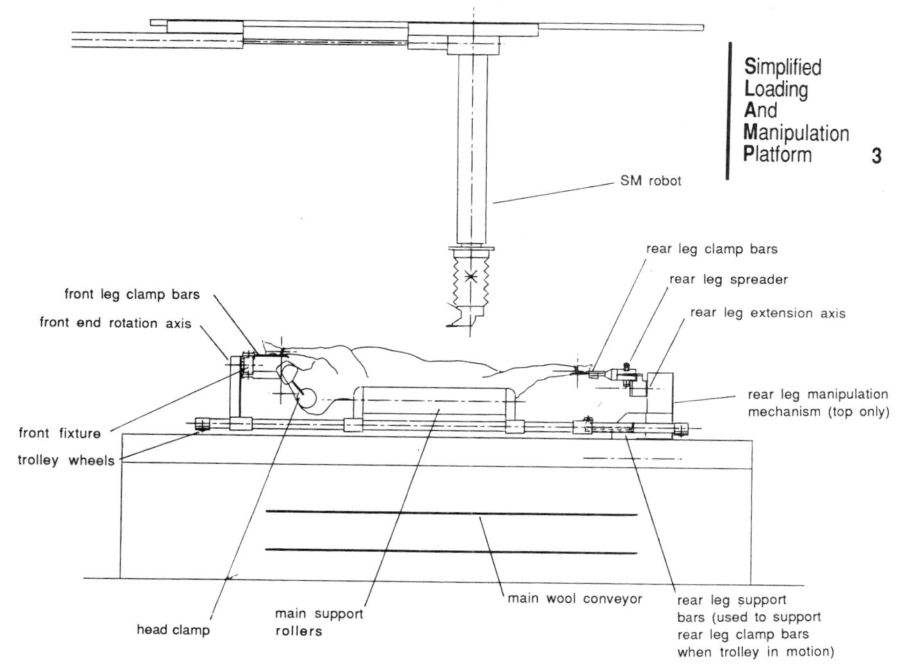

Figure 9.10 SLAMP—A roller cradle concept adapted for automated shearing. The diagram shows the arrangement for the laboratory prototype of the SLAMP manipulator. The diagram illustrates the trolley on which the sheep is transported—the only part of the SLAMP manipulator visible is the rear leg manipulation mechanism and the main platform on which the trolley rests. Actuator drives for the front fixture, the main support rollers and deployment of the rear leg support bars are contained within the platform and engage with drive components in the trolley when the trolley is located on the platform.

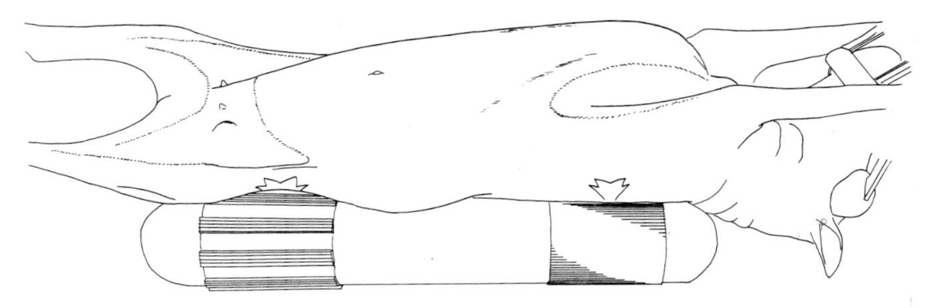

Figure 9.11 SLAMP shearing position—belly. The animal is supported on its back by the two main support rollers and leg and head clamps. The two rollers are contra-rotated to stretch the skin of the sheep in the region of the stifle joint at the rear end the brisket at the front end.

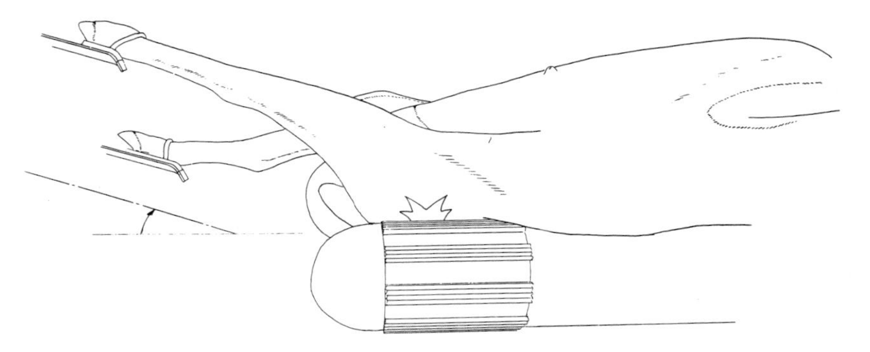

Figure 9.12 SLAMP shearing position—rear legs and side. The rear legs are rotated through about 60° and then pivoted to the side away from the rear leg being shorn by about 20°. The near side roller is rotated to stretch the skin. The position is symmetrically reversed to shear the other side.

reasonable comfort for the animal (figure 9.14). One of the front legs remains attached to the front end post to restrain the sheep longitudinally.

We have found that manipulation of the head and the upper front leg on each side into the positions shown takes place along an approximately circular arc centred inside the brisket.

In order to achieve this kind of manipulation a special front end fixture is required to hold the front leg clamps and head clamp. The fixture is rotated as the sheep is turned on rollers. At two positions where the sheep is lying on its respective sides the upper leg clamp and the head clamp can be removed from the fixture by swing arms which move the end of the leg and the head to achieve the neck shearing position.

A height adjustment is also required to lower the front leg into the most advantageous position for shearing. The end of the front leg needs to be lowered relative to its position in the front end fixture, to help stretch the skin of the neck.

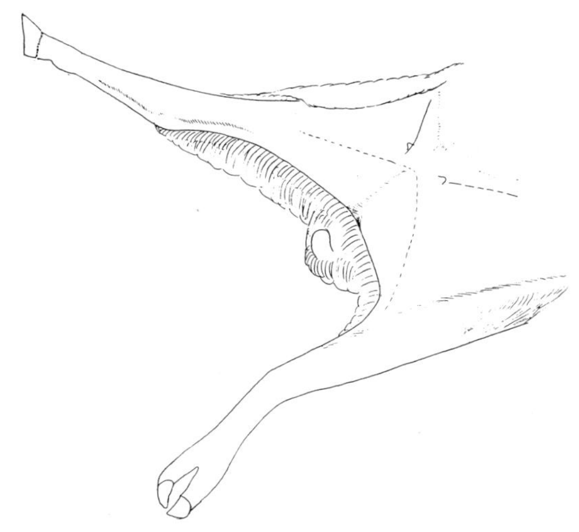

Figure 9.13 SLAMP shearing positions—crutch. The rear legs are spread apart with the sheep in the belly position lying on its back. The legs are then rotated by approximately 20° to improve access for the robot.

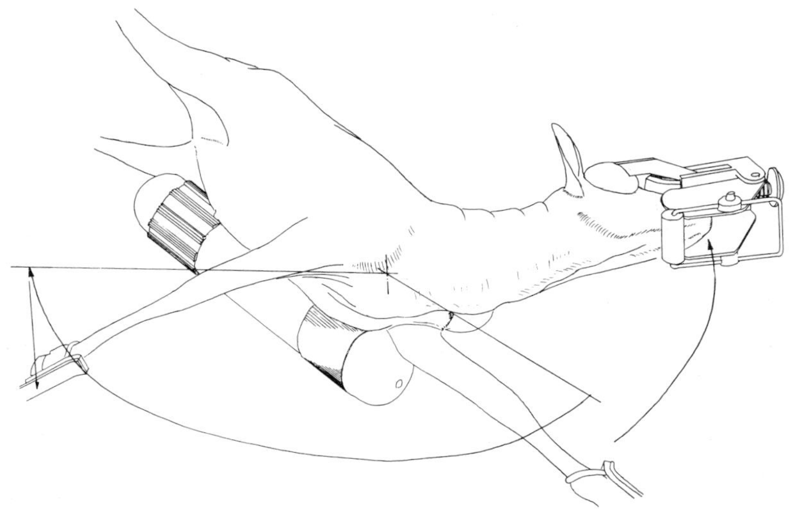

Figure 9.14 SLAMP shearing positions—front leg and neck. The upper front leg is rotated about the brisket through an angle of approximately 120° and then lowered as shown. The head and neck are rotated through an angle of approximately 60° as shown. The head and leg are returned to their starting positions to shear the upper half of the neck and the sides.

When the shearing of the neck has been completed the head and front leg are rotated back through their respective arcs and returned to their starting positions in the holding fixture. The sheep can then be rotated to the next shearing position.

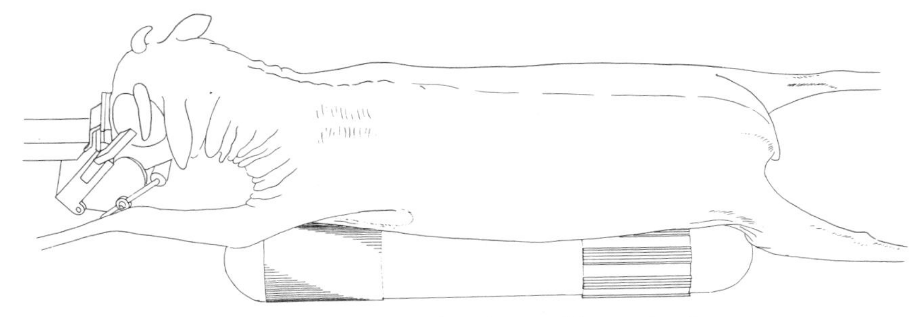

Figure 9.15 SLAMP shearing positions—back.

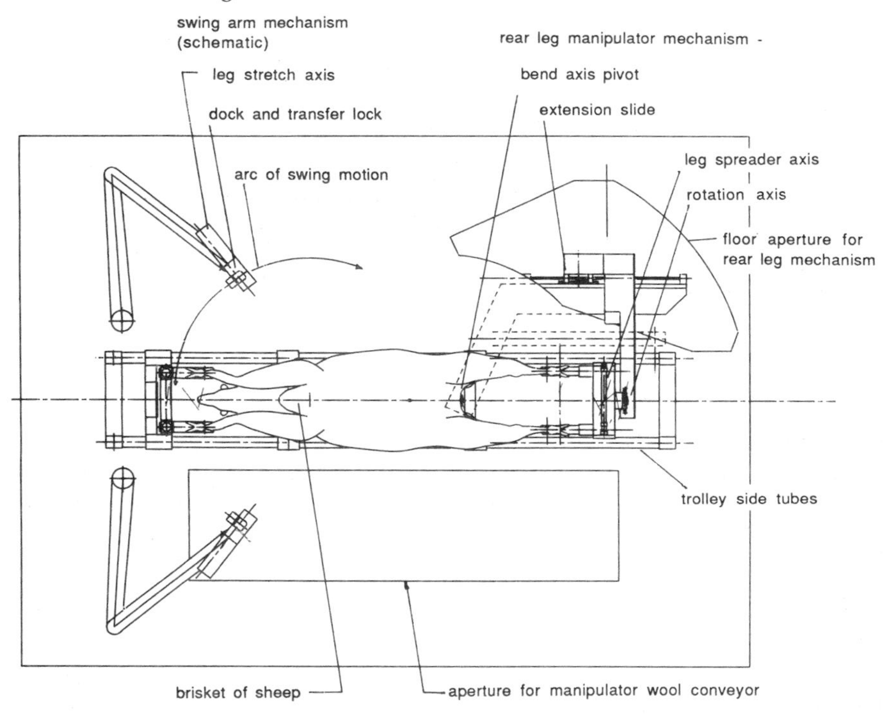

Figure 9.16 SLAMP manipulator—plan view. This diagram illustrates the components of the manipulator which are built into the manipulation platform as well as the trolley. Note that the longitudinal position of the brisket of the sheep is fixed regardless of sheep size, therefore, the extension slide for the rear legs has to accommodate variation in rear leg length in addition to variation in body length.

Effects of size variation

Any automatic manipulator must be capable of handling a wide range of sheep sizes and we have chosen to design the cradle so that it is capable of handling sheep in a weight range from 25 kg to 85 kg live body weight. Figure 9.1 (p. 251) shows the extent of size variation which we anticipate.

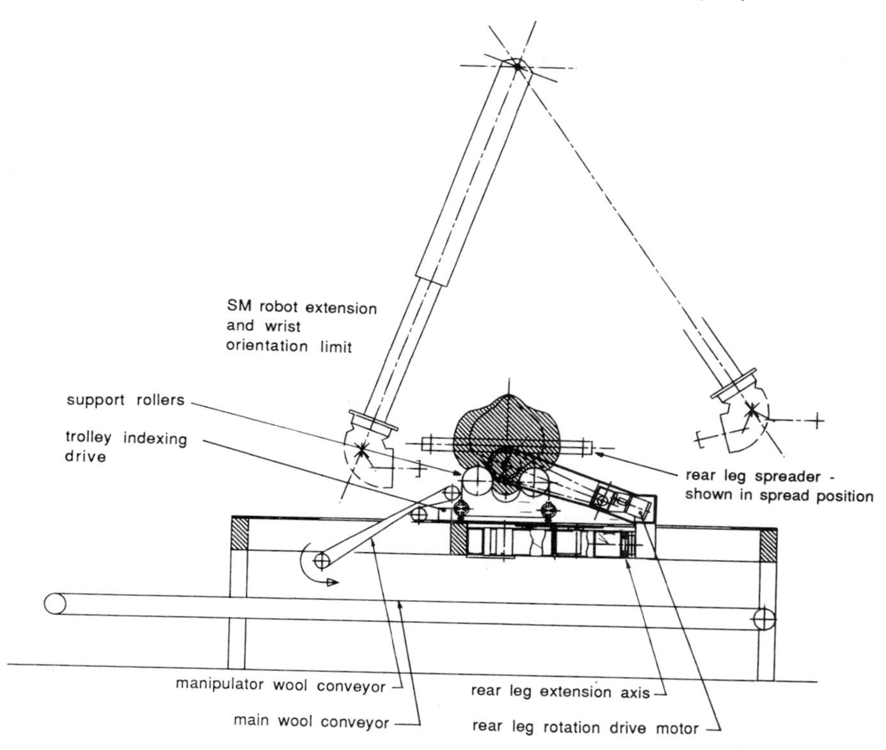

Figure 9.17 SLAMP manipulator—end view. Details of the manipulator and main wool conveyors are shown. The fleece peels off on to the manipulator wool conveyor during shearing. Once shearing is completed the entire fleece is discharged at an appropriate moment on the main wool conveyor which runs under all shearing stations. The diagram illustrates the laboratory prototype arrangement.

For most of the manipulation sequence size variation can readily be accommodated by longitudinal adjustment of the rear leg clamps on their rotating fixture and of the front fixture holding the leg and head clamps. The requirement for head and front leg manipulation for neck shearing creates some special problems as the variation in leg length is typically not in proportion to the variation in body length.

We have chosen to locate the sheep longitudinally so that the front of the brisket is at a known position and therefore the effective centre of rotation for the head and front leg in the neck shearing position is practically the same for all sheep. The major remaining difficulty concerns the length of the neck. The neck length does not vary in proportion to that of the front legs and so the position of the head clamp requires some adjustment in relation to the position of the front leg clamps in the rotatable fixture.

Head clamp

The other key to successful shearing of the head and neck of the sheep lies in the design of the head clamp. It is highly desirable to clamp the head in a way which does not obscure any wool which needs to be shorn and yet achieves a rigid connection between the clamp and the skull of the animal.

We have devised a head clamp which satisfies this criterion and which also effectively blindfolds the sheep and protects the eyes. It is well known that a blindfolded sheep which is lying on the ground will remain immobile for several hours without any further restraint. We have found that blindfolding significantly quietens the animal, even during shearing and manipulation movements which are imposed on the animal.

The skull of the sheep is clamped by the two eye cups, each of which has a small protruding lug which engages with a depression in the animal's skull just behind the raised eye sockets. The two arms holding the eye cups can be moved in to engage with the skull and a mechanical locking mechanism prevents them from being separated until the animal is finally released. A third device consisting of a pivoted locking bar is placed under the jaw of the animal to prevent rotation of the head about the eye cups (figure 9.14).

With this design of head clamp there is a small patch of wool between the eye sockets and the ears which is difficult to access for shearing. However, this can be left on the sheep until shearing has been completed and then can either be chemically removed or removed by hand at the unloading station.

Automatic loading and unloading

A system that is well designed for automation needs to satisfy the following requirements:

1. Maximum utilization of the automated plant.
2. Separation of human operators from the workspace of automatic machinery, but clear access for visual monitoring.
3. The jobs which require hard physical work are performed by the machinery.
4. The human operator, if required, needs to have sufficient scope to exercise his or her skill to maximize the productivity of the automated equipment so as, at the same time, to maximize job satisfaction and earnings.

A recent innovation in crutching machines is depicted in figure 9.19. Sheep are led up a race to two work stations where shearers roll them out on to crutching cradles. Unlike most earlier innovations, these machines are being used commercially and it is reasonable to expect that the practical problems of handling sheep through them will be overcome by progressive refinement. This development shows how machines can be built around people and their skills.

We chose to remove the complex and uncertain task of restraining the sheep to a loading station, similar in principle to a commercial crutching machine. An operator rolls sheep (with mechanical assistance) on to a standardized trolley consisting of the two parallel rollers, rear leg clamps and front end fixture incorporating both head and leg clamps. The sheep is inverted from a device similar to a Gun Crutcha (Young, patent, 1979) by the human operator who first applies the head and leg clamps. In doing so there is an opportunity to shear parts of the head and the crutch by hand if the sheep is a particularly difficult one to shear in those areas. The operator will also be responsible for assessing the type of sheep and feeding information into the robot control system to adjust the shearing strategy. Typical information would include the extent of grass seed contamination of the fleece, presence of stained areas of wool, the sex of the sheep, the condition of the sheep and the shearing qualities of the wool, and could be input to the system control computer by a limited vocabulary speech dialogue.

It is also feasible for the operator to take certain measurements of the wool and its characteristics which may be useful for sorting the wool automatically

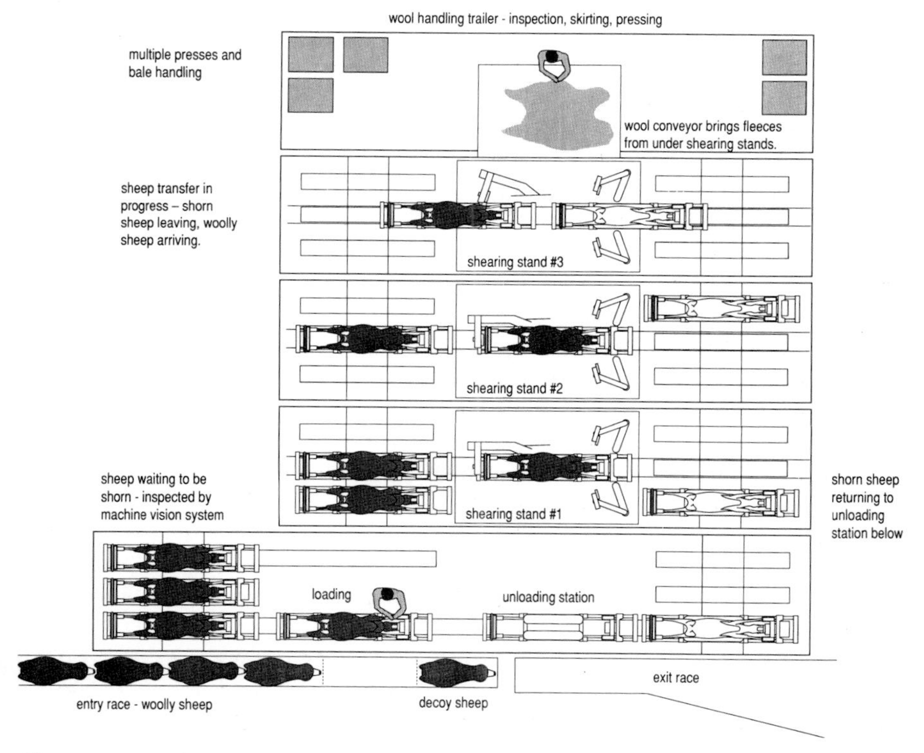

Figure 9.18 Shearing system plant layout. Each large rectangle represents a standard container sized semi-trailer. Though they have been drawn slightly apart for clarity, they would be locked together on-site. The sheep trolleys move around the system clock wise, driven by a mechanical indexing drive. A control computer ensures that each shearing station has a woolly sheep ready to be loaded and an empty slot for the shorn sheep to be discharged into. Loading and unloading operations can be performed on a batch basis and do not have to be exactly synchronized with the operation of the automated shearing stations.

Figure 9.19 Mobile crutching rig. The rig is operated commercially by shearing contractors in Western Australia. (Roof, sun shades, tow hitch and other details have been omitted for clarity.) A third operator removes crutchings from the far side and helps to feed sheep from yards nearby.

after shearing. The measurements would be taken by electronic instruments connected to the system control computer.

The overall arrangement of the proposed plant is shown in figure 9.18. After the loading station the sheep is automatically transported to one of several shearing and manipulation modules. After shearing, the shorn sheep is transported back to an unloading station where the operator removes the clamps, applies anti-lice and other treatments and releases the sheep. At this stage the operator can also remove any pieces of wool left unshorn by the robots. The system has some holding capacity between the load and unload stations and the shearing and manipulation stations so that woolly sheep are always available for immediate transfer once a shorn sheep has been ejected from the shearing and manipulation module.

The sheep transport system required to move animals between the loading and unloading stations and the shearing and manipulation modules is mechanically very simple because the sheep are all transported on a standardized fixture, which is moved through constant distance steps through the system. The system control computer ensures that sheep are moved in response to the varying cycle times of each shearing robot, and that information pertinent to each sheep is downloaded into the appropriate robot controller.

Description of manipulator

The manipulator which we have designed consists of two major components. The first is a removable trolley consisting of the two parallel rollers and two end fixtures to which the legs and head are attached. The second part consists of the platform to which the movable trolley is secured, a mechanism to manipulate the rear end of the sheep and two swing arms which manipulate the front legs and head. Figure 9.10 (p. 265) illustrates these features.

The removable trolleys consist of three assemblies all linked together by two parallel steel tubes which run the length of the cradle. In the middle there is the roller assembly consisting of the two independently driven parallel rollers. The rollers are capable of being driven in either direction so that they can be used either for rotating the sheep (when they rotate in the same direction as each other) or for stretching or bunching the skin (when they rotate in opposite directions). The rollers themselves are replaceable and we anticipate that three different sizes will be required to deal with the full range of sheep sizes.

At the back end of the trolley there is a rear end fixture which holds two leg clamps to which the sheep's legs are secured using a rope passed through pinch cleats.

At the other end there is a rotatable fixture holding a head clamp and two leg clamps (similar to the rear leg clamps).

When a trolley holding a sheep arrives at the manipulation platform, pneumatically operated clamps secure the trolley, and the roller drives and front end fixture drives are engaged mechanically (the trolley does not contain any actuator drives—all movable components are locked in position while the trolley is away from the manipulation platform).

A mechanism at the rear end of the platform engages with the two rear leg clamps so that the rear legs can be stretched, turned with the sheep, spread apart for shearing the crutch or swung to either side through about 20°. Two swing arms can be moved through circular arcs centred on the sheep's brisket position; these arms engage either with a front leg clamp or the head clamp. A rotating key-lock inside each clamp is turned when the swing arms engage with the clamp so that the clamp becomes locked to the swing arm instead of the front end fixture. The swing arms can be raised or lowered, or moved radially to stretch the legs or neck. For neck shearing the front leg is rotated through approximately 120° and the head and neck through approximately 60°.

The remaining component of the manipulator is a wool removal conveyor. As shearing progresses the sheep is turned and the fleece peels off the sheep on one side. The wool removal conveyor supports the weight of the fleece so that it does not pull out delicate folds of skin which could easily be cut during the next shearing blow. Once the entire fleece has been shorn the full fleece is discharged downwards on to a main wool removal conveyor which runs under all the shearing stations.

Actuators

The manipulator is actuated by pneumatic cylinders and electric stepper motor and lead screw drives. The ARAMP manipulator was almost entirely hydraulically actuated. However we have found that the finished cost of electric actuation

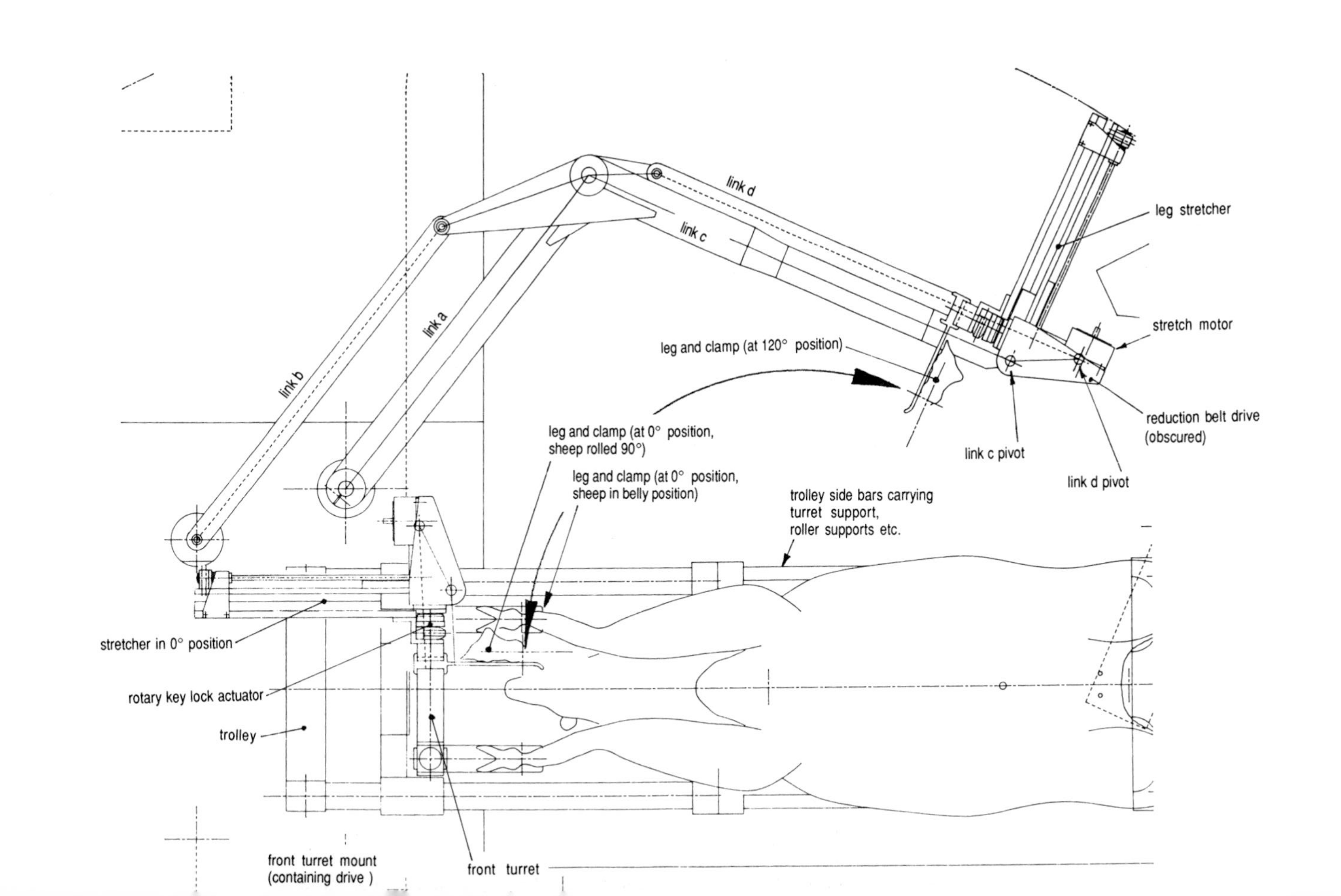
link d
link c
link a
link b
leg stretcher
stretch motor
leg and clamp (at 120° position)
reduction belt drive
(obscured)
link c pivot
link d pivot
leg and clamp (at 0° position,
sheep rolled 90°)
leg and clamp (at 0° position,
sheep in belly position)
trolley side bars carrying
turret support,
roller supports etc.
stretcher in 0° position
rotary key lock actuator
trolley
front turret mount
(containing drive)
front turret

arms manipulate the head and front legs into neck shearing positions.

The arm consists of a double pantograph virtual centre mechanism similar to the ORACLE wrist. It is constructed from tubular main members 'a' and 'c' and flat tracking links 'b' and 'd'. A leg stretch mechanism is carried on the end of the arm.

The leg clamp bars and the head clamp are normally locked to the turret which rotates as the sheep rolls on the rollers. To transfer a leg or the head to the arm, the sheep is rolled 90° from the belly-up position. The arm moves the leg stretcher to the transfer position. A locking key carried by the rotary pneumatic actuator engages the leg clamp bar, at the same time disengaging a key in the turret, releasing the clamp. In this way, the clamp bar is transferred from the turret to the arm. Turning the actuator in the opposite direction releases the clamp bar from the arm and relocks it to the turret.

The arm is raised and lowered (through 300 mm) by a counterbalanced ball screw lifting actuator under the platform. The 120° swing movement is obtained by a linear (ball screw) actuator working a lever on a sliding collar, also under the platform.

The longitudinal position of the turret is initially unknown. When the cradle is mounted on the platform the arms position the stretchers over the turret and the stretch motions are used to search for magnetic markers on the turret mount. A calibrated offset value then provides the exact location of the leg clamp bars. Different offset values are stored for each cradle in use.

devices has fallen dramatically when compared with equivalent hydraulic actuation for this application.

Force sensors and overload protection devices are fitted to all movements which could injure the sheep in the event of uncontrolled actuator movements.

Controls

SLAMP 3 is computer controlled, as ARAMP was, though with quite a different approach. Although the approach is largely conventional, the effort justifies more than a passing mention. A later review of our work criticized our choice to build our own control system on the basis that our research resources should have been directed at sensing instead. I do not intend to refute the criticism as I believe it is justified and I have already explained the background to our decision. As is so often the case, the overriding need was to integrate aspects of the manipulator control system with the robot control software. First the manipulator stretches the sheep, and the positions of the drives provide the location of the extremities of the sheep—the ends of the legs and the head. Second, in the event of a problem, we need to *pause* both the robot and the manipulator with the option of continuing the sequence.

The decision to use electric motors brought new problems which we had hitherto been able to ignore. Most robot users have to contend with electromagnetic interference (EMI) from the motor drives and this decision introduced EMI to what had been an electrically 'quiet' environment. Jan had taken strict precautions with the SM robot electronics in any case, but there have been significant problems with its sensors and lesser problems with controls.

Figure 9.21 illustrates the manipulator control system. The interfacing with the A900 computer provides, in essence, memory mapped access to registers in the controller; a program can read from or write to a shared common block in the A900 computer in the knowledge that this data will appear at the respective places in the controller within the SM robot I/O cycle time (5 msec). The choice to communicate through the SM robot interface was one of expedience and saved both hardware and software effort. It was also expedient to incorporate the digitizer encoder electronics within the manipulator controller—three 5,000 pulse incremental encoders provide rotation transducers for the arm shown in figure 3.9 (p. 65), and switching logic is used to obtain a four to one increase in pulses.

The optical connection reduces pathways for EMI to penetrate the SM robot control circuits.

Limit switch position sensors are built from Sprague Hall effect integrated switching circuits. These are robust to EMI effects but were surprisingly sensitive to mounting stresses, particularly those arising from differential thermal expansion effects.

Manipulator motion sequences are generated by an interpreter which passes groups of simultaneous commands to a small execution monitor which is destined to operate in parallel with the robot motion control software. SM will ulti-

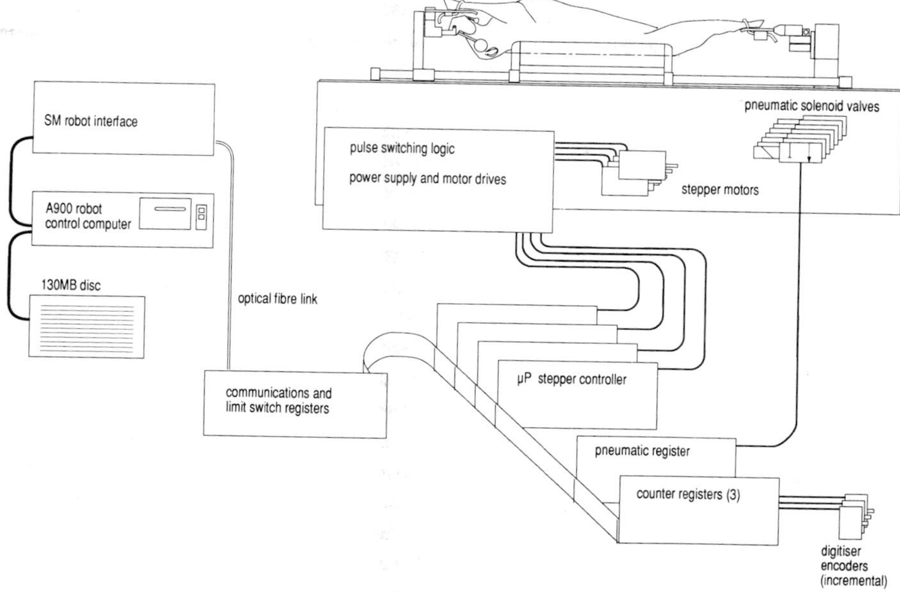

Figure 9.21 SLAMP manipulator control system—refer to text.

mately need to shear sheep while they are being manipulated to achieve a reasonable shearing cycle time, and some close coordination will be essential. With one or two exceptions, though, the robot is working on one end of the sheep while the *other* end is being rotated or stretched. We do not anticipate control difficulties—rather the problem will be programming and sequencing difficulties, particularly with breaking and resuming otherwise automatic integrated sequences of operations.

The interpreter provides a small set of elementary instructions.

Electric drives:

Start a drive moving to a position (absolute or relative displacement);
Start a drive moving to a reference position marked by limit switch sensors;
Start a drive stretching the sheep (move till force sensor switch activated);
Wait (time delay);
Wait for a given drive to finish moving;
Wait for all active drives to finish;
Read drive positions.

Pneumatic actuators:

Switch air on;
Switch air off.

Some pneumatic movements require careful sequencing with reference to limit switches—special routines were built into the interpreter to perform these 'macro' operations.

About 30 files of sequence commands provide instructions for manipulating a sheep from one shearing position to another in a defined sequence. A further program provides a 'safe' user interface to select the right sequence of command files, knowing the current manipulator state, and the new desired shearing position. This program carries an internal network (a directed graph) so that it can calculate the most appropriate sequence of moves, if one exists.

Thus the command which the CLARE programmer uses is 'go to shearing position n4' which we expediently encode as

run 'SLMAP n4';

so that CLARE does not have to know anything specific about manipulator movements.

Experience

I do not think that we would choose stepper motors again, given the opportunity. We chose them originally to avoid the need for feedback encoders, saving wiring complications. We thought that we could predict the torque requirements such that we could choose motors which would never miss a programmed step. What we have found is that sheep struggling and mechanical 'glitches' provide sufficient uncertainty to violate these assumptions and that cheaper alternatives can be used for several movements. Where positioning accuracy is needed we would now prefer to use DC or AC servo drives.

There is one particular situation which underlines the need for feedback. The manipulator is a complex machine, and with the variations in sheep which we encounter, there are several possible faults which occur. Most of these are caused by motors missing steps. When a sequence has failed part way through, we are usually faced with two alternatives.

1. A skilled manipulator programmer operates the interpreter in interactive mode, moving one drive at a time, to restore the manipulator to a known state (a given shearing position).
2. The sheep is removed by hand, and . . . a skilled manipulator programmer operates the interpreter in interactive mode, moving one drive at a time, to restore the manipulator to its initial starting position.

As yet, there is no way for an unskilled operator to extricate SLAMP from a failed state.

For us this is not a problem. Steve Ridout, who wrote the control software, is usually available to rescue SLAMP when it fails or another researcher can follow Steve's excellent documentation, given 30 minutes or so. But this is a problem for the future which needs to be resolved for a practical system. It could provide an interesting research topic.

Conclusion

Since 1984 our principal technical aim has been to develop equipment for shearing an entire sheep in a time which is likely to be economic. We have devised a system comprising separate transportable modules which can be joined together on site to form an integrated automated shearing system. Each module consists of a single trailer which can be transported by a prime mover or form part of a road train.

Our SM robot and SLAMP form the heart of the system as the shearing module. A number of these shearing modules can be set up side by side all served by a single module for loading and unloading sheep.

The third type of module is a wool sorting module where wool can be automatically or manually sorted, and pressed into conventional or jumbo bales.

This shearing system represents an elegantly simple concept which meets all of the important practical requirements:

(1) a modular system which is fully transportable, quick to set up before shearing and to mobilize afterwards, and adaptable to different shearing conditions;

(2) highly productive and fully automatic shearing stations which produce an entire fleece in one piece on to a conveyor for automated inspection, sorting and packaging;

(3) a high degree of safety and reliability through the use of inbuilt computer monitoring of hardware and software operations;

(4) operating costs (to the farmer) comparable to manual contractor shearing rates.

We are confident that this type of shearing system will provide a practical alternative to traditional manual shearing and will be readily accepted by wool producers.

Project control issues

The renewed vigour, energy and money injected by the Corporation to support the manipulator research and development did not come without strings attached. As Technical Director I was soon aware that there was to be tighter control and more accountability. Russell Flack stayed on with the project and visited the project every month to monitor progress and to make sure we had the resources we needed. For a few months this worked well. We had monthly contact with a Corporation representative who could make executive decisions on their behalf. We had a sympathetic ear for our successes and failures and we were responsible to a research committee who understood what engineering research was all about.

Yet by the end of 1987 I was concerned once again. My own research output had been fragmented by building a new team and coordinating technical work on half a dozen different fronts. As our problems mounted the frequency of Russell's visits increased making still further demands on my time. In July 1987 the sub-committee supported my suggestion to appoint a full time Project Manager and to move away from the constraints of annual funding yet there was no progress by the end of the year. At the beginning of 1988 I promoted Wai Chee Yao, our accountant since 1982, to a new position of project administrator. Her role was to spend time with the engineers to help them keep track of the myriad of technical details and to free me for much needed research and trial work. She was to spot delays before they became too serious and to provide the detailed progress data for Russell. She had enough experience for the job and quickly earned the respect of the engineers.

When our increased funding was first approved in 1987, the Corporation believed the manipulator would be completed in time for a demonstration in February 1988. They were persuaded that this was quite impractical and settled for February 1989 instead. The timing was dictated by budget decisions which were made between February and April each year and the February 1989 date was to be immutable.

Anxiety levels were rising fast by the end of 1987 when David produced his final breakthrough. This was compounded by delays to robot shearing trials needed for shearing speed measurements to confirm SLAMP 3 productivity predictions. Even though Wai Chee's appointment freed my time, the trials were well short of completion by the 1988 budget review in February.

It was an uncomfortable meeting. Peter Kovesi, Graham Walker (who joined us as a graduate at the end of 1986 to work on surface modelling and vision), and Darryl Cole (who joined us as a shearing expert on Allen White's recommendation) had worked hard to prepare our first demonstration of vision guided shearing. The demonstration consisted of six shearing blows, but the sub-committee members were not impressed. Resources had been lavished on the project for the twelve months, yet the year's results seemed disappointing. The sub-committee took the unprecedented step of offering a 7.5% bonus as an incentive for a complete shearing demonstration twelve months later. I was privately unhappy about this, though the rest of the team were enthusiastic. And somehow, an expectation emerged for a 12 minute shearing time, though I cannot recollect how.

From the start of his involvement Russell had insisted that all of the work undertaken by the team be monitored by project management and controlled methods. The work was broken down into individual tasks which were assembled into a critical path network and a computer package was used to analyse progress each month. This was a frustrating exercise during 1987 as the programme dates receded into the future month after month. It became more meaningful once the design details of SLAMP 3 were settled. As the engineers responsible for construction David and Jan prepared detailed work estimates for all the manufacturing and assembly tasks. Russell's computer predicted that SLAMP 3 would be completed by mid August 1988 but I had my doubts, yet I could not alter the date without overriding the estimates which Jan and David had committed themselves to.

The team settled into a more predictable routine. Parts gradually accumulated in the workshop and other parts of the shearing system gradually came together. The machine vision system was commissioned and a complete revision of the motion control system (chapter 11) was completed by July.

At the beginning of July the computer was still confidently predicting completion in mid August. Privately I thought that November was a more likely date but that would still leave two months to prepare for the big demonstration in February. By the middle of August it was obvious to everyone that all was not well, but our biggest mistake still lay hidden.

David and Jan had decided in March to switch from hydraulic to electric actuation—they chose high-power DC stepper motors for the 12 positioning movements. It was a sensible choice based on David's observation that electric actuators were becoming cheaper and easier to maintain whereas hydraulic actuator devices often leaked and were not decreasing in price. Jan provided David with motor performance curves based on a commercial motor controller. By May several factors were pushing us towards making our own controllers. At up to $5,000 each the commercial controllers represented a major budget item which we had not allowed for. The Corporation would be far from happy to have such a big increase in the budget right at the start of the financial year. Our second reservation was concerned with sequence control. Until we could be fully confident about safety we wanted to be able to pause the manipulator and the robot at any time if the sheep seemed to be distressed. Experience had shown us that most of the time it would be safe to resume. However the only controllers available from local suppliers provided an emergency stop; after an emergency stop the motors could not be restarted without first moving to a reference position. There were simply no controllers which could be purchased through suppliers and which incorporated this very necessary 'pause and resume' facility. In June, July and August Jan was hard pressed designing and building his controllers and amplifiers. Packaging limitations forced him to halve the planned operating voltage for the stepper motors to 40 Volts. The motor characteristics were by now very different from the ones which Jan had supplied David back in March. Delivery problems delayed the first motor tests until the end of August leaving little time to manoeuvre. By now the tenuous performance margins which had been allowed in March had completely disappeared with the de-rating of the electrical drives and much lower mechanical efficiencies in transmission components than we had expected. With completion dates rapidly receding once again there was no choice but to accept slower manipulator movements.

The project management and control system had been much more thorough and complete than the one I had operated while we were building the SM robot yet it was no more accurate. Instead of the five months predicted in March 1988 it took nine months to complete SLAMP 3. The manipulator was moved under the SM robot on Christmas Eve 1988 with less than five working weeks left before the big demonstration. Three weeks were needed to finish commissioning leaving only days for final practice. However, without some form of project management, it might not have been completed at all.

Demonstration day

February 21st was to be the climax of three years' hard work. The date had been fixed 18 months before and our reputation was 'on the line'. Often it seemed unfair that our work over three years was to be judged on just one trial. With endless trials behind us, with all the practice, the preparation, the anxiety, all directed at one trial; was it all a necessary diversion?

I wrote the following account at the time, though I have edited it slightly for publication.

Saturday February 18th

Preparations have taken most of our time and energy for weeks now; a scarce summer for relaxation, but I am pleased overall. The only disappointment is that there has not been enough time to refine the shearing to keep within the target time for the whole sheep of 12 minutes. The shearing pattern is identical on each side and this is the root of the problem because a non-symmetric pattern could be much more efficient. Yet it is the only practical approach because we have been able to save half of the preparation time. The manipulator has only just started to work, so there has been no time for the steady refinement that I had hoped for. Even so, the robot shears well and we remove the entire fleece in one rather ragged piece.

The entire system works, but slower than we hoped. There are extra shearing blows to be as sure as possible that no pieces of wool are left behind. Wherever we tend to miss a small bit, we have added a blow.

The manipulator is slow, and all the movements are sequential; with more time and confidence we could program them to operate in parallel.

The vision system is slow too, and it has to be closely watched. It demonstrates possibility but not reliability!

And all the sequencing to link the multitude of procedures together takes time too.

We are all thrilled to have reached this stage. Apart from being able to show shearing on the whole sheep, the behaviour of the robot is more predictable and it can shear very fast when set up properly. Of course, it has not been possible to do that in time for this demonstration, where the emphasis is on the manipulator, and shearing the whole sheep in one operation. That we have managed to get that working is amazing. In the last week, we have had four major break downs and lots of minor problems to deal with. We are very good, just now, at fixing problems!

We achieved the first complete shearing just ten days ago, though with breaks and hitches. SM first sheared a sheep on SLAMP 3 in mid January, just the belly and the rear legs. It was not until the beginning of February that all the sheep positions could be reached, and then only by dint of hard work and patience. The manipulator mechanisms needed many minor adjustments and good humour was often a prized commodity. Mechanical parts are still being added, even today.

We are all very tired, and looking forward to Tuesday when we can relax. We have all we need now—the demonstration works, given all the necessary attention to detail. We have a high-quality video filmed last Thursday; if we have a major breakdown, the committee can see it.

We have rehearsed our roles; the machinery works yet the demonstration is more theatre than technical show. How *we* act is much more important than how the machine acts on the day! When . . . if . . . we have a breakdown or a problem we have to just relax and fix it, appearing calm and confident. If we're calm, the review committee will be too. So now a rest and a short practice on Monday. On Tuesday they will be gone by midday, on to their meeting in town, so we can all relax by the pool and celebrate.

Wednesday 22nd February

A pilots' strike disrupted our programme. Half the Wool Harvesting Research Advisory Committee arrived on time at 9 am; we were to mount a second demonstration for the other half when they arrived. Performing two demonstrations is always more than twice as hard as one.

Monday's practice was not a good start to the week. The first sheep started well, but a small cut stopped the robot at the same spot repeatedly on both sides. I could not find any reason for this behaviour! We tried a second sheep, with much better results, but I remained uneasy.

I was particularly concerned about an increasing tendency for the robot to wobble alarmingly and testing at the end of the practice revealed an alarming increase in wrist actuator overshoot errors. This problem has appeared and disappeared over the last few months. But with a recent servo valve failure due to contamination, a report from a specialist confirming the presence of contaminants in the hydraulic circuit, and the increase in wrist overshoot error, I thought that the chances of outright failure on the day were rather alarming. Particularly as we would have to do four sheep—four times the normal daily run for routine experiments. I advised Russell Flack that it would be a good idea to call off the full demonstration, or at least warn the visiting committee of the situation, but he decided to proceed.

We started well on Tuesday. I had slept well, as had other team members, since we all knew that everything that could be done had been done. During final preparations, there was plenty of good humour—a good sign.

The first half of the committee arrived on time, and it was clear that the more influential members were among them.

The sheep were selected of course. Darryl Cole had set aside the best shearing sheep and then selected from these for the demonstration day; the remainder were used for practice. The first sheep was prepared and loaded while the committee was briefed and shown to the viewing area in the passage outside the laboratory. The blind was down while they took their seats. The sheep was struggling alarmingly—more than the norm—and if anything was going to cause problems, that was. We received the signal to start.

I pressed the *return* key to start the automatic sequence. Off went the lights under computer control, the light stripe projectors flashed and seconds later the blind was raised just as the robot started to shear . . . perfectly. Prof. Stone and Russell provided the commentary in the viewing area; they both had radio earphones so I could keep them informed without the committee being able to hear. Steve Ridout sat beside me watching SLAMP, Peter Kovesi and Graham Walker watched the vision system monitor in a darkened corner of the laboratory, David and Darryl watched from the far end to me, and the rest of the team watched on closed circuit TV in the next laboratory.

The robot continued smoothly, repeating a bit when the sheep did struggle at one critical moment. And on, and on till the sheep was finished without another hitch. John Silcock, the chairman, was beaming from ear to ear. He had been a staunch supporter of the project for ten years. I did not have a good view of the shearing from my controlling workstation, but both Darryl and David were smiling.

Off went the first sheep, and on came the second without a break—a routine rehearsed only once. The projectors flashed again and the robot was shearing the second belly. I missed the gentle knocks on the door, and the clatter as Ian Hamilton climbed in the window—the only access out of sight of the committee. Steve and I had forgotten to connect a sensor cable to the second sheep cradle, and fortunately the other team members next door had noticed. But they had no way of warning us after knocks on the door had gone unanswered. (We had unplugged the telephone to avoid an unwanted distraction while shearing.) So Ian climbed out of their window and into ours to carry the vital message. Steve calmly stepped forward to connect the cable seconds before its absence would have aborted the run!

The second sheep was almost finished when the sheep flicked its head and

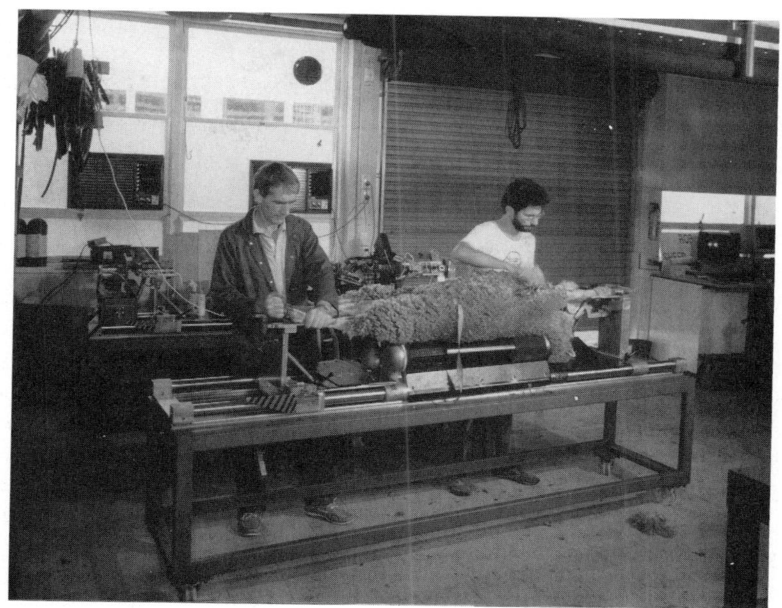

Figure 9.22 Demonstration sequence—1. Darryl (left) and Peter prepare a demonstration sheep, attaching the legs to the clamp bars and clamping the head. This and other photographs were taken during a dress rehearsal.

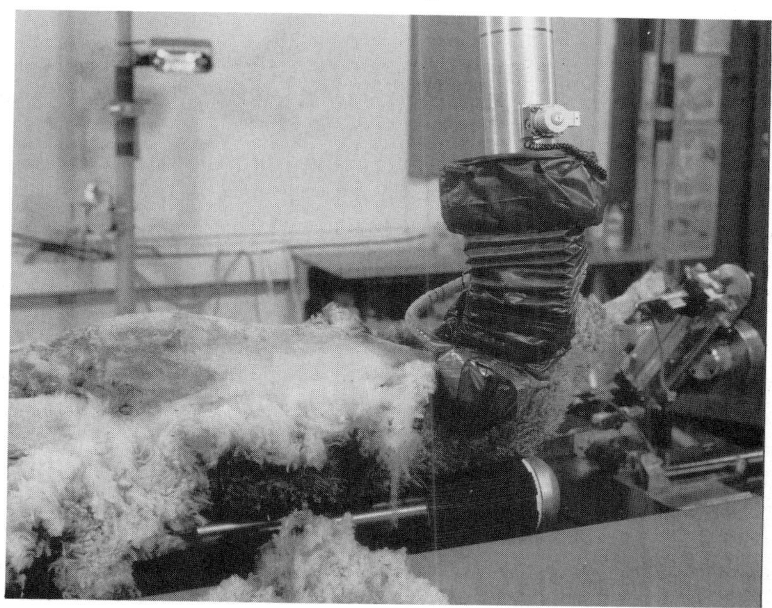

Figure 9.23 Demonstration sequence—2. The belly has been shorn and the robot is pressing down the skin flap near the rear leg while the roller turns to pull the skin taut. The separated belly fleece can just be seen on the conveyor in the foreground.

Figure 9.24 Demonstration sequence—3. Shearing the front leg, preparing for neck shearing. The sheep has been rolled 90° and the front leg has been manipulated by the swing arm; the stretcher mechanism can be seen in the foreground.

Figure 9.25 Demonstration sequence—4. Shearing the neck—the robot has to shear along the straightened skin folds and heave the wool up and over the head.

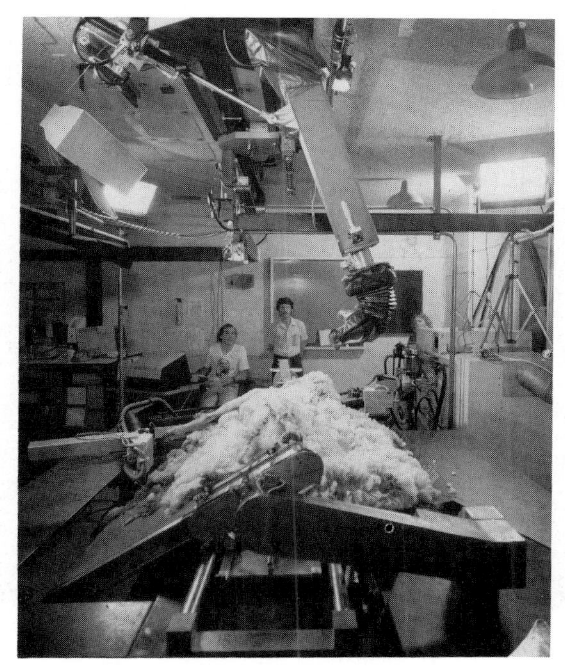

Figure 9.26 Demonstration sequence—5. Two thirds done. The front leg is being returned to the turret before the sheep is rolled over to shear the other side. In the background the author at the operators terminal, Steve Ridout checking SLAMP and Lynette Lynn filming the shearing on video. Note the main machine vision camera (just to the right of the SM upper arm) and the spotlight next to it. Extra lights have been erected for high-quality video filming.

the clamp released. I paused the robot and Darryl calmly mounted the platform to secure the sheep; the robot quietly resumed once he was clear and a minute later two fleeces lay on the conveyor belt under the platform. The committee were thrilled, almost as much as we were. It was certainly our best result.

The committee members were briefly shown around the laboratory and introduced to the team before being given morning tea. David, Peter, Jan and I joined them and I told them how proud we were to have achieved and surpassed their long standing objective 'to shear 95% of the wool under automatic control'. I also announced my intention to step aside from directing the project. There was little reaction to this; they pressed me with questions about reactive wool, shearing time, and the effort needed to reduce it to four minutes—our target. I explained the ideas we wanted to pursue—the hybrid follower, more efficient shearing, a better cutter and so on. They decided to keep to their planned schedule, and left behind a lunch we had arranged for them.

We shared their lunch among us and for the first time in months it was difficult to fill in the time while we waited for the stragglers[2] to arrive.

[2] Also means sheep missed during main muster for shearing.

Figure 9.27 Demonstration sequence—6. Sheep being rolled to right side to complete the shearing. Standing in the background are Darryl Cole (left) and Allen White.

There were just six and they looked weary having spent more time in airports than on the long flight to Perth.

We were relaxed as we settled into the routine once again, I was almost bored. The wrist oscillation was worse though the shearing seemed fine for the first sheep. The changeover was completed without a hitch and the robot started the last one. As the belly was shorn, the oscillations became more pronounced and when the robot turned to shear the side and rear leg, I stopped the run. I checked the wrist and the status display showed a growing servo error. The second group seemed a little disappointed—somehow they were never warned that we would be so far off the twelve minute shearing time but they were soon off to join the meeting at a city hotel. At 4:30 we opened the beer.

Today (Wednesday) we celebrated with a lunch and a relaxed afternoon around the pool. At long last it's all behind us, and although I feel dazed, there is a long overdue sense of exhilaration and achievement.

Thursday February 23rd

Russell Flack visited us to announce the committee decisions. They were disappointed that the time was 25 minutes instead of 12 but they were still pleased and would pay a bonus of 5%. They approved the next budget and requested a plan with a five year time horizon. There would be no more big demonstrations.

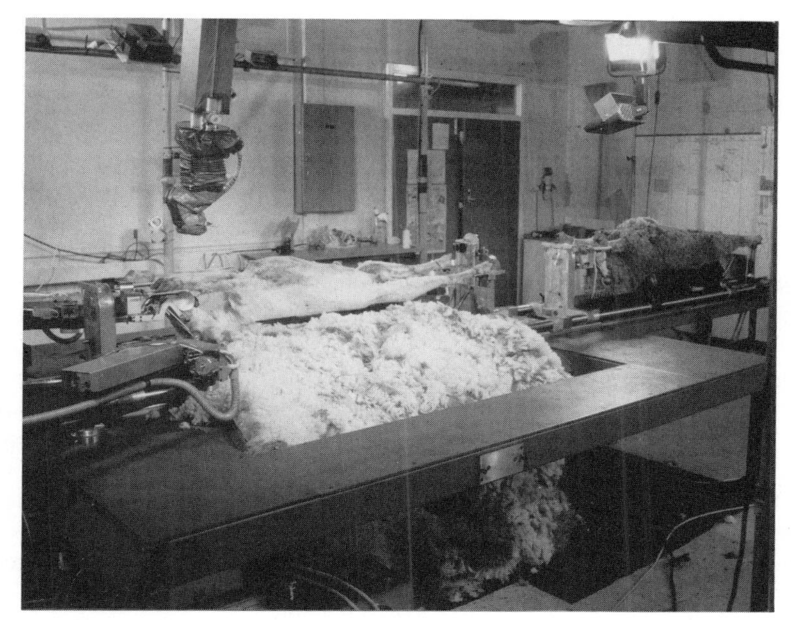

Figure 9.28 Demonstration sequence—7. Change-over. One sheep has been shorn and another waits to be loaded. The fleece is deposited on a large conveyor under the cradle, which was still to be fitted at the time of this rehearsal. A light stripe projector and two halogen lamps used by the vision system can be seen on the overhead bar behind SM.

It seemed that as I was stepping aside, they were making all the planning changes I had long wished for, but that's life!

Epilogue

The next day I discovered how lucky we had been. Just as I started to demonstrate the robot to a student, it suffered a cardiac arrest—an uncontrolled shaking—and I slammed my hand on to the emergency stop button. Shortly afterwards, the robot was stripped for a complete overhaul, in the course of which we discovered the source of metal fragments which we first noticed in 1985, and which were still emerging from its innermost plumbing. An adjustment screw on a commercially manufactured cylinder scraped metal from the extend actuator piston on each stroke and a pile of metal shavings had accumulated at the bottom of the cylinder. Any sudden movement could create enough swirling of oil to pick up fragments and carry them into the delicate servo valves. If one of the larger fragments had found its way to a servo valve . . .

10
Seeing the sheep

Surface models and shearing patterns encoded in command files provide the working knowledge for SM to shear sheep. The surface model provides a reference surface for robot movements. The command file defines the movement sequence, the location of the movements with respect to the surface, and parameters which define the style of movement and operation modes for the sensors. We achieved an enormous simplification when we arranged the surface models such that each part of the sheep can be represented by constant coordinate values; the surface models account for the shape and size variations between sheep. This greatly simplifies the structure of our command files. We calculate a new surface model for every sheep in a way which preserves invariant surface coordinate values as far as possible.

In the early years of the project we discovered that we could accurately predict surface models with size and weight measurements. The extremities of the sheep retained much the same shape; it was sufficient merely to stretch models to obtain a reasonably accurate representation of legs, head and so on. To predict the shape of the body sections we devised statistical techniques. We measured the length, weight and width of the sheep at several points and found that we could predict the shape of the body sections to within approximately 20 mm and the surface slope to within six degrees.

There was one critical assumption underpinning all of this; that the sheep could be repeatably positioned on the manipulator. This was a cornerstone of the design for ARAMP. Each sheep was registered lengthways against a neck rest, and then the sheep was mechanically restrained such that the same parts of the same sheep would always manipulate to the same positions in space. Thus the position of a given part of another sheep would depend only on the intrinsic size and shape of the animal and not on how it happened to be placed in the manipulator.

When we designed the SLAMP manipulator we deliberately relaxed this constraint. We did this on the assumption that we would use machine vision to locate the sheep as well as measure its size and hence make up for variability in positioning. This change allowed us to contemplate the roller type of cradle that we see today. While we are now concerned about sheep positioning variability on the SLAMP manipulator, it is easy to forget these decisions which were made years ago and without which the elegance of the SLAMP design would never have been conceived.

At the beginning of 1986, we assumed that machine vision could provide us with a simple solid state ruler—a device to measure the size and position of sheep on the cradle for predicting surface models. It looked easy; we chose off-the-shelf equipment. We recognized that some new image processing methods might be needed but that we would be able to make a lot of progress using conventional methods.

Machine vision

In our culture, vision is our primary sense and we take it for granted. Our eyes receive enormous quantities of visual information, yet our brains can reduce this to simple understandable shapes and concepts.

In a typical machine vision system, a TV camera provides a computer with images; for each image brightness values for 250,000 or more points (called pixels) are stored as vast arrays of numbers. We do not understand how the human brain manages to analyse images so quickly, but we can analyse machine vision images with computers, rather slowly by comparison. If we design the images carefully, we can write simple programs to analyse the images and take useful measurements reasonably quickly.

In comparison to human sight, machine vision images are blurred, and lack detail, and computers have great difficulty recognizing image features which are quite obvious to a 1-year-old child.

It is worth the effort. With no moving parts, machine vision provides a simple non-contact measurement technique which can be extremely reliable and robust under field operating conditions. Any alternative technique will involve equipment which is more prone to breakdowns and is likely to be more expensive to operate and maintain.

Peter Kovesi spent six months with SRI on a kind of self-funded sabbatical leave in 1986 and learned much about practical machine vision techniques. Meanwhile, Jan Baranski investigated alternatives for vision hardware.

On Peter's return in mid 1986 we purchased an IBM PC-XT computer, a colour camera, a frame grabber[1] and some peripheral equipment. We chose a PC based system because it was cheap and there was no comparable vision equipment for Hewlett Packard computers. The frame grabber was supplied with a library of image processing software subroutines written in C and Peter started work on building a basic machine vision system. In late 1986 Graham Walker joined the team and was soon spending much of his time helping Peter prepare software for the vision system.

Jan was particularly interested in colour image processing but we found that several factors made this look rather unattractive. Firstly, sheep are essentially colourless objects; any colour they do show depends more on wool contamination

[1] Converts the TV signals to digital form and stores one or more images in memory.

than any intrinsic colour. Equipment for analogue preprocessing of colour video signals was expensive and hard to obtain at the time.

There was one more interesting use for colour processing. We wanted to distinguish the outline of a sheep as seen by a camera above the manipulator. Fleece wool is quite dark on the outside, but can be brilliant white on the inside. Scraps of shorn wool lying on the manipulator base could confuse a view of the sheep outline. Even if the base was only indirectly illuminated by reflected light, the white fleece scraps could reflect as much light as brightly illuminated unshorn fleece. We thought of flooding the background with coloured light, and then performing a hue separation to isolate the relatively uncoloured foreground. We also thought of image subtraction techniques—taking a picture of the manipulator before the sheep is loaded and another after loading—subtracting the images should provide a silhouette of the sheep.

Alas, what seems obvious to the human eye is far from obvious to the camera. Neither technique showed much promise. We found that bright primary colours in the background produce noisy luminance signals on composite video, and colour distinctions needed considerable processing to resolve. Coloured light from the background spilled on to the sheep! The image subtraction failed because the presence of the sheep on the manipulator changed shadows and lighting effects around the manipulator—most of the image changed significantly!

We saw more promise in carefully controlling lighting with computer controlled shielded linear halogen lamps and spotlights.

It took longer than we thought to put the system together. There were complications in calibrating the cameras and obtaining suitable lighting. In February 1988 we demonstrated vision-guided shearing for the first time and apart from minor enhancements the vision system has remained unchanged since (figure 10.1). We demonstrated that machine vision could be used as a solid state ruler for measuring sheep and confirmed that the statistical surface prediction techniques would work, given vision measurements (Walker 1987). Many of the measurements could be made reliably with simple computations in a very short time. Although we were able to demonstrate that it works, we failed to demonstrate reliability. We have found more variability in sheep images than we expected and it has been much more difficult than we expected to locate even apparently obvious features in sheep images.

Priorities shifted; first the vital demonstration of fully automated shearing in February 1989, better manual sequence control software, improved programming techniques and yet another round of demonstrations and a further review of the project in 1990. Three years after the original vision system was commissioned, we can think about overdue refinements.

Research priorities have now been directed at surface model generation for the entire sheep population and, to do this effectively, we need more reliable vision measurements. Presently, image processing is the weakest element of the system; as I shall describe in the following paragraphs, we use very basic tech-

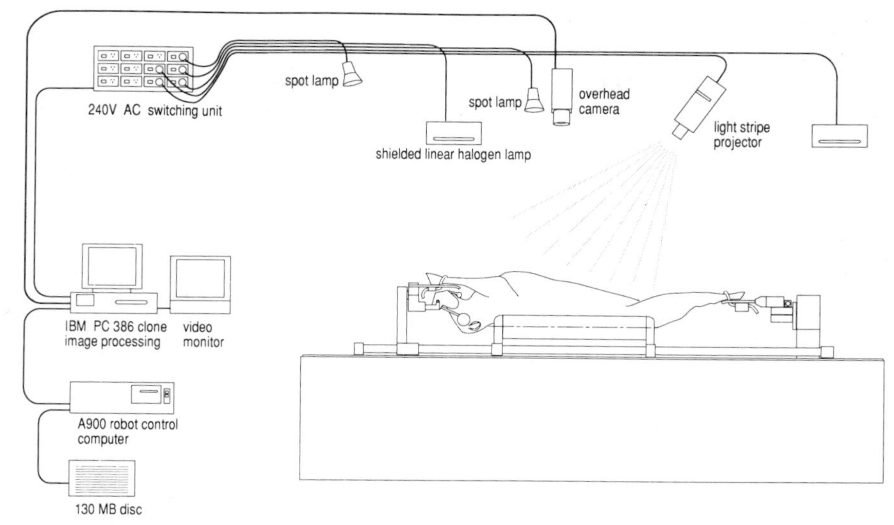

Figure 10.1 Arrangement of vision system equipment—refer to text.

niques. Yet to overcome our present limitations, we will have to break new ground in machine vision research. We have tried the best and most respectable image processing techniques on images of sheep and all have been found wanting. We need to develop our own techniques if we are to demonstrate that our solid state ruler is truly reliable.

The vision task

Compared with many research projects in machine vision we have comparatively modest objectives. All our sheep images look similar—we use the term 'semi-structured' images. Robotics researchers use the term 'unstructured' to refer to the cluttered everyday, natural environment. Many vision research projects are aimed at operating in such environments, typically for mobile robot navigation. At the opposite extreme a factory inspection station provides a highly structured environment. Notionally identical parts are placed one after another in front of the camera with constant lighting conditions at a constant distance. Our aim lies half way between the extremes in terms of difficulty. All sheep are different to a surprising extent on close inspection. The erratic texture of the fleece confuses conventional edge-finding techniques and loose pieces of wool lying on the manipulation platform can confuse visual analysis. Yet all sheep have four legs and a head, they are covered with fluffy wool, they will always be the same way up, more or less, and roughly in the same position.

It is important to note that we had no intention of using the vision system to guide the cutter directly. A shearer uses sight for guiding the handpiece, but a good shearer can shear almost as well blindfolded. For much of the time, the cutter is hidden beneath the wool and vision is only used indirectly. We have always planned to use vision to predict or adapt surface models.

To predict surface models, we need to be able to locate surface features, preferably in three dimensions. We will know the approximate thickness of the wool and the weight of the sheep in advance. From this information we can predict roughly where to look for the features we need to measure and their orientation. We can control lighting and to a lesser extent background, and we can sometimes create our own visual features (figure 10.2).

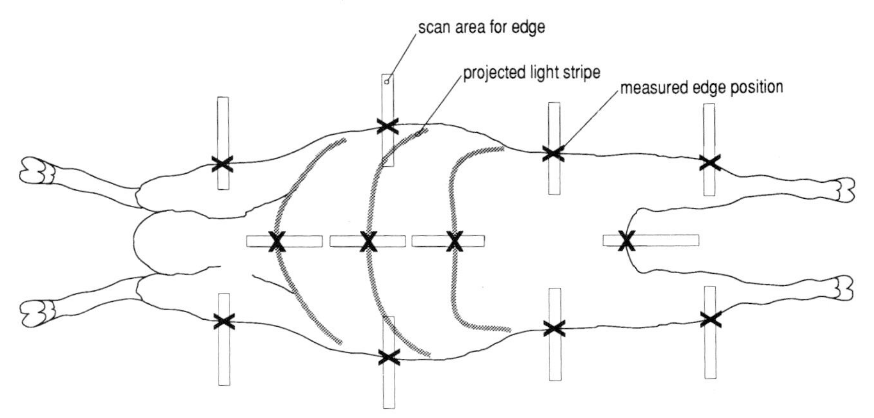

Figure 10.2 Typical vision measurements. Edges of the sheep are detected along image search bands approximately normal to the edge and about 5 pixels wide. Stripes are detected within comparable search bands.

For three-dimensional measurements we can use structured lighting —stripes of light projected across the sheep's body. Once we know the plane of each light stripe we can convert a two-dimensional measurement made in the image plane to a three-dimensional location. This relies on accurate calibration of the cameras and the light stripe projectors—the results are extremely sensitive to small calibration errors. We also need to correct for lens distortion.

In this second phase of machine vision research we wish to provide reliable measurements; that is, measurements with a known and high degree of reliability. We also need to know when measurement processes have failed so we also need to calculate our confidence in each measurement. This provides a safety margin—if the confidence level is too low then the measurement data is discarded and replaced with other information.

Equipment resources

The hardware and software arrangement is shown in Figure 10.1. The vision system is presently based on an IBM 386 compatible personal computer (PC). A Matrox PIP 1024 video digitizer board (frame grabber) is installed in the PC. This provides inputs from three black and white CCD TV cameras and the frame grabber can store four images at a resolution of 512 x 512 pixels. The processed images can be displayed on a TV monitor with graphics overlaid.

We first used Sony CCD colour video cameras (DXC-102P) though the images are digitized in black and white. We have since worked with purpose-designed black and white cameras which appear to supply images with less noise. We use Sony auto-iris VCL-08Y lenses, fitted with Sony wide VCL-0746 conversion lenses, giving an effective focal length of 5.6 mm.

A 16-bit digital interface controls lights and projectors mounted around the sheep through a purpose-built switch box. The shielded linear halogen lamps provide overall illumination with a sharp cut-off to keep the manipulator platform as dark as possible—for these we used Phillips Apollo Kombi 500 W outdoor floodlights mounted upside down.

The PC is connected to the robot control computer via an RS232 serial link operating at 9600 baud.

The major part of the vision software acts as a command interpreter. Commands can be accepted either from the keyboard of the PC or the robot control computer and the replies go to the PC display or the computer respectively. Most of the dialogue between the robot control computer and the PC is in terms of camera plane coordinates—integer values—to reduce communication overhead. Three dimensional calculations are mainly performed by the robot control computer.

The interpreter software is primarily based on a library of image processing functions. We use the library routines supplied with the frame grabber for access to the hardware and have created our own routines for image processing operations.[2] All the image processing software is written in C and runs under the control of the DOS operating system.

Feature detection and location

We have tried to restrict image processing to the detection and location of easily found features—retroreflective markers, edges and light stripes. We have been able to locate these features using fixed thresholds in each processing operation, though not with sufficient reliability.

Marker detection

We have created bright features in our images using small pieces of highly retroreflective material—3M Scotchlite 7615 which is used for front projection screens. This material reflects light directly back towards its source with a very narrow spread, approximately 1 degree. To enable a camera to see these spots a 100 W domestic flood lamp is placed beside the camera lens. While most of the image is dark, light from the retroreflectors saturates the camera and can easily be detected.

We attach these markers to parts of the manipulator which we need to locate using the vision system. Most of the manipulator parts are moved by stepper motors. It is easier to obtain the positions of these manipulator parts from the

[2] Commercial image processing software is almost always aimed at processing the entire image—a tediously slow process which is needed if the processed images are to be interpreted by people. So far we have managed to avoid this necessity.

manipulator controller. However, some parts of the manipulator, such as the head clamp are free to move in pivots and need to be located by the vision system. A retroreflective marker is mounted on the outside of each eye cover—part of the head clamp. Provided they are not obscured by wool these can be used to measure the head position.

We calculate a search region which defines the expected position of the marker. We use fixed thresholding to decide whether each pixel in the search region is part of the marker or not. We calculate the u and v image coordinates of the centroid of the marker pixel. The threshold level can be just below the saturation level of the camera.

The attraction of this method is that the retroreflectors are passive elements. We do not need any energy supplies for the light spots, they can be turned on or off by the light source next to the camera. With adequate care they can also be used with full background illumination although normally we turn off the background illumination before measuring spot locations.

Edge detection

Most of the image features we need to measure are parts of the sheep and we know roughly where to expect edges and their orientation. Instead of scanning the whole image for edges we need only scan in regions where we expect edges, perpendicular to the expected edge direction. It is not necessary to find the entire outline of the sheep because we only need to know the boundary at a few significant points. In this way we can greatly reduce the amount of computation needed.

To find edges in the image we look for sharp changes in the intensity (i.e. large derivatives/discontinuities). This is done numerically by convolving the image with a mask. The mask that we have used with most success is a rescaling of the derivative of a Gaussian (Canny 1983) given by

$$F(u) = -u/\sigma^2 \exp(-u^2/2\sigma^2).$$

The rescaling ensures that the peak response of this mask to a step edge is independent of σ. Very broad masks are usually employed to filter out noise and small clumps of wool. Typically a σ value of 5 pixels is used; the effective width of the mask is about six times this, making it 30 pixels wide. This corresponds to about 100 mm in the 'world' with our current camera viewing distances. This mask is directionally sensitive; for it to be able to detect an edge it must be applied to the image so that it crosses the edge at roughly 90°. In an arbitrary image where one does not know the rough positions and orientations of edges beforehand one would have to apply the mask to all points in the image at several orientations. Typically two orthogonal directions are used, but sometimes more. Using our prior knowledge of the scenes we are able to reduce this to a one-dimensional problem. We have also experimented with the Morrone-Owens edge detection technique with some success. In this method edges are assumed to be points of phase congruency in the image rather than points of rapid intensity change (Morrone and Owens 1987).

We have implemented a modification to the standard edge detection technique in an attempt to allow us to emulate some of the behaviour of the human eye. We justify trying to imitate human visual performance in a machine vision system by noting that the performance of a machine system is generally judged to be 'good' if it finds features considered to be significant by the human visual system. The human eye is very much more sensitive to edges in low luminance conditions than in bright conditions. Thus two edges which might be marked as having equal intensity gradients numerically, may be perceived to be quite different by the human eye. Edges that we 'see' to be important may not be 'calculated' to be important, which can be a source of frustration! We have implemented a behaviour of this sort by logarithmically rescaling the raw digitized intensity values as shown in figure 10.3. Note that intensity differences at the low end of the scale are exaggerated and at the high end they are suppressed. In some applications one might wish to do the opposite. In this way the sensitivity to dark and light edges can be manipulated as required. In our case there are many 'bright' edges in the texture of the wool that we wish to suppress. Figure 10.3a shows a 'slice' of digitized intensity values moving from the dark background across the boundary of the sheep on to wool. Figure 10.3c shows the rescaled intensity values. Note how the intensity transition is made 'squarer', making the detection of the sheep boundary easier. Figure 10.3d shows the result after convolving with a first derivative of a Gaussian mask. We have found this technique to be very useful in our application, improving the reliability of edge detection.

Light stripes

To provide three-dimensional information we project stripes of light using 35 mm slide projectors with 200 W halogen lamps. Slide projectors were cheap and readily available; the slides were hand cut from plastic or cardboard.

We need to use broad stripes—a fine stripe from a device such as a laser projector breaks up in the cluttered terrain of the woolly fleece.

We hoped that stripes could be detected with a fixed threshold. Unfortunately even the broad stripes from the slide projector can be broken and erratic in the places where they are of most interest. Further, light from the stripes can penetrate gaps in the fleece and diffuse in lower layers of wool. While the surface reflection is brighter, diffused light can exceed a threshold low enough to guarantee detection of surface reflection off the darkest wool.

Stripe detection tends to be most erratic on top of sharply curved parts of the sheep, such as legs, where the wool flops down and large cracks confuse the stripe image (figure 10.4).

Locating features in three dimensions

The process of taking an image with a camera results in three-dimensional world (x,y,z) coordinates being mapped into two-dimensional image (u,v)

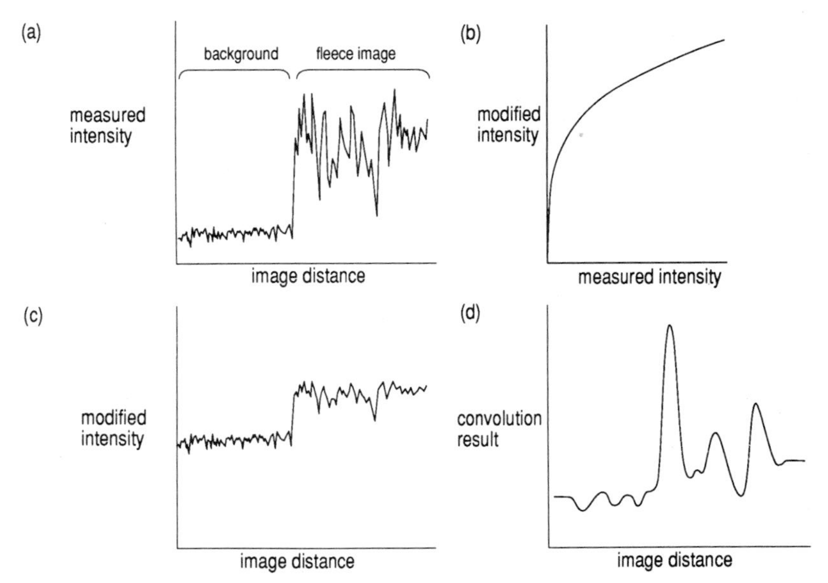

Figure 10.3 Edge detection. Logarithmic modification (b) of measured intensity values (a) produces a result (c) which emphasizes edges in darker image regions in comparison to texture 'noise' in the fleece. A wide convolution mask is applied to detect the edge.

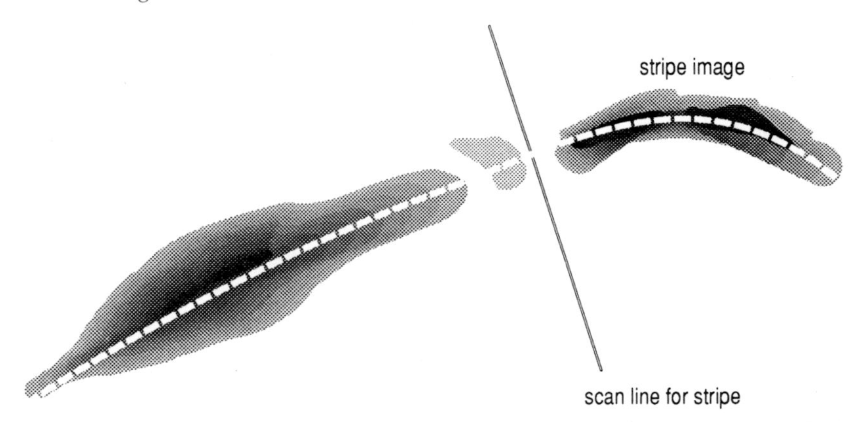

Figure 10.4 Problem with stripe detection—a broken stripe image is missed by the scan line which would normally intersect the stripe.

coordinates. Ignoring lens distortion, we can use a linear model to represent this process mathematically as

$$\begin{vmatrix} su \\ sv \\ s \end{vmatrix} = \begin{vmatrix} c_{11} & c_{12} & c_{13} & c_{14} \\ c_{21} & c_{22} & c_{23} & c_{24} \\ c_{31} & c_{32} & c_{33} & c_{34} \end{vmatrix} \begin{vmatrix} x \\ y \\ z \\ 1 \end{vmatrix}$$

where s is a scale factor.

This 3 x 4 matrix is known as the camera calibration matrix. Given this matrix and given an expected position of a feature in the scene in world (x,y,z) coordinates, one can calculate where one should search for it in the image.

The reverse transformation is also fairly straight forward; given the position of a feature in the image, finding where it is in world (x,y,z) coordinates. We use two techniques, the constraint plane method, and stereo.

Use of a constraint plane

If one knows that the feature of interest lies in some fixed plane, then from the equation of this plane and the calibration matrix one can generate a 'sensor matrix' which transforms image coordinates to world coordinates (Bolles *et al.* 1981). This method is typically used in 'light stripe' sensing; we also use it for reflective markers which are mechanically constrained in a known plane.

$$\begin{vmatrix} sx \\ sy \\ sz \\ s \end{vmatrix} = \begin{vmatrix} s_{11} & s_{12} & s_{13} \\ s_{21} & s_{22} & s_{23} \\ s_{31} & s_{32} & s_{33} \\ s_{41} & s_{42} & s_{43} \end{vmatrix} \begin{vmatrix} u \\ v \\ 1 \end{vmatrix} .$$

Stereo

Given the position of a feature in two (or more) images taken from different positions, the calibration matrices of the cameras can be combined to produce four (or more) equations in the three unknowns x, y, and z. The pseudo-inverse method can be used to solve these overconstrained equations. We are only able to use stereo measurements on simple features that are readily found in each of the images. This is because our cameras are very widely separated, and hence generate very different views of the scene, preventing the use of classical feature-matching techniques.

Calibration

The aim of the calibration process is to generate the calibration matrix for the camera. The method we use is that of Bolles *et al* (1981). Basically the process is to take an image of the scene, pick out a number of points in the image for which we know the world (x,y,z) coordinates, and hence establish the transformation from world to image coordinates. There are 11 unknown values in the camera matrix (the 12th value is an arbitrary scaling factor, set to one). For each point in the image for which we know the world coordinates, we can generate two constraint equations, one each for the u and v coordinates. Thus we need at least six such points in the image for which we know the world coordinates to be able to generate enough constraint equations to solve the 11 unknown values. In practice more than six points are used to provide greater accuracy and some protection against degeneracy in the constraint equations.

Accuracy

Before making allowance for lens distortions, the location accuracy was poor. Using the constraint plane method positions could be determined to ±15 mm. Stereo measurements using two cameras provided ±10 mm error (perhaps because we are solving a set of overconstrained equations). The errors came from several sources.

1. Quantization error—each pixel in the image corresponds to approximately 3 to 4 mm of the 'world' with our camera arrangement.
2. Numerical problems in determining the calibration matrices. Initially these problems were severe but after some judicious rescaling of some of the columns of the matrix to be solved these problems were reduced considerably.
3. Lens distortion—we use wide angle lenses to obtain sufficient field of view within a restricted working space.

The quantization problem could be reduced by exploiting knowledge of the shape of the features being detected, and using floating point arithmetic, to obtain sub-pixel resolution.

Moving the robot to a position specified by vision measurements was initially a source of great frustration—errors of up to 10 cm were encountered at times! This problem appeared to be caused by accumulation of errors in the vision system and the robot.

For our early calibration tests we adopted a conventional approach using a calibration frame structure placed in the camera field of view. The location of the structure was measured, and the positions of retroreflective markers on the structure were known. The frame was large and awkward to handle. There were error accumulations in measuring the structure position and orientation, and in locating the robot coordinate frame.

We eliminated several sources of error by using the robot to position a marker at known positions in the field of view of each camera. This way, we derived the relationship between where the robot '*thinks*' it is and image coordinates. We use 12 robot positions to provide reasonable accuracy and the procedure is convenient to perform. While we initially used a retroreflective marker, we have switched to a battery powered LED fastened in a known position on the cutter. This procedure has proved to be accurate and convenient. It is interesting to note that Brooks *et al* (1987) use motion of their mobile robot to calibrate its vision system directly in 'robot distance units'.

The planes of the light stripes are calibrated by placing known plywood target planes on the manipulator. Once the image stripes have been located, we can calculate the plane of each stripe, knowing the position of the camera.

Once the cameras and stripes have been calibrated, a simple test procedure checks each aspect of the system during an initialization sequence. If fixed manipulator features do not appear at the correct position in a given camera image, we can deduce that the camera has moved and needs to be recalibrated. We check the light stripes by a similar method.

Lens distortion correction

We use wide angle lenses in order to scan the image of a full sheep almost 2 metres long lying only 1.6 metres beneath the top camera. The lenses which we use introduce some distortion so that a large rectangular object appears distorted by up to three or four pixels at the corners. This corresponds to errors of the order of 15 mm at the sheep.

We compensate for this distortion using the method outlined by Bowman and Forrest (1987), using third-order polynomials. While Bowman and Forrest suggest second-order, we have found that in practice third-order corrections are needed.

$$u' = a_0 + a_1 u + a_2 v + a_3 u^2 + a_4 uv + a_5 v^2 + a_6 u^3 + a_7 u^2 v + a_8 uv^2 + a_9 v^3$$
$$v' = b_0 + b_1 u + b_2 v + b_3 u^2 + b_4 uv + b_5 v^2 + b_6 u^3 + b_7 u^2 v + b_8 uv^2 + b_9 v^3$$

where u' and v' are the corrected image plane coordinates,
u and v are the original image plane coordinates, and
a_i and b_i are coefficients to be determined by least squares approximation during the camera calibration sequence.

Polynomial lenses are uncommon, though. Fourth-order functions provided more accuracy, but were also more susceptible to small calibration errors. The third-order functions provide usable accuracy within the image area we use. We hope to switch to a radially symmetric non-linear compensation function as described by Tsai (1986). However, this is complicated by the fact that we use 512 x 512 pixels from an image sensor designed for the 625 line PAL television standard. Our image area is therefore offset from the lens centreline.

Programming

Peter Kovesi and Graham Walker developed a compact programming technique for our vision system based on an interpreter and a library of image processing functions. The interpreter has two versions: the first version can be used interactively and the second operates via a serial RS232 link to the robot control computer. The dialogue in both circumstances is identical except for a simple error detection protocol adopted for machine to machine communication (Kovesi and Walker 1989).

Examples of operations which the interpreter performs range from basic functions such as:

SNAP ***camera buffer gain offset*** (snap an image from a camera to a given buffer number)

FWRITE ***filename buffer*** (write an image from a buffer to a disc file)

to more complex functions such as:

EDGE u_1 v_1 u_2 v_2 (find an edge scanning along the line from u_1 v_1 to u_2 v_2).

The A900 robot control computer is better equipped with software and special hardware to perform three-dimensional computations; it also has access to surface models and associated data. Therefore, we restrict the processing within the PC computer to image analysis in two dimensions. This has the side benefit of reducing communication between the two computers. The relatively slow communication speed is not a major difficulty while we are in our research phase.

We implement three-dimensional computations within special extensions to the CLARE interpreter. For example:

set pt: spot([s, *r, ns*]);

searches for a spot which should be within distance *r* of the point **s**. *ns* is a sensor matrix number; the matrix is obtained by combining a known plane with a given camera calibration matrix, and is defined with an earlier command. The coordinates of the resulting position are placed in the vector **pt** which can then be used to adjust a surface model.

By programming vision operations at this level we avoid having to modify measurement sequences after minor changes to camera positions or recalibration. All the required adjustments are automatic.

Results

We routinely use the vision system to take measurements for predicting most of our surface models, albeit under close supervision. However, the belly of the sheep has so far defeated vision measurement. Fortunately there are some promising new methods to try.

There are two principal difficulties which we frequently encounter—image variability and sheep variability.

Image problems

We would like to use the vision system for predicting the belly patch: one of the typical views of the belly appears in figure 10.5. Here we encounter our most

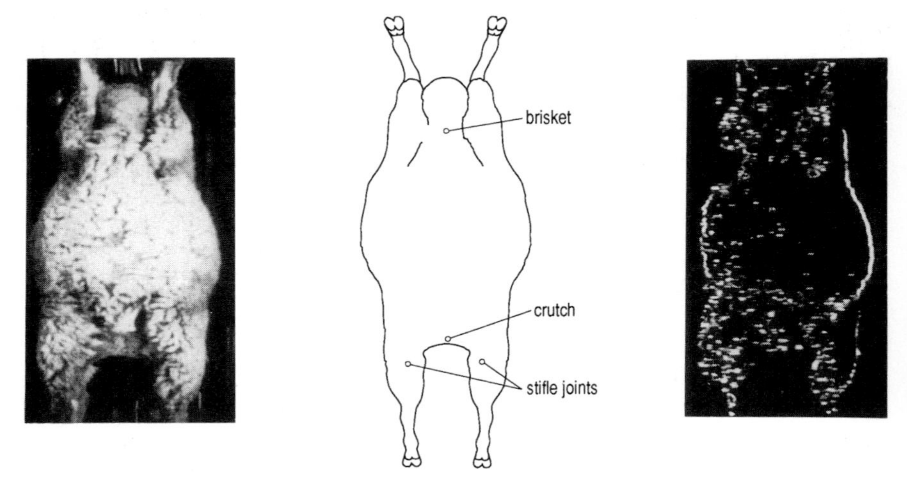

Figure 10.5 Experiment with adaptive edge detection—refer to text.

severe problems. Some of the measurements are easy e.g. the crutch x coordi nate. The most important measurements are the locations of the elbows, the brisket and the stifle joints. In these locations, gaps in the wool have opened up with inversion and stretching of the legs. Stripes projected on to these areas have gaps and variable intensity. We would also like to know the brisket location, but here there is a great amount of variation in the appearance of the sheep. Notice how there are substantial apparent intensity variations along the illuminated edges of the sheep.

We provided a copy of this image (which was selected at random) to a research group refining the energy method for edge detection (Morrone and Owens 1987). They have developed an as yet unpublished technique for adaptive thresholding which operates more successfully in the presence of noise than other edge algorithms. Members of the group were quite confident that their technique could handle the image. Operating with a right to left scan, the edge detector results appear alongside the original image. Notice how the right-hand edge has been lost along the front and rear legs. Their method works better than most of ours, yet it has been easily defeated!

Figure 10.6 illustrates a further trial with the same technique, this time on the

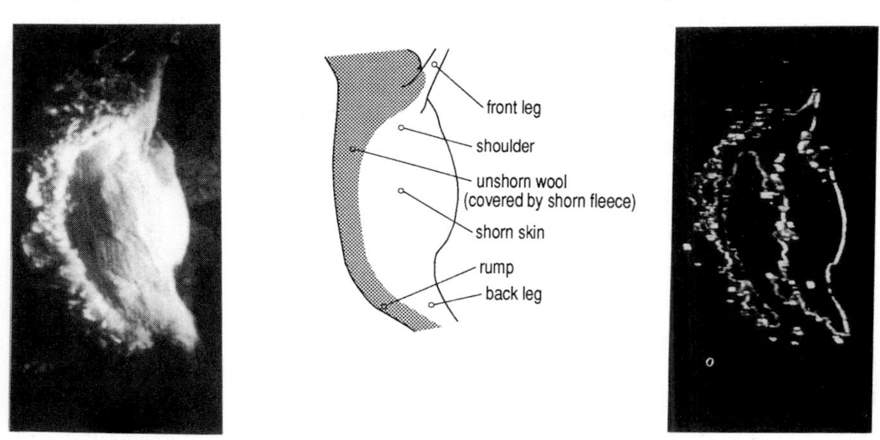

Figure 10.6 Experiment with adaptive edge detection—refer to text.

shorn side of a sheep. We are principally interested in the location of the edge of the shorn fleece which curves along the left-hand side of the lower part of the image. While we perceive this as a clear continuous curve, it has been lost in the 'spaghetti' produced by the edge-finding algorithm.

We have concluded from these and other experiments that a workable feature detection must make use of prior knowledge about the images—texture, shapes and variability, and may have to work in two-dimensions. It may not be sufficient to use one-dimensional edge finding methods and simply apply them in several different directions, as has been the case in nearly all machine vision systems so far.

Sheep variability

We use stripes to measure the z (vertical) coordinate of the brisket. However, finding the y coordinate is more difficult. Not only is the stripe often broken, but the wool profile may not represent the sheep shape accurately enough to locate the brisket underneath. Sometimes, even human operators can fail to locate the brisket in the image.

Sheep positioning varies to the extent that we often misplace the search areas for important features. Referring again to figure 10.6, in another phase of measurement we project a light stripe across the shorn area near the shoulder blade to measure the height of the shorn skin. We scan the stripe within a predetermined rectangle. Sometimes, though, the sheep is able to wriggle, and the shorn wool heaps higher so that the search area is obscured by shorn wool. For the time being, the operator checks that this does not occur and a manual stripe measurement can be taken if it does occur. In future, we must improve our measurement methods to detect this automatically and prevent such errors.

Fortunately, with such a simple measurement technique as we have used so far, with fixed thresholds and search areas, there is plenty of room for improvement.

One promising technique is based on the energy minimizing splines (called snakes) suggested by Kass *et al* (1987). These are simulated flexible wires which are made to settle on weak or ill-defined image features using a variant of potential field techniques. At the time of writing, we are developing ideas which extend the concept of a snake to build in knowledge about the expected shape of a feature.

11
Sequence control

In chapter 4 I presented a coherent account of motion control computations for sheep shearing but I carefully deferred revealing the details of software structure until this chapter.

In 1986, I followed the advice of Mark Nelson, a computer science graduate student, and separated the geometric issues of motion control from the sequencing and structural issues. Mark suggested a formal framework for sequence control and laid the groundwork for the structure of our current motion control system. Peter Kovesi helped me write the software to make it all work.

We soon discovered that we needed easier methods to develop and test shearing software. And the main requirement was for easier ways of executing parts of the shearing sequence without having to start from the beginning—in short improved manual sequencing. After completing the interpreter which controls the SLAMP manipulator, Steve Ridout rewrote the syntax for CLARE and the routines which perform the lexical analysis. Peter, Steve and Graham Walker then developed an 'interactive programming system' which provided the ease of programming we needed.

The first part of this chapter describes automatic sequence control and the second part manual sequence control. Both issues involve safety and I will begin with an account of the only serious accident with a sheep in 12 years of trials on live animals.

Learning from an accident

Trial 377: November 6th 1984. Michael Ong, Peter and I were testing new shearing software, when smoothly, quietly, and deliberately, the cutter descended relentlessly inside the chest of the sheep. By the time we realized what had happened, it was too late for the sheep. We stopped the robot and lifted it clear using the manual controls. The sheep died quickly.

After we recovered from the shock, we sat down to take stock of the situation. Was it a hardware or software failure? Was it operator error? Why had the sensors not stopped the robot moving down? There were many unanswered questions. There would be no more shearing tests on live sheep until we had some answers.

In a way, I was surprised that it had not happened before. In nearly four hundred shearing tests with experimental hardware and software, ORACLE had inflicted a mere handful of skin cuts requiring any attention. (A Merino sheep has remarkable self-healing abilities. Its soft skin is easily cut by thorns, fencing wire and twigs, quite apart from shearers. All shearing causes some cuts—robot shearing causes fewer cuts, and less serious ones. Thanks to its healing abilities, quite apparently serious cuts can often be left to heal without treatment—though we always applied an iodine spray.)

Peter and I substituted our metal practice sheep for the real thing. Repeated attempts to reproduce the failure were fruitless. There was no evidence of a hardware fault, and the symptoms suggested a deliberate programmed manoeuvre. The software had recently been modified to plan reposition movements to recover from skin cuts automatically, and this seemed to be the most likely source of the problem. But there was no data recorded which could have confirmed this.

A week later, after some close analysis, I found a possible, but extremely unlikely combination of events which could have led to a deliberate movement to the point [0,0,0] which lay inside the sheep, close to where the robot ended up.

We learned some important lessons.

First, the robot has to stop itself! Since we have a robot which is capable of planning its own movements, we cannot predict its behaviour in advance. So much so, that we cannot recognize an inappropriate movement or potentially dangerous movement until it is too late to stop it, even with quite slow movement speeds. This is compounded by the difficulty of seeing the cutter—the surrounding wool screens it from view. So we cannot expect an operator to be able to stop the robot in time. A vital lesson to learn.

Since we cannot absolutely guarantee the correctness of the software it follows that we must build in automatic checks to prevent unintended or dangerous movements. Even if the software code were absolutely correct, we cannot be sure of the software *design* which may be faulty. Fault detection software must counteract the effects of software errors, even in itself! It sounded for a moment like teaching the insane to diagnose themselves.

Secondly, the operator cannot react to a hardware failure quickly enough. Therefore, automatic checking for hardware faults is equally important.

Finally, there must be enough data recorded automatically to find out what happens just *before* a failure. Assuming that most problems originate in software, at least in our research environment, a record of the behaviour of the software is essential. We need the robot equivalent of the aircraft crash recorder. Yet recording the results of hundreds of thousands of computations every second would simply not be practical. Even a summary—20,000 results per second—would fill our disc capacity after just a few minutes of shearing.

I devised a solution which combined the idea of the crash recorder with a continuous loop tape recorder. First, details of each commanded movement are recorded automatically all the time. This data is valuable for analysing trial results, and shearing performance, quite apart from any benefits in tracing software faults. But commands alone are not sufficient to pinpoint the causes. So, I hit on the idea of recording about 20,000 summary computations per second, in a ring buffer like a continuous loop tape recorder. There is enough memory to ensure that at any given instant the last two seconds of recorded results are available. If a fault is detected, the ring buffer is immediately copied to the disc, and a trigger is set so that the ring buffer will be copied to the disc when it is full again. This way, the disc recording will always reveal what happened before the fault and how the system responded just afterwards. We can also have continuous recording for a limited time if we need it.

It took three months to take the software apart and build in fault detection with four of us working on it.

We used three methods to detect software faults, none of which involved a significant overhead in time or memory space.

1. Each conditional branch statement was changed so that the value of the condition flag had to be exact. Any unexpected value would trigger a fault indication. For example, the following code

```
if(state.gt.0) then
 do operations for state = 1
endif
```

   was replaced by

```
if(state.eq.1) then
 do operations for state = 1
else if(state.ne.0) then
 call Fault( . . . record error and take appropriate action . . . )
endif
```

   In this way, any value other than 0 or 1 (the expected values) would trigger a fault indication.
2. Many variables were subjected to range checks. For example, if the displacement of an actuator exceeds a preset limit over one motion step, a fault is triggered.
3. Marker values are written into memory locations between arrays or blocks of data, and checked regularly. If the marker value changes, it indicates that data has been misplaced (array subscript out of range, for example).

Once a fault has been detected, extra data is automatically inserted into the ring buffer to provide detailed data for analysis and an appropriate response action is chosen from a table. For most internal software faults, the robot is either paused (the operator can decide whether to continue), or shut down completely.

The results far exceeded our expectations. When we first started running the robot again, the automatic fault detection system reported many errors. Errors which had been in the software all along, but which we had never noticed. The adaptation mechanisms built into the robot control program had automatically compensated for them! Later, our data recording method was to become a tool for verifying the correct operation of the software on-line.

Many times, we wished we had been able to stick the wool back on a sheep and have another go. But once shorn, it was impossible to repeat an experiment.

There was only one chance to record data, and the method of recording devised at this time became an invaluable development tool.

SM safety

ORACLE was decommissioned two months after trials resumed but safety was a major issue in the design of SM. ORACLE was floor mounted, large, and cumbersome. While there may have been problems for the sheep, the operators were relatively safe except during maintenance operations. We had had a near accident when Jan Baranski was inspecting wiring between the upper and lower links. The wiring connected the pressure transducers back to the servo cards. He touched one of the cables and the robot lurched up without warning, then lurched back again. The power was on because the fault was intermittent and only occurred when the robot was near a particular position. Jan thought it was a broken wire, but he could only confirm that by touching it with the robot in position. Maintenance workers on robots are particularly at risk, as accident records suggest. And robot accidents are more likely to lead to death or serious injury than accidents with other types of industrial machinery.

Unlike its predecessor, SM was to be small, compact and ceiling mounted. With a much larger workspace, it posed a new level of hazard. It could move fast and silently without warning, at up to three metres per second. Safety regulations required a screen enclosing the robot workspace but this would have caused many problems. We needed to be as close to the sheep as possible to see and understand what was happening, for the safety of the sheep as much as ourselves. We used the laboratory as the safety enclosure—there are windows along one side which provide safe viewing from the adjacent corridor. Microswitches on the doors were wired into the automatic shutdown circuit, and we adopted a rule that at least two team members had to be in the laboratory during any operations with the robot powered.

When the shutdown circuit is interrupted by pressing an emergency button, or loss of power, a valve immediately shuts off the hydraulic supply to the robot, and check valves on the main actuators close, locking the arm in position. The check valves can be reopened manually so that the arm can be pulled clear of the sheep if necessary.

The automatic shutdown circuit is activated by a watchdog timer which responds to a failure of the computer or electronics power supplies. The timer is a simple monostable circuit which has to be explicitly reset every few milliseconds as part of the normal input/output process controlled by the computer. The integrity of the multiplexed communications is monitored by a single test channel which transmits a signal to the wrist electronics and carries the same value back to the computer on the input line. Random voltage values are sent to the channel, and the voltage inputs are checked by the computer. Two successive discrepancies trigger an immediate hydraulic power shutdown. Any break in the communication cables will be detected by this process.

The computer monitors the input values for each actuator potentiometer.

If any signal jumps too fast, or lies outside the normal working range, or departs too far from the set actuator position, the hydraulic power is shut off immediately.

One proposal we seriously considered was to place an operator's stand near the sheep, and use our obstacle avoidance techniques to guide the robot around the stand, keeping the operator safe. This was discarded because we needed to see what was happening from different directions. Many stands would be needed. So we fitted pressure-sensitive mats to the floor around the sheep platform. Each mat produced a coded signal. If the robot moved into the space above the mat which was occupied, it stopped automatically. The mats were arranged so that a break in the cables connecting them caused a fault condition stopping the robot.

This worked well for a while, but the pressure mats were unreliable. They became contaminated with oil, dirt, wool and urine after some time and had to be regularly replaced.

Four years later, we were ordered to comply with new safety standards for all industrial robots. A light curtain guard was connected to the automatic shutdown circuit to detect a person entering the robot workspace. The software was arranged such that the robot could only be restarted after an emergency stop (either operator or software triggered) by an explicit operator command. A two-character confirmation code is randomly generated by software—the robot only restarts if the operator types in a matching code. Warning lights and signs were also needed, with comprehensive emergency controls and a training programme for all team members.

Fortunately, by then we had overcome most of the problems requiring close observation of the sheep during shearing. If we had been forced to comply earlier, some of the problems may have taken longer to solve. To promote safety awareness we keep a tally of accidental incursions by team members through the light curtain. When a team member exceeds a points limit, he or she buys the trials team lunch!

There have been no more serious accidents: we know that our precautions have protected us and the sheep from several software faults. The safety software led me to explore new ideas for automatic sequence control. For I soon realized that abnormal shearing conditions could be detected by the same range tests on motion control variables that we use to guard against software errors.

Automatic sequence control

The best laid plans . . .

. . . often go astray. This applies particularly to shearing and several kinds of operating problems arise during robot movements, necessitating automatic replanning.

Walking along a city footpath provides good analogies. Most of the time you can continue in a chosen direction by veering gently from side to side to avoid fixed obstacles and other people walking the opposite way. Occasionally it will be necessary to slow down for a moment or two. However, if a sudden downpour of rain occurs, you might deviate to take shelter, and only continue after the rain stops. You may also find an unexpected barrier, and have to abandon your chosen route, and plan a new one. Most shearing adaptations take the form of repeated minor path and speed adjustments; these are classified as continuous adaptations. When these fail, however, a movement has to be interrupted and may need to be completely replanned. Operating conditions for a shearing robot vary considerably and even skilled shearers have to interrupt shearing and restart blows, particularly on difficult sheep.

All the adaptations described in chapter 4 can be classified as 'continuous'. The path of the robot is adjusted at each computation step so that errors are progressively reduced at subsequent steps. However, for each adaptation mechanism there is a limit to the extent which can or should be provided. When these limits are exceeded some form of recovery is essential. We have called these conditions 'operating errors' as they often indicate a fault with the shearing pattern or surface model.

The shearing pattern is described in terms of a sequence of blows. Figure 11.1 is part of the standard instructions on issue to all shearers in Australia and

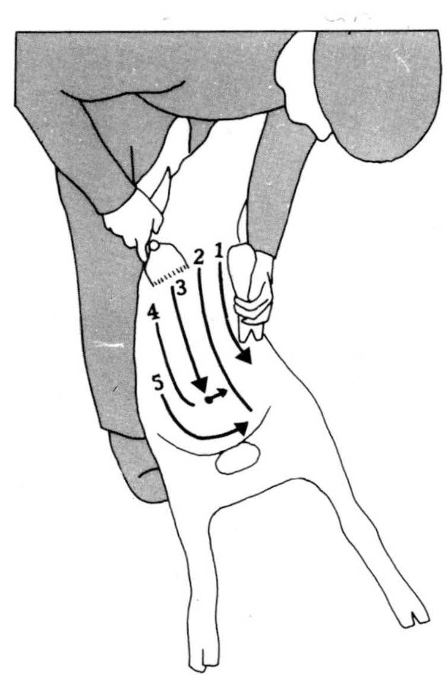

Figure 11.1 Shearing instructions for belly on issue to shearers throughout Australia. (*'Tally-Hi'* shearing)

shows the blows for belly shearing. The programming technique described in chapter 4 is similar, except that it is coded symbolically and defines more detail. Yet there are also implied instructions which the shearer uses if a blow cannot be completed in one continuous movement. If the skin is cut, or wool clogs the cutter, or the comb snags a skin fold, the shearer has to stop and move the cutter to clear the problem and then resume shearing the rest of the blow.

In doing this the shearer has:

(1) detected an abnormal condition;
(2) diagnosed the condition;
(3) selected an appropriate corrective action;
(4) performed the corrective action;
(5) resumed the originally planned sequence.

The ability to take corrective actions and to diagnose and decide which corrective action should be used is part of the skill which has to be learned from experts and refined by practice. Considerable experience is required before a shearer can correctly diagnose most problems and recover quickly.

Many manual skills are described in this form. First the basic procedure is described which is executed if nothing goes wrong. Secondly, there is a series of statements which take the form 'if . . . then do . . .'

Error recovery

The term used for this aspect of robotics is error recovery. Several papers have been published since 1985 (for example Lee *et al* 1985, Harmon 1986, Smith and Gini 1986). Most papers focus on replanning object manipulation tasks after one step in a whole sequence has failed. What to do, for example, if a part is found to have slipped from a gripper during an assembly task.

Two types of error recovery have been described by Smith and Gini (1986) citing Randell *et al.* (1978). *Backward recovery* consists of going back to a pre vious system state that is known to be correct. In this context the recovery from a skin cut described from above is seen to be a type of backward recovery. *Forward recovery* consists of performing alternate actions with the robot that lead it to a state which it would have reached had it been able to continue with the planned actions. The planned actions can then be resumed from that state.

It is not appropriate to attempt to implement the equivalent software techniques in the context of an adaptive robot. Backward error recovery typically involves the notion of rolling back the system to a known correct previous state. However, it may be neither possible nor desirable to move the robot exactly in reverse through previous states. Instead, we need a sequence of actions (including movements) by which the robot can reach some known correct previous state. Before restarting the robot in this previous state it may be necessary to alter some of the parameters which govern the robot's behaviour, based on a diagnosis of the likely cause of the problem.

Smith and Gini (1986, p. 285) define forward recovery as a determination of the difference between the actual state and the desired state of the robot and then developing a sequence of operations to achieve the desired state. It is implied from their paper that the notion of 'desired state' is the successful completion of a sequence of movements, each of which is explicitly defined in the robot programming language. If a robot has to pick up a part and turn it over, the means by which the part is manipulated is not so important; reaching the final position determines success. In shearing, most of the movements are *process steps*; it is necessary to perform shearing along the path of movement to remove wool, and it is not sufficient simply to reach the end of a shearing movement by some alternate route.

It follows that error recovery involves the *resumption* of a shearing movement, or its replacement by an equivalent movement. And planning the recovery actions is complicated by the need to be able to respond to errors *anywhere* along a planned shearing blow.

Let us define the term *interruption point* as the point (or state) at which motion of the robot was interrupted in response to an error, and *recovery point* as the point at which normal motion can be resumed. The recovery point could be between the start of the blow and the interruption point—this loosely corresponds to backwards recovery. If an alternate shearing style or extra blows are used, then the recovery point could be further along the same blow, or even later in the original sequence—this corresponds to forward recovery.

Examples of operating errors

The term 'operating errors' is introduced to describe situations in which the adaptation of robot motion by path modification, wrist reorientation, movement speed or cutter drive speed is inadequate, or exceeds preset limits. Some examples follow.

Model error

A typical operating error is a situation in which the position of the skin under the cutter measured by sensors deviates from the predicted surface model by more than a certain distance, say 60 mm.

Response: the sensor calibration should first be checked to ensure that the sensors have correctly detected the skin of the sheep. This can be done by moving the cutter away from the sheep and measuring the sensor outputs in free air. Next, the shearing is recommenced further back. If the error fails to recur, then it is possible that wool characteristics may have caused the sensors to measure the distance to the skin incorrectly, and the sensor characteristics may require modification. If the error does recur, the shearing is still possible, but at a slower speed. On-line adaptation of the surface model can be used to reduce the error.

Excessive drag force

A simpler kind of operating error can be detected by measuring the drag force required to push the cutter through the wool. If the drag force is excessive we can assume that exceptional wool conditions are being encountered and an alternative shearing action is required.

Response: we can repeat the last part of the shearing again—this often clears the problem. If we are shearing parallel blows, we can program the robot to lift and shovel the wool to one side before resuming the blow. Another possibility is to shear the rest of the blow using more overlap and then resume—effectively using two blows instead of one. If the cutter is tunnelling under the wool, we can lift the cutter up and forwards to break open the fleece.

Skin cut

A skin cut is a further example; it can be detected by monitoring the level and smoothness of the resistance sensor signal (figure 4.11, p. 93). The normal adaptive response of the robot to a high resistance sensing level is to lift the cutter away from the skin; however if forward motion of the robot continues this action will make the cut worse.

Response: the robot must be temporarily moved backwards to withdraw the comb teeth from the skin and then shearing can be resumed from a recovery point some distance behind the interruption point, with a slower shearing speed until the site of the cut has been passed.

Diagnosis

The concept of signatures (Lee *et al* 1985) defines an error in terms of a variable being outside a range of values in one of a group of machine states, for which the range limits apply. We have extended this concept to introduce frequency counting since a single instance of an error may only indicate a transient 'deviation' which can be tolerated.

We have chosen to calculate two kinds of signature; an alarm and an event.

Alarm signatures

An alarm signature defines a machine state 'to become alarmed about'. In other words, a state in which further (restorative) action may be needed. The signature is encoded with the following information:

alarm description
variable to be monitored
lower limit
upper limit
pattern of machine states

The last item specifies a pattern of machine states in which the range limits are valid. The machine states reflect the different operating modes in which variable limits may be important:

shearing	S
repositioning	R
stopped	.
moving forward	^
moving backward	v
close to, or on skin	_
away from sheep	/
executing recovery	X
landing	L

sensor contact	#
hydraulic power on	=
hydraulic power off (dry run)	+
under manual control	B

Each state is defined by a single character code. Two sets of states can be defined. The first set defines the states in which the variable range limits are to be tested. The second group defines states in which the variable range limit must not be tested. For example:

ALARM Cutterslow 'Cutter running slow'; Acrpm between 0 and 300; State SL; Notstate +; end;

Acrpm is the variable name; the condition is tested during landing and shearing but not when the hydraulic power is off.

A special set of alarm variables is provided for testing. These are calculated by different parts of the software and are written into predefined locations in a reserved shared common block. Examples are:

Perr	pitch error relative to surface model,
Herr	height relative to surface model,
Acrpm	actual speed of cutter drive motor (rpm),
Drag	drag force on cutter, and
FLegForce	filtered stretch force on rear legs (from manipulator).

Events

These are more complex than alarms; events introduce time or space dependency and a Boolean combination of alarms or other event inputs. By this means we can set up a decision tree which implements a crude method of diagnosis. Events at the ends of the decision tree have actions associated with them which will trigger recovery procedures.

Event signatures are defined as:

event description
Boolean combination of alarms and/or other events
incidence rate
forget rate
accumulator
rate selection (time, distance)
pattern of machine states (as for alarm signature)
action code

The 'incidence rate' and 'forget rate' parameters form a frequency counting facility. The accumulator value (initially zero) is incremented at the 'incidence rate' at every motion step at which the Boolean combination is 'true', and decremented (to zero) at the 'forget rate' at every step when it is 'false', as shown in figure 11.2. The event 'fires' if the accumulator value exceeds a critical value. Different types of operating errors can be detected using this facility by selecting different incidence and forget rate values. The accumulator can be incremented or decremented in proportion to elapsed time, or distance shorn. If

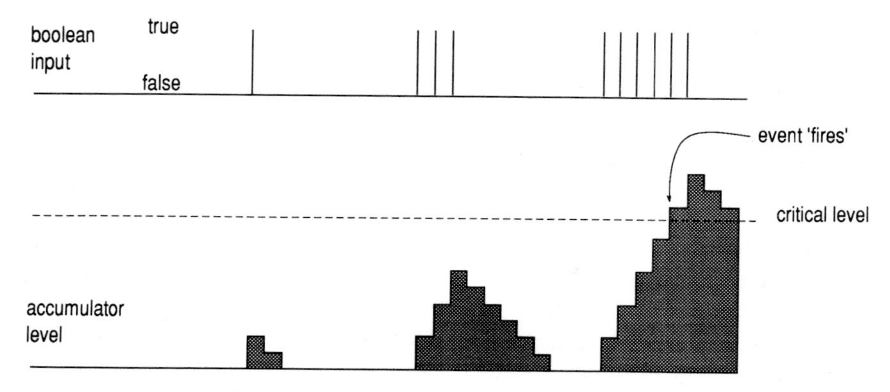

Figure 11.2 Forgetting algorithm behaviour with intermittent faults—refer to text.

incidence rates are not specified, then the event fires immediately the input expression becomes 'true'. The algorithm was originally developed to handle intermittent faults which might otherwise generate unacceptable false alarms.

The action code value determines the response to an event firing (if any). The principal option is to request an as yet unspecified recovery procedure. The safety software provides a series of options for dumping data recorded in ring buffers to a log file—all of these are available too and most events are programmed to use this. Fault response actions can be requested, such as a pause.

Can this relatively simple scheme reliably infer typical operating errors such as unusual wool properties (which affect sensor behaviour), surface model errors, sheep struggling and skin cuts? Is the absence of inference rules a problem? The experimental evidence collected so far on several hundred sheep suggests that this scheme is adequate. It must be noted that the monitoring of variables which are derived from the higher levels of the continuous adaptation mechanisms does represent a form of inference. Most of the problems listed cannot be observed simply from sensor data. However, they can be inferred by simple observations of the continuous adaptations made by the robot.

Having a method for diagnosing problems is only part of the solution. The selection and implementation of a suitable recovery is the next requirement.

Selecting recovery procedures

All events which trigger a recovery procedure lead to the same recovery selection procedure. There are two reasons for this.

First, the event which triggered a recovery being attended to may fire again before the recovery has been completed. In this case, repetition of the recovery may be pointless—instead it may be better to select a different recovery.

Second, several events may be triggered at the same time—there may be conflicting alarms or different diagnoses. Experience has taught us that problems occur in 'showers'—more than one problem tends to occur at the same time. If the cutter snags a skin fold, there may be a high drag force. Yet by the time recovery selection starts (albeit on the same motion step) a skin cut may have been registered as well.

Recovery procedures are specified as hypotheses of the form:

recovery procedure *i* will result in problem *j* being overcome

Each hypothesis has a single primary selection criterion (PSC) corresponding to problem *j*, and a number of secondary selection criteria (SSC). The PSC is the event representing the problem which triggers the need for a recovery procedure and which the selected recovery is intended to overcome. Each SSC is an alarm or event. Hypotheses are grouped in families, each family sharing the same PSC and SSC's.

Each PSC is assigned a score, representing its priority. Each SSC is also assigned a substantially smaller score corresponding to a certainty factor. SSC's are used to ensure that the most appropriate recovery is chosen in given circumstances. For example a family of hypotheses to overcome drag problems can use shearing styles as SSC's to select appropriate recoveries for tunnelling, peeling (parallel blows), wrinkly skin shearing or leg shearing.

Once a recovery has been commenced, it is monitored for its effectiveness in overcoming the problem indicated by the PSC.

Each recovery procedure includes a test command—e.g. test time:0,5. This means that the PSC is monitored; if it recurs in the interval 0 to 5 seconds (centimetre distance units may be selected instead) then the recovery has failed. Otherwise the problem can be considered to have been dealt with.

If the PSC fires during the testing interval, the score of the hypothesis is reduced by a failure penalty and a new choice is made. The same hypothesis may still score enough to be selected again. The recovery in progress is abandoned and the chosen replacement is commenced.

If a different and more serious problem occurs during the recovery then a new hypothesis is selected and the current one abandoned.

Some recoveries may be strategic—in the sense that they represent an alternative shearing strategy for dealing with a known problem. This kind of recovery can be interrupted and later resumed if other more serious problems occur. Other shorter recoveries are called 'local'; these are simply abandoned if a more serious problem arises.

Programming examples

Normal shearing is programmed by specific coordinate values, though these may be defined by variable names:

blow [cp1, cp2, cp3]; move [restpose];

This is a simple example of a shearing sequence. The 'blow' command specifies a shearing movement along a path specified by the points cp1, cp2 and cp3. The reposition to the start of the blow is planned automatically—no specific command is needed for it. After the blow, the robot returns to a rest pose (position and orientation specification). The terms cp1, cp2, cp3 and restpose would need to be assigned suitable values before executing the sequence, of course.

Signatures

```
ALARM   DragA 'Drag adaptation limit'; var Dlevel above l.6;
        state '^'; notstate 'L'; end
EVENT   Dragprob 'Drag problem'; qual DragA; action 100; end
```

Hypotheses

```
FAMILY  Dragprob;        ! A family of hypotheses will follow all
                         ! of which will share Dragprob as the
                         ! PSC. No SSC's are listed.
HYPO Drag1 'Drag recovery 1'; call RbackP; score 220; fail 30; end
HYPO Drag2 'Drag recovery2'; call RposTP; score 180; fail 30; end
HYPO Drag3 'Drag help'; call OpHelp; score 170; end
```

The alarm specification specifies an adaptation limit, to be tested while the robot is moving forwards (state '^') but not during landing (notstate 'L'). This alarm is the input to the event 'Dragprob' (qual DragA) and triggers recovery selection (action 100).

The family of hypotheses uses this event as the primary selection criterion, and no secondary criteria are specified. The first hypothesis scores 220, reduced by 30 each time the procedure 'RbackP' fails to prevent the event 'Dragprob' recurring during the testing period. It will be allowed two attempts; next the procedure 'RposTP' would be tried, and if that fails, operator help is solicited through the procedure 'OpHelp'.

This is a simple example of problem diagnosis and recovery selection. We have found that about 60 signature rules (alarms and events) and about 40 recovery hypotheses are adequate.

Recovery procedure

The recovery procedure is defined implicitly in terms of the interruption point location, though the coordinate values can be explicitly referenced by special variable names.

```
Procedures RbackP; sing '3CFCFCFEDC___'
vary shvel: shvel*0.75;
set shyof: –0.3;
set shvel: 0.05;
test shdist: 0.01, 0.10;
blow rpath (–0.07, 0.03);
end;
```

This recovery procedure reduces the shearing speed for the remainder of the blow by 25% (vary command) by changing the plan for the partially completed blow. Each recovery procedure 'sings' a characteristic song so that the operator can recognize recovery decisions without having to watch a screen. The shearing speed for recovery is set at 0.05 m/sec (set command); this command modifies the parameter value inherited from the interrupted blow. The test command specifies that a recurrence of the PSC after 0.01 metres and before 0.10 metres of shearing distance indicates failure. The blow command makes use of the 'rpath' function which generates a recovery shearing path from the inherited plan. In this example, the recovery blow path runs from 0.07 metres behind the interruption point to 0.03 metres in front. The reposition movement needed to move back to the start of the blow is planned automatically using inherited parameter values. The recovery point is then 0.03 metres in front of the interruption point.

```
Procedures RposTP; sing '3CCGGAAG___';
startmove [(–i*.03 + k*.05), 0.2, 3, rot(As,j,0.3)];
movetowards (ppos + kp*.15 + ip*.03); ! control point defined in
                                      ! terms of interruption point
                                      ! pose elements
                                      ! ppos, ip, jp, kp
movetowards (ppos + kp*.15 + ip*.03); ! repeated point.
                                      ! Lift wool clear
test shdist: 0.01, 0.10;
blow rpath (–0.07, 0.03);
end;
```

This alternative procedure uses an explicit reposition (referring to the interruption point coordinates) to clear the wool from around the cutter.

```
Procedures OpHelp; sing '4EFEFEFEFEFEF___';
move (p–i*0.05 + k*0.02);
cutter off; sensor off; air off;
run 'key *opint';   ! display program operator's menu keys
                    ! ABORT / SKIP / REPEAT / TRY AGAIN
tr 1;               ! transfer to operator at console
cutter on; air on; sensor on;   ! get ready to resume—will only be
                                ! needed for 'try again'
blow rpath (–0.12, 0.03); test time:0,0;
end;
```

In this last example, the operator has to select a menu key option. The operator can also use any normal CLARE command to clear the problem, or manual intervention at the sheep. The testing command is no longer relevant but a dummy command is still needed.

Apart from some additional housekeeping commands which I have omitted, most recovery procedures are of comparable simplicity. Notice that recoveries are almost completely independent of the context in shearing command files. This means that we can refine recovery procedures independently of shearing command files and vice versa.

Software structure

At the end of 1984, we had already started to provide elementary recovery actions for the robot. In the event of a skin cut, we stopped shearing and moved the robot backwards (a regression) by about 80 mm and resumed the shearing blow. In fact, the accident which prompted serious work on error detection occurred while testing this feature and revealed a fundamental structural weakness in our software.

The recovery motion was controlled by a separate motion module within TOM. We had separate modules for shearing and repositioning; we simply added a cut-down version of the reposition module which would move the cutter backwards (figure 11.3). We also needed to modify the shearing module so that it could restart a blow after a recovery. The weakness was that each different recovery needed a separate module. Since some recoveries would require extra shearing blows, we would need a duplicate shearing module. And further, some recoveries might encounter further problems which could only be overcome by further recoveries. I soon realized that a recursive approach was needed but this was too difficult to contemplate at the time. In 1986, Mark Nelson worked with us for 6 months and developed a more formal approach to software structure (Nelson 1986, 1990). For the next 18 months my attention was diverted by problems with sheep manipulation, finding new research staff, project reviews and shearing demonstrations. Finally at the beginning of 1988, Peter and I

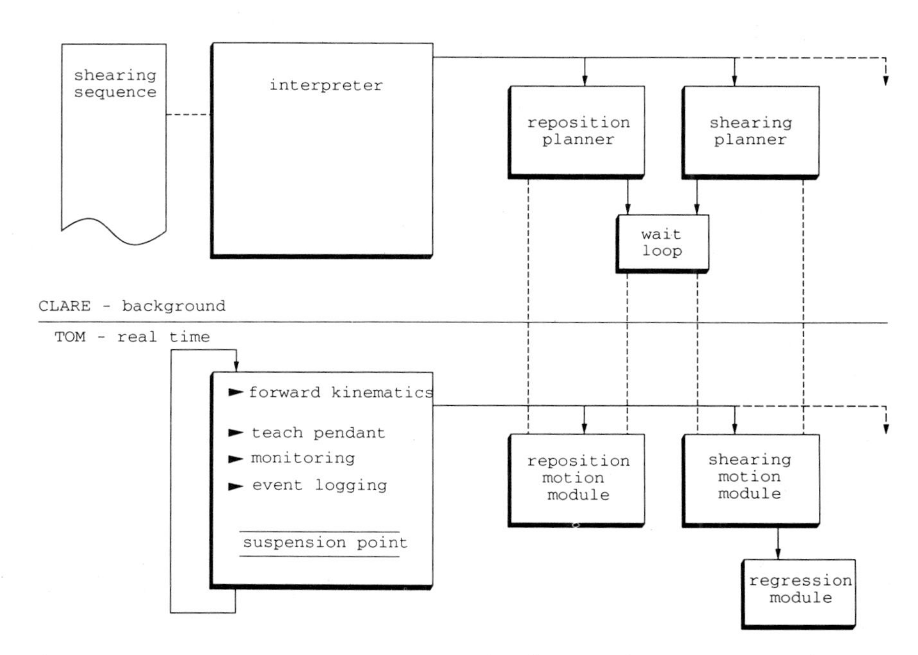

Figure 11.3 Non-recursive motion control software scheme in use until 1988. Refer also to figure 4.8. Execution control follows the solid lines with arrows. Data is passed via the dashed lines. Several other motion modules were used—they have been omitted for clarity. Also, the shearing planner called the automatic reposition planner, and hence initiated reposition motions as well as shearing motions.

started work on a recursive sequence control system and by August it was ready for trials. Turning now to figure 11.4 we look in more detail at the software structure needed to implement this scheme.

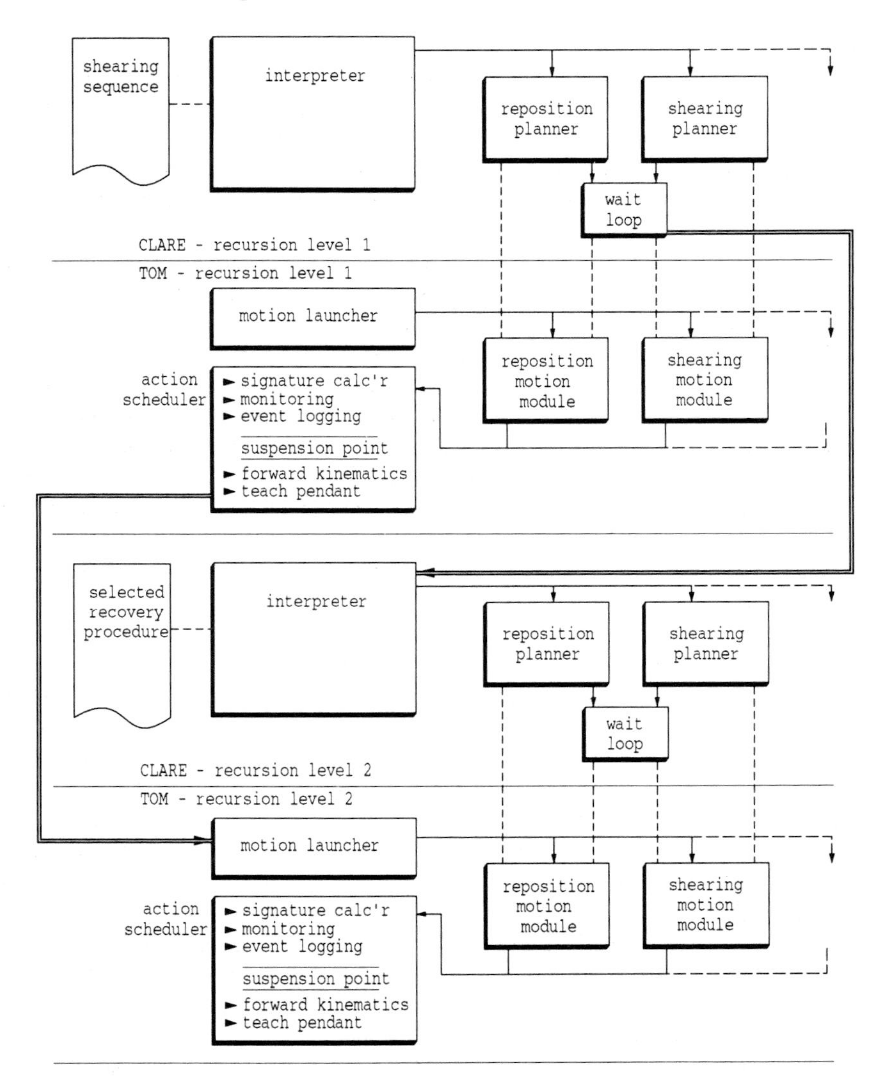

Figure 11.4 Recursive software at level 2. The interpreter has been called recursively from the level 1 wait loop to execute a selected recovery procedure. This causes the action scheduler at level 1 to call the motion launcher recursively to initiate level 2 motions. As in figure 11.3 control passes along the solid lines with arrows and data passes along dashed lines. Refer to text.

The input to the sequence interpreter consists of a normal shearing sequence and a set of procedures which define what the robot might do to recover from operating errors or abnormal shearing conditions. The script encoding the alarm and event signature definitions is also an input to the interpreter, though in a different syntax.

The normal sequence is executed by the sequence interpreter which generates, with the help of the path planners, a sequence of movements which are executed by TOM. When a movement has been initiated the action scheduler calls an appropriate motion module to calculate the robot motion required. There are different motion modules for different styles of movement, such as repositioning, shearing, landing, etc.

A motion module calculates the cutter motion, one step at a time, and calls the action scheduler as it contains the suspension point at which TOM pauses while the calculated motion for each step is interpolated. When TOM reawakens, the action scheduler returns control to the motion module which calculates the motion required for the next motion step. When the movement has been finished, control passes back to the motion launcher which sets a flag to indicate that the motion has been completed. The sequence interpreter then breaks out of its wait loop and returns to the planner. The planner can initiate another motion or it can pass control back to the interpreter for the next command in the sequence.

As well as containing the suspension point, the action scheduler calls routines to calculate the pose of the arm at the start of each motion step (forward kinematics), to monitor the condition of the arm actuators and sensors, to calculate the signature rules and to control event and data logging.

The signatures are designed to react to abnormal shearing conditions. When an event requiring a recovery procedure fires, the hypothesis evaluation routine selects an appropriate recovery procedure. This triggers a lock which prevents the action scheduler returning to the motion module; instead it is kept ready to launch a new motion while it waits for the sequence interpreter to issue recovery motion requests. This, in effect, results in an interruption to the original motion being executed at level 1. All temporary variables needed by the interrupted motion module are retained on the stack. Figures 11.5 and 11.6 illustrate the internal structure of the action scheduler and motion modules in more detail.

When a recovery request has been made the sequence interpreter branches out of its wait loop and calls itself recursively to execute the chosen recovery procedure. Now the sequence interpreter is at level 2.

The recovery procedure is now executed by the sequence interpreter. New motion requests are carried out in exactly the same way as has been described above except at level 2. When the planner issues a motion request, the action scheduler launches the motion at recursion level 2. When the recovery procedure has been completed the action scheduler, at level 1, is unlocked and allowed to return control back to the original motion module to resume the interrupted movement.

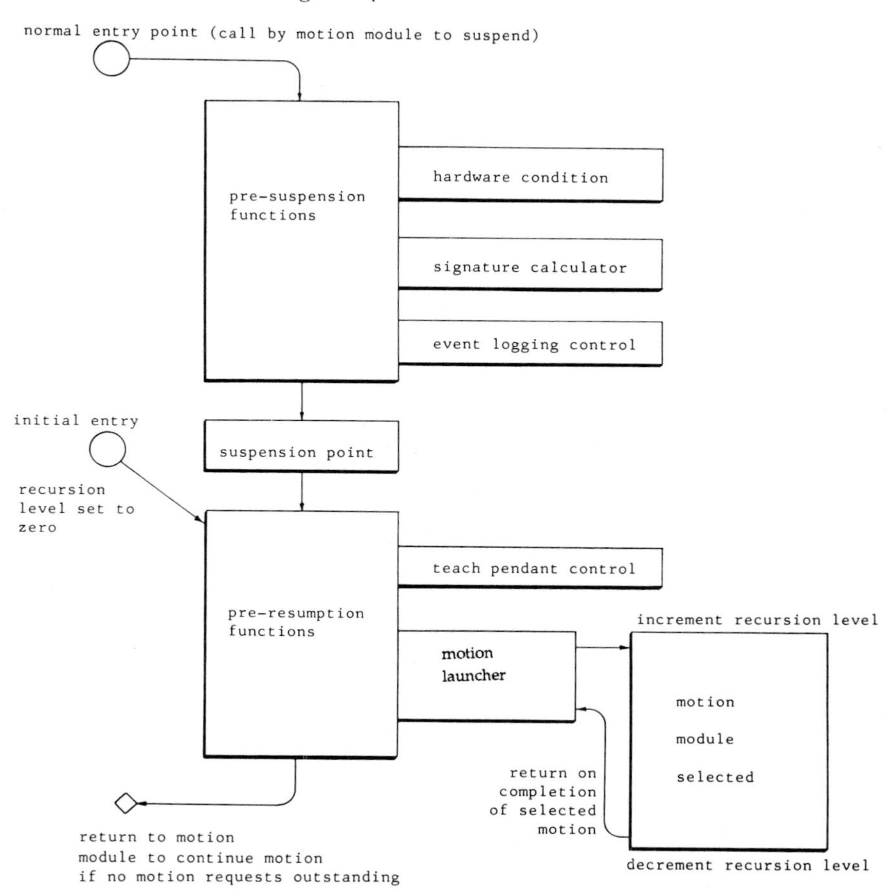

Figure 11.5 Internal structure of the action scheduler. Refer to text

This imposes an important requirement on any motion module. Before computing the next movement step, the motion module must check to see if an interruption occurred and must reset temporary variables if necessary. For example, the cutter may have been moved backwards along the shearing path from the interruption point. The shearing module checks for this, and resets internal pointers and trajectory parameters appropriately.

Each movement is specified by an 'action data block'. Some 1200 bytes in length, this block specifies the motion type, parameter values, control points (for trajectory curves), orientation frames, etc. It also contains blanks which are filled in with data collected during the course of the movement: the windrow which might be used for the next shearing blow is an example.

When designing this scheme, I devised a message-passing sequence which would allow the action data block for each movement to be passed between the real-time and background modules. I discarded this idea eventually because

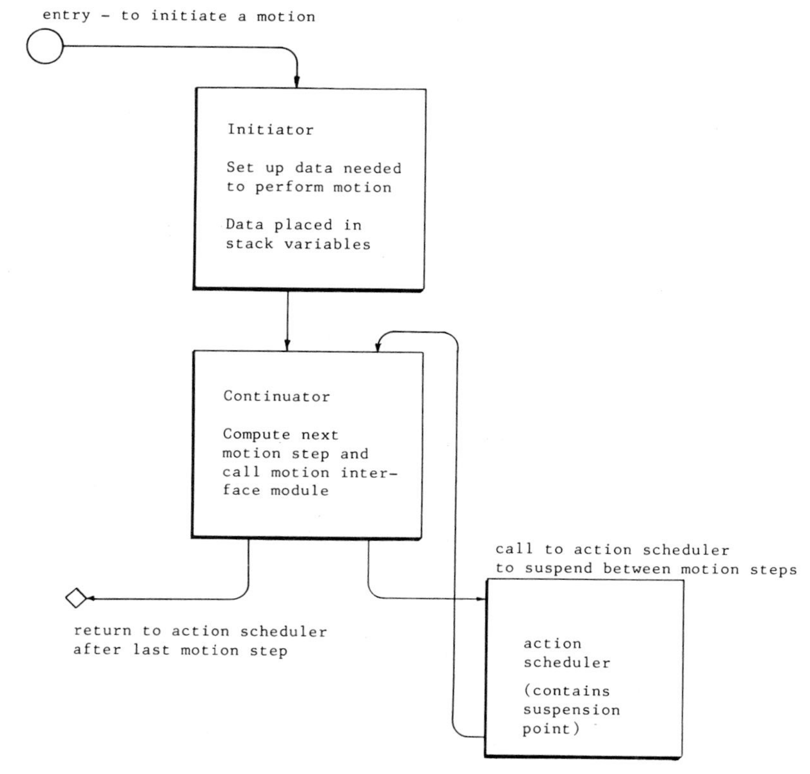

Figure 11.6 Internal structure of a motion module showing linkages to action scheduler.

copies of the action data are needed in several places, and I could not guarantee that all the 'copies' of the action data would be correctly updated during the recoveries.

The action data block describing the partially completed plan for a given movement must be available to the interpreter when a recovery is requested. This allows several important facilities to be provided. First, recovery movements inherit the same parameter values as the interrupted motion unless the values are changed by commands in the recovery procedure. In addition, or alternatively, parameter values in the partially completed plan may be altered so that the remainder of the interrupted movement is executed with the altered values after resumption. Third, recovery movements can be planned in terms of the state of the interrupted plan. These facilities are illustrated in the programming examples.

Apart from storing and updating the necessary copies, I also became concerned about the time overhead needed just for passing the data, and verifying the integrity of the sequence.

Instead of message passing, I opted for a 'blackboard' approach; an array of action data blocks is stored in shared memory—one block for each recursion level. Controlled access routines help to prevent disasters from coding errors.

We have chosen to implement a fully recursive arrangement because it allows the full range of robot motions and other programming facilities to be available for recovery procedures. Recovery procedures may be interrupted if more important problems arise. Recursion also simplifies the handling of the plan data which defines the motions. Temporary data for the movement being performed at each level of recursion is simply stored implicitly on the stacks associated with TOM and CLARE. No special provision needs to be made in the structure or data storage in the motion modules to permit interruptions to occur, although some modifications may be necessary to motion calculations just after an interruption has taken place. Thus the use of recursion leads to an elegantly simple arrangement of software which is highly versatile and easy to tailor to a particular application.

There is one further subtlety which needs to be explained at this stage. If at any time during the execution of a recovery procedure it becomes apparent to the system that the recovery itself needs to be interrupted the system can choose between interrupting the recovery (and going to yet another level of recursion) or simply abandoning the recovery already in progress and starting afresh with a new one. In the former case the sequence interpreter will plan a new recovery based on the plan inherited from the interrupted recovery and go to a higher level of recursion. In the latter case the interpreter will plan a recovery based on the plan inherited from the originally interrupted movement at the first level of recursion. In the same way a recovery procedure, once completed, can specify that the originally interrupted motion be aborted and that the sequence interpreter be resumed at the next command following that motion. In this way recovery motions can either take the form of interruptions to normal motions or substitutes for normal motions.

The perceptive reader will understand that the use of recursion is not the only way to achieve this end result. A conventional non-recursive interpreter and motion control system will suffice. However, the experienced programmer will appreciate that the use of recursion greatly simplifies the handling of temporary data in the system without, in this instance, leading to any significant increase in computation overheads.

Operator intervention

In our older motion control system, the routines for operator intervention were separated from the normal motion routines; thus one could safely use these routines even if the normal motion software seemed to be faulty. This was achieved at a price—complication and vulnerability to elusive timing faults.

I was determined to avoid this problem when designing our new motion control software. I resolved to make the control structure safe enough to use for operator intervention. For if it were not, how could one ever trust it for fully automatic operation?

I nearly had cause to regret my decision because I found that the provisions which I had to make for operator intervention complicated the entire scheme. It seemed simple enough; the 'pause' button triggers a lock which prevents control being passed back to the motion module after TOM is reawakened. Subsequent button selections then trigger one of the following.

red button—a top priority recovery procedure which backs the cutter away from the sheep using elementary straight line movements. After this has occurred, a motion control status flag is set to shut down all motions at all recursion levels, triggering a collapse back to the base (zero) level of recursion. The interpreter collapses in parallel, returning control to the operator.

operator interaction—a recovery procedure which passes control to the operator console until the operator types a 'TR' or 'EX' command to continue with the stored commands. The operator can type commands (using programmable menu keys) to repeat the interrupted movement, continue it, finish it at the interruption point and skip to the next movement, or abandon the shearing sequence completely. We use a similar arrangement for operator assistance recoveries—control is passed to the operator keyboard. When the operator has fixed the snag or problem, the robot can resume shearing.

manual escape—a procedure built into the software which does not rely on recovery procedures. A motion control status flag is set to shut down all motions at all recursion levels, triggering a collapse back to the base (zero) level of recursion. The interpreter collapses in parallel, returning control to the operator. Once this occurs (instantaneously on a human time-scale) the operator can use the motion buttons on the teach pendant to move the robot away from the sheep.

The complications arose when I discovered that the vital red button recovery, or even the pause request, could be ignored under special conditions. But as this applied, in theory, to any recovery request, correcting the scheme was worthwhile. In any event, the need for safe operator intervention was a non-trivial problem to solve.

Operating system requirements

Our software structure demands certain important operating system facilities. It is not often appreciated that robots impose some of the most demanding requirements on operating systems for response time and reliability.

Language

We have used Fortran-77 for all our robot control software, with the exception of certain hardware and operating system dependent interfaces. We are fortunate in having a version which supports recursive subroutine and function calls—without this we could not have contemplated this scheme for recursive motion control.

The compiler is not fast, nor does it optimize code. However, a little discipline and careful observation of the assembler listings of emitted code help to achieve fast computation times.

In future, we will be moving to C or C++ for new developments. But the Fortran core may survive for some time to come. Perhaps it has been most clumsy for implementing the CLARE interpreter. Yet we see the interpreter as an interim step to some kind of compiling approach. Either we will compile shearing (CLARE) instructions to some intermediate or machine code, or we will translate shearing instructions automatically into Fortran or C code for later compilation.

Real-ime control

Efficient real-time operation is essential. Naturally everyone has a different opinion about what constitutes 'real time'. In our situation, a typical 500–1000 μsec response time for a memory resident software process to be started after an interrupt or other program request is sufficient speed. In addition, the easy incorporation of user code into operating system interrupt handlers is essential for the fastest software processes which require response times of the order of 100 μsec or less.

The recursive structure of TOM imposes a special need. At the suspension point, TOM stops executing while CLARE tests its wait loop until TOM is reawakened by a request from HEATHER. While suspended, TOM must be kept in a state of suspended animation with all its internal memory structures intact, so that when it is reawakened, it continues executing from its internal suspension point. It is not sufficient for the operating system to merely restart TOM on a request from HEATHER, or for it to execute at preprogrammed intervals.

Simplicity is another important requirement. A simple operating system is more likely to provide a predictable response than a complex one.

Networking of the robot control computer raises some difficult operating system issues. Network access can disrupt operating system response to 'real-time' requests and can seldom be accomplished with simple software. For this reason, we have avoided network access.

Observations

This arrangement has evolved over several hundred shearing trials with live animals during the last few years, and several important conclusions can be drawn which may be relevant to other applications for adaptive robots. The first is that most of the abnormal conditions are diagnosed from internal states of the robot control system which arise only indirectly from the sensor inputs. Direct sampling of a sensor signal to diagnose a problem is the exception rather than the rule.

The second observation is that the processes of adaptive robot control and sensor fusion are necessarily intertwined at different levels of software. The hierarchy of operations performed is derived more from the time-scale of the intended computations than from a philosophical separation of the processes of perception and action. Perception and action in this system are irrevocably intertwined. There is no actual or intended separation of these two processes.

The third observation is that we have chosen to implement simple rules to diagnose problems and select appropriate recovery procedures. The necessity for rapid automatic decision making was the primary reason for this. We have since realized that neither the sensing means nor our understanding of the process of shearing is in any way adequate enough to contemplate maintaining an internal model of the robot environment which would permit a suitable recovery procedure to be inferred. Our recovery selection techniques are more analogous to instinctive reactions. A skilled human shearer seldom needs to pause to see what is wrong. A skilled robot shearer will know its limitations and if recovery procedures fail to resolve the problems then it will ask for human assistance.

On the negative side, the method of specifying hypothesis selections is clumsy and difficult to alter, once set up. Testing and verification poses special problems. We plan to take another look at these aspects when time and resources permit.

Manual sequence control

About 4000 lines of text encode the CLARE commands for shearing a complete sheep and calculating surface model adaptations in response to manipulator and vision measurements. In ideal circumstances, we could expect the operator merely to press a start button for the whole system to commence automatic shearing. It would stop only when it ran out of sheep.

We run two kinds of shearing experiments—trials and demonstrations. In trials, we often (in the earlier days mostly) shear only part of the sheep—the rest may be hand shorn on the manipulator or finished by a shearer later. In demonstrations we aim to shear the entire sheep automatically. Yet in both situations, reliable and flexible manual sequence control is essential, particularly during demonstrations when the unexpected happens. Fellow researchers will understand how many faults somehow hold themselves in reserve just for demonstration days—they never occur during routine trials. In our case, we find that sheep seem to react to our nervousness when we have an important or influential audience to satisfy; operators, machinery and sheep all are affected by stage fright. When something goes wrong, we need to be able to resume from near where we left off with a minimum of delay and fuss.

As yet we do not have a perfect system by any means. Yet it does offer some features which are worthwhile in a research environment and are not hard to implement with a little forethought.

Command files

Each command file consists of commands in a known syntax. The CLARE interpreter maintains a symbol table which permits variables to be defined: these can be either scalar or vector quantities with a predefined number of elements. New variables can be defined at any time as local variables—their values are stored within the interpreter. It is also possible to define locations within certain shared common blocks as variables. In addition, motion parameters which are part of the action data block are also accessed as variables.

On entry to a command file, the current action data block (containing values of all motion parameters) is saved in an explicit stack. The interpreter calls itself recursively to execute the file and the choice to use an explicit stack was made to conserve machine stack space in a limited address space. (The A900 computer has a 16-bit architecture.) As well as the action data values, the status of all symbols is saved.

New variable definitions within the command file are valid for the extent of the command file. Redefinition of existing names is not permitted as it represents poor programming practice in this environment. On exit from the command file, the symbol table status array is restored, erasing memory of variables defined within the command file. The action data block is also restored. The values of local variables (defined outside the command file) and shared common values are retained. This dual system may seem to be inconsistent, but it is convenient and avoids the need for explicit parameter passing between command files. It is also simple to program and to reconfigure when needed.

```
def a:4.0;           ! define a local variable a with an initial value 4.0.
adef b{21,4}:6.0;    ! define a shared common block location as a
                     ! variable (block 21, offset 4) with value 6.0.
set shvel:0.4;       ! change value of (predefined) motion
                     ! parameter shvel .
tr belly.c:          ! transfer to command file belly.c
show shvel,a,b;      ! display values of shvel, a, b. Assuming that
                     ! belly.c changes all of them, only the value of
                     ! the motion parameter shvel is restored on exit.
                     ! Thus local variables can be used to pass back
                     ! values from command files.
```

The loss of temporary local variables can be avoided by chaining from one command file to the next, but when the interpreter finally finishes the chain sequence the temporary definitions disappear.

Command file calls can be conveniently nested in this way; we impose a depth limit of 5.

A **tr 1** command transfers control to the operator's terminal—when the operator types 'EX' or 'TR' control passes to the next command. Once again, though, the interpreter is called recursively to do this with the same sequence of saving and restoration.

Command file internal structure

Each shearing movement in a command file is labelled; it has been our intention to carry this label down to all levels to ensure that data recorded during the movement is appropriately labelled. We have yet to complete our intentions in this respect. The real purpose of labelling is to permit shearing to be restarted from selected points in a command file. The selection of the restart point is left to the operator, though some machine assistance could be provided.

To achieve this, we impose some internal structure on the command file:

```
! Command file for shearing belly patch
   lp bellya;                 ! load belly surface model
   air on;                    ! turn on compressed air
   cutter on;                 ! start cutter
   def AsF*9:Eframe([45,0,35]);   ! define an orientation frame as
                              ! a local variable
EndInit;                      ! end of initialization section

   blowdef b1;                ! label for first blow
   movement;                  ! define start of this movement
   startmove[. . .            ! start first repositioning movement
   set shvel. . .             ! define parameters for first blow
   blow[. . .                 ! first blow

   blowdef b2;                ! label for 2nd blow
   movement;                  ! define start of this movement
   startmove[. . .            ! start next repositioning movement
   set shvel: . . .           ! define parameters for 2nd blow
   blow[. . .                 ! next blow

   blowdef b3;                ! label for 3rd blow
   movement;                  ! define start of this movement
   startmove[. . .            ! start next repositioning movement

Entry b3; movement; park[t,u,h, vel, frame ]; . . . other commands . . . ;
EndEntry;

   set shvel: . . .           ! define parameters for 3rd blow
   blow[. . .                 ! 3rd blow

   blowdef b4;                ! label for 4th blow
   movement;                  ! define start of this movement;
   startmove[. . .            ! start next repositioning movement
```

Normally, we would invoke this command file (assumed name belly.c) with:

```
tr belly.c
```

The initial part of the command file performs preparatory operations such as sensor calibration, opening log files, starting the cutter and defining local variables. It is terminated with an 'EndInit' command. CLARE continues with the next command in the sequence and commences shearing at the first blow.

However, by using the invocation:

```
tr belly.c b3
```

CLARE ignores code following the 'EndInit' command until it finds the 'Entry b3' command. Then command file execution resumes at this point with a 'park' command which will move the robot to a point above the surface coordinate values *t,u,h* with a given speed and (optional) orientation frame. The park procedure can automatically plan a safe movement clear of obstacles. The robot is then positioned ready to land for the next shearing blow in the sequence.

In the case of a normal entry (and for entry points later in the file if entering at 'b3') the code between each 'Entry' command and the succeeding 'EndEntry' command is ignored.

The one discipline which is needed here is to ensure that all parameters needed for a given blow are defined immediately before the 'blow' command. We can make exceptions to this—some parameters only need to be set once for a given patch. This is good practice for two reasons.

First, by using a standard sequence of parameter definitions (though CLARE does not require this of course) the programmer needs to think about their values—are they appropriate for this blow?

Second, it is easy to make changes to the command file by inserting new blows or deleting old ones *without* having to check that parameter values for succeeding blows will be affected.

The command 'movement' following the 'blowdef' commands serves as a place marker. If a shearing movement needs to be repeated, CLARE returns to the place marked by the preceding 'movement' command.

Supporting features

Given that attention will be focussed on a particular piece of code, the system (screen-based) editor can be invoked directly from CLARE to present the text pertaining to a particular blow or movement:

edit belly.c b3

. . . or simply

edit, b3 . . .

once the command file name has been used in a 'tr' or 'edit' command.

We also have (crude) facilities for displaying the blow paths on a graphic display of the surface model.

During shearing code development, the robot can be run in a 'dry' state, without power. The movements are displayed by an animated stick drawing of the robot on the surface model, or successive robot positions can be painted on the screen, such as that shown in figure 4.3 (p. 83).

Programming by showing

In contrast to most industrial robots, we have relatively limited facilities for 'teach mode' programming. Team members and visitors have often suggested that we program the robot with an instrumented shearing handpiece which would allow shearing movements executed by a skilled shearer to be recorded and replayed on subsequent sheep. I may have missed something obvious to others; maybe the complications which I have seen could easily be overcome. I think it is worth noting them to make sure they are recognized and either overcome or bypassed in future.

First, any movements recorded for one sheep need to be generalized for other sheep before they can be used; a surface model is required for this. A model could be derived from the movements, in combination with feature locations obtained by (improved) machine vision or digitizer. Until now, we have had to shear the wool off first, and then measure a surface model. By the time the model has been measured, the sheep has often moved, or sagged, or relaxed. The cutter forces deflect parts of the sheep too.

There is a further complication; the finished sequence must allow for surface modelling errors since we cannot exactly predict the shape in advance. The senses of vision and touch, the skill and experience which guide the hands of an expert shearer cannot be reproduced. Therefore the shearing pattern needs to be tailored to compensate for sensing and prediction errors.

It would not be easy to build the instrumented handpiece either. Accurate measurement of the handpiece pose (position and orientation) is needed, at high speeds (up to 1 m/sec). The forces exerted by the handpiece on the skin and wool need to be measured too. All this in a fully operating lightweight handpiece which has the same shape as the robot cutter.

We developed a restricted 'edit by showing' facility which generates some of the code automatically. The robot is moved under manual control over the sheep and a sequence of positions is recorded. These are then converted to surface coordinates so they will be adapted automatically for different sheep. We learned that our conventional 'teach pendant', even with movements available in surface or tool coordinates rather than Cartesian or joint coordinates, was far from satisfactory. We thought about using a 6-axis joystick (e.g. the commercially available *Spaceball*). Yet after we solved the control problem, there would be significant software problems which are beyond our resources to solve at this time. Finally, there are safety and regulatory implications in using non-standard controls.

Although it can be tedious to encode shearing sequences, the trial and refinement of sequences takes much longer. The coding effort is not significant. The part that seems to take most time is thinking: how can one use the programming techniques available to produce the desired effect, and, more difficult, *how will it behave in all the different shearing conditions?*

As I mentioned in chapter 4, we would like a 'nice' graphics interface for programming, with icons, pull down menus, good robot animation and display of the whole sheep as a solid (inviolate) body. This would allow more people to program and refine shearing procedures without having to learn coding techniques. Those experienced with robot programming will recognize a major software development project here, quite beyond our present resources, but becoming easier all the time as the capabilities of commercial CAD workstations improve almost monthly. For us there are other problems requiring more thought and less software engineering.

12
Skilled work

Shearing is skilled hard work. For thousands of years, we have designed machines to take the hard physical component out of manual work, to act as extensions of our hands and feet. Yet with shearing, that is not enough. The machine needs to have shearing skills to work effectively.

After 15 years of research, we have learned that replicating manual skills is not easy, particularly shearing. Manual skills are part of everyday life, but the neural and physiological processes which support them are not easily appreciated, any more than apparently intellectual skills. I once attended a lecture by a famous art critic, an internationally known expert on the work of Picasso. I listened carefully as he speculated on the reasons why Picasso had incorporated various characteristic elements in his paintings, his colours, the chequered red and yellow designs, the seemingly dismembered and disfigured human forms. By the end of the lecture, I was just a little irritated by what seemed to be academic speculation which could have easily been resolved by asking Picasso before he died. I asked the obvious question, and was surprised by the answer. 'Yes I did ask, and after many many conversations he was quite unable to tell me.'

Manual skills are similarly obscure. In his magnificent review of the intellectual problems confronting artificial intelligence research, Penrose (1990, p. 490) describes the apparent role of the cerebellum, the primitive part of the brain responsible for coordinated skills. The flowing grace of a dancer, the precision and timing of a juggler, the quiet relaxed movements of a champion shearer, the accuracy of a professional tennis player: all these skills are acquired by practice and refinement. Conscious thinking about the skills can temporarily disrupt performance! The true relationships in the skill are buried deep within the brain, far away from access by analysis.

The hope that we can somehow replicate human neuron elements with high-speed electronic circuits and hence build machines which can learn coordinated skills for themselves has inspired a new generation of artificial intelligence research. Artificial neural networks may one day achieve this, but my instinct suggests that this is a long way off.

Whitney (1986) has argued strongly that robotics researchers should focus on *what* people do rather than *how* they do it. He was interested in replicating surface grinding skills used for deburring and finishing metal products. For industrial processes like grinding, his arguments have a great deal of merit though much of the intellectual stimulation of artificial intelligence research comes from the quest to understand ourselves.

Dickmanns (1988) has shown how motor vehicles can be driven automatically by computer control at high speeds along highways. Once again, he has focussed on what the role of the driver is, rather than how the driver does it.

A human shearer moves the handpiece forwards, pressing the comb teeth down on the skin. The mechanical cutter severs the wool fibres as they pass between the comb teeth, and the shorn fleece peels away to the side. This much is apparent to a casual observer. Holding the sheep still is part of it too, of course. Yet as any novice will tell you, it's not at all easy. Somehow, the handpiece just doesn't move easily, sometimes not at all. Sometimes, ragged pieces of wool are left behind, and other wool needs two, three or four passes to shear cleanly. It is not always easy to find out how a skilled shearer does it.

Physical strength and endurance are part of the skill too; a novice stands sweating and panting after shearing the first sheep. After the first lesson, muscles and joints will ache for days. Later, years of hard physical strain leave permanent partial disablement for far too many shearers: it is this factor which is one of the strongest motivations to devise shearing robots.

What we do know is that a typical novice shearer needs to shear 10,000 or more sheep to achieve a reasonable level of competence. And our experience with shearers confirms that accessing human skills is a tortuous process, full of surprises, as each contributing factor becomes apparent.

Looking back, one can perceive the route we have taken towards a skilled shearing robot. First, we attempted to replicate the movement of the cutter; we built ORACLE for this and demonstrated a basic competence and the possibility of robot shearing. ARAMP demonstrated that sheep could be automatically manipulated into different shearing positions, though the design was constrained by the political requirement that it be seen working with ORACLE. SM represented a considerably refined robot with more delicate touch, dexterity, speed and strength. SLAMP represented the next step; holding the sheep and presenting each section of fleece in near ideal condition.

Paradoxically, SLAMP represents the Whitney approach. In essence, we started by defining what the shearer does when holding the sheep:

(1) restraining the sheep to keep it still,
(2) moving it from one shearing position to the next,
(3) stretching the skin when necessary, lessening the cutter drag and reducing any tendency to clog the cutter,[1]
(4) holding the body with the legs and left hand so the skin to be shorn is firmly supported by the flesh and is flat or slightly convex in the direction of shearing, and easily accessed by the handpiece held in the right hand, (hands can be reversed);
(5) positioning the sheep so the shorn fleece rolls cleanly off on to a growing pile on the floor; and
(6) balancing the weight of the animal to minimize stress on the shearer.

[1] The opposite sometimes applies. For some blows, the sheep is 'dumped'; the animal is positioned to compress the wool so that a long, slow blow removes more wool.

We then set out to build a machine to do this for the SM robot—not an easy task as chapter 9 relates. SLAMP performs quite differently from a human shearer; it was designed to suit the robot's workspace and to use machinery which is simpler than human legs and arms (coupled with human sight and other senses).

While the SLAMP concepts were emerging, we gained access to first-rate shearing experts for the first time. This was appropriate though; in the earlier stages, our gaps in shearing skill were apparent even to casual observers much of the time. Shearers visited us, and their observations were useful, but we could easily understand where we needed to make improvements. By mid 1987 though, we were reaching the point where deeper understanding was vital for progress. We were by then fully committed to replicating human shearing skills, not necessarily by duplicating a human shearer, but certainly by using an almost identical tool.

Before describing our interactions with the human experts, I will review what we knew then about the task, and the mechanical skills we had accumulated at that stage. I have used the present tense; our understanding is now deeper rather than being different.

The task

As with any application of industrial robots the first requirement is to thoroughly understand the task to be performed. Yet this understanding only comes in stages. At each stage the attempt to apply one's understanding reveals the gaps in that understanding and it is only recently that we have closed our gaps sufficiently to make the task of the machine seem as effortless as it does to a skilled shearer.

A typical Merino sheep carrying 12 months wool produces a fleece weighing 5 kg. The fleece tends to hang together like a piece of thick pile carpet so shearing is much more like peeling this carpet away from the sheep than mowing grass. Much of the time the shearing cutter is out of sight under a thick layer of fleece and a human shearer relies much more on his appreciation of the shape of a sheep and on tactile sensing than on vision. Tactile sensing here is enlarged to encompass its widest meaning, that is the sense of proprioception, force sensing, sensing of vibrations from the tool, as well as direct contact with the skin of the sheep.

As well as the obvious variations in size, sheep vary in shape and in the relative proportions of each part. The wool characteristics vary enormously from tight fleeces which require a great deal of physical strength to tear apart, to loose open fleeces which require almost no effort to shear. The skin also varies from being taut all over the sheep to a mass of folds and wrinkles. Merino sheep usually have too much skin for the size of their bodies; as well as forming folds and wrinkles the skin is highly mobile and slides easily over the flesh underneath.

The shearer or manipulator holds the sheep and manipulates it into the best possible shearing position for each part of the fleece. Legs, head and neck are manipulated so as to stretch the skin to reduce its mobility as much as possible and to flatten out the folds and wrinkles.

Not only is the sheep positioning and manipulation important—the shearing comb is a critical part of the shearing tool. The tool has evolved over the last 100 years through a process of trial and error stimulated partly by the fruitless search for a better method of cutting wool (chapter 5). Shearers and manufacturers of combs now have a much greater understanding of how the comb works. As shown in figure 5.5, (p. 135) the undersides of the comb teeth run along the skin and the depth of the comb teeth therefore determines the thickness of stubble remaining after shearing. The action of combing the wool ahead of the severance mechanism keeps the comb in contact with the skin so minor positioning errors can be tolerated. However this also means that lifting the comb in response to sensor signals does not necessarily move the comb away from the skin—the skin can be lifted at the same time by the wool which is still unshorn. The extent of this effect varies enormously of course between different parts of the animal. Where the skin is tight and the underlying flesh firm, there is very little tendency for the skin to be lifted.

The shearing comb, about 85 mm wide (figure 12.1), is large compared with

Figure 12.1 Plan 'photocopy' of a shearing comb. The grinding marks can be clearly seen.

the curvature of much of the sheep. When shearing legs and other sharply curved sections only part of the comb is used. This means that the sensing devices behind the comb will provide very different readings for these parts of the sheep from those registered on the main body areas where the whole comb lies against the skin. Thus we see that the use of the sensors varies with different shearing conditions. The knowledge of how to use the sensors under different shearing conditions must be built into the software.

Robot skills

For the purposes of this discussion, I will define manual skill as a human action, performed primarily by hand movements and perfected by a combination of learning and refinement through practice. Adaptation is an essential element of any manual skill, so that variations in working conditions can be accommodated.

Robot skill

A skill incorporated into a robot confers a specific ability which can be used to perform a task. Whereas a human can acquire the skill through explanation and then refine it by practice, the robot needs to be programmed. Thus the skill needs to be expressed in a formalism or design which can be implemented in the robot controller. Some skills can be wholly or partly represented by the hardware of the robot.

We normally think of skill as having an intellectual embodiment alone. However, we have built an embodiment of the skill of manipulating and positioning the animal as a computer-controlled machine. This machine has been designed on an understanding of human skill, just as the robot movements encoded as software are based on a similar understanding. Therefore, I would argue that a robot skill can be just as much a hardware embodiment as a software embodiment.

Delicate touch with fast reactions

A delicate touch with fast reactions is the first requirement for robots working on live animals. We achieve this by a combination of sensing, mechanical design and adaptive motion control software. Conventional factory robots cannot perform this type of task because they are optimized for point-to-point positioning moves, with high repeatability and precision.

Both the control system and the structure of the SM robot have been carefully designed with adaptation requirements in mind. The follower axis provides a means of adjusting the cutter height relative to the skin with a very fast response time—between 2 and 4 msec. This fast movement takes care of adaptations which the main arm is too slow to follow; the dynamic requirements for arm control are reduced by providing the follower axis.

Dexterity

The second important attribute of a shearing robot is kinematic dexterity. Not only must the arm have sufficient reach to access the sheep on the manipulator but it also needs to be able to align the cutter with the skin and have sufficient freedom of movement to perform the complex coordinated movements required for shearing, repositioning and pushing the wool out of the way.

For this we designed the SM arm which does not have any singular positions within its usable workspace. We had to design a completely new wrist mechanism—both are described in detail in chapter 8.

We chose to mount the robot overhead to provide the best possible reach to a horizontally disposed workspace with the sheep supported from underneath. The pivoted telescopic arm design provides a simple mechanical solution for the upper end of the arm and the unique wrist design provides the dexterity required and the freedom of movement to orient the wrist mechanism. A result of this development is a robot of unparalleled kinematic dexterity in which mechanical simplicity has been achieved in both appearance and actuality.

Sensing

Control of the arm is ultimately dependent on the quality of the information received from sensors. The most important is resistance sensing in which a minute electrical current flows from the comb to the sheep which is connected to ground. As the contact pressure of the comb increases so does this current. If the skin is cut at any time there is a sudden increase in the current. For this sensing arrangement to work the conductivity of the wool must be substantially less than that of the skin of the sheep.

Skin proximity sensors located under the cutter are equally important. Our capacitance sensors measure the distance between the cutter and the skin to a maximum of 80 mm and with a frequency response in excess of 1 kHz. The sensors are affected by variations in wool properties. The pneumatic shielding device keeps wool sufficiently compressed against the skin so that proximity sensor readings are largely unaffected.

The resistance sensing provides an intermittent and noisy signal from the tips of the comb teeth. The capacitance sensors, on the other hand, provide a much more consistent distance indication but one to which a certain degree of uncertainty is attached because the sensors are located well behind the teeth of the comb. The robot control system is able to use the best quality of each type of signal with impressive results. Mostly, the robot presses just firmly enough to keep the cutter working with a force of about 2 to 6 N. Although there is a large tolerance of different mechanical surface characteristics, the moisture levels near the skin affect performance. The robot presses too firmly on dry calloused skin and stability can be marginal on some hard, boney parts of the sheep.

Surface models—a sense of proportion

The surface model performs two important functions for the robot. First it provides information to help interpret sensor signals. As explained above the capacitance sensors are located some distance behind the front of the comb, therefore, knowledge of the surface model tells the robot how to allow for the surface curvature between the front of the comb and the location of the sensors. The surface model also provides advance warning of changes in surface curvature. Second, the surface model serves as a reference for planning robot movements.

Every sheep is different yet they all look alike. We use surface models to represent the differences between sheep so that much of the remaining software required to control the robot remains the same for all sheep. The surface model introduces two important simplifying steps.

The first is that each point on a complex curved surface presented by the model is defined by just two coordinates in the same way that a point on the earth's surface is defined by latitude and longitude. Shearing blows can be planned as one or more straight line segments across the surface. Thus as few as four coordinate values can specify the line of the blow on the surface. These values can be translated automatically into the mass of data required to move the cutter to the beginning of a blow and guide it along a path on the surface.

The differences between sheep are accounted for by calculating a new surface model for every part of every sheep to be shorn. We have methods for using measurements of the sheep to predict the position and shape of each part of the surface. Many parts of the sheep, particularly the extremities vary significantly only in size; the shape remains largely the same. Thus surface models for the legs, for example, can be generated from models for a standard sheep, suitably adjusted to take into account the leg lengths.

The shape variations which occur in the main body sections of the sheep are more difficult to deal with. The most successful method we have developed for this is statistical prediction. We measure a large number of sheep, collecting surface shapes together with sets of size and weight measurements. The two sets of data can be statistically related so that by measuring the size and weight of a new sheep we can predict a relatively accurate surface model, even though the surface is hidden by up to 100 mm of fleece.

These methods of prediction lead to the second important simplification. The position of a particular part of the sheep's body will always appear at the same surface coordinate values no matter how large or small the sheep may be. This arises from the choice of normalized surface coordinates—the coordinate values range from 0 to 1 in each direction. Thus a given spacing in surface coordinates varies with the size of the sheep.

This allows us to write programs for shearing the sheep in surface coordinate values which are the same for every sheep. In this way we exploit the similarity between sheep; surface models provide a powerful conceptual device for separating those parts of the process which are similar from those parts which differ between sheep.

Coordination—motion control and sequencing

The motion control software links all the elements together and provides the means of programming shearing. Naturally, the motion control system is specific to shearing though it could be used for a wide variety of other applications.

The CLARE interpreter provided a means for programming shearing style. The placement of each shearing blow is defined with respect to the surface model, as described in chapter 4. We can specify parameter values to control shearing speed, cutter attitude, sensing parameters and overlap offsets. In other words, we can display the level of motion skill which might be apparent to the casual observer mentioned earlier.

In mid 1987, B-spline repositioning and blow paths were still in the 'ideas' tray, ideas for recursive motion control were emerging, and the hybrid follower was latent. All these developments were stimulated by our interactions with shearing experts which were about to commence.

Acquiring skill

Our first step was to organize a shearing school for the entire research team to learn the basic skills of manual shearing. Many team members did not have the physical fitness to achieve much skill or confidence but it was nevertheless a valuable exercise. We learned much about the importance of manipulating the sheep to obtain a shearable surface—near flat with a stretched skin in a convenient position to reach with the handpiece.

Our contacts with professional shearers had already alerted us to the differences in perception between experts. A significant factor according to one expert could safely be ignored according to another. And we came to realize that there is no recognized coherent description of shearing skill in a form we could understand without having to learn the skill ourselves.

We had expected some resentment from shearers, particularly the Australian Workers Union which represented them. Yet we never encountered any from the shearers who visited our laboratory. Many were sceptical and most left with little fear of being replaced with robots in their working lifetime (10 years on average). Most were surprised that a machine could shear at all, and one or two were excited by the possibility that a machine could do the hard work for them. Most of the shearers who worked with us for a day or so every few weeks were awed by the machine's capabilities and were unable to make comments which were useful to us. They could watch *us* shearing and correct our technique, but passed little comment on the machine's style. Perhaps it was so terrible to them that they felt that any criticism would offend us. Or perhaps the level of technology seemed so far from their experience that they felt unable to offer useful suggestions.

A chance encounter brought us into contact with a professional shearer who had the rare combination of expertise, lucid communication, and a gift for translating his skills into a new setting. Allen White had already gained a reputation both as an expert shearer and for his ability to teach others. He joined the team for two months full-time in May and June 1987 during the lull between shearing seasons.

Working with Allen was quite a different experience. He listened quietly for hours while we explained how the robot and manipulator worked. He intently watched the robot shearing and was soon asking us to try different ideas. He carried an air of confidence and assurance, inherited from a background as the son of a Maori king, his tall figure and powerful physique somehow at odds with his quiet modesty. Allen realized that he had to transform his shearing skills into a new framework of machinery and mechanisms, and appreciated the possibilities and limitations of computer control. Perhaps this came with his teaching style which was to understand the learner's framework and show how improvements could come by gently extending it rather than by dismantling it and replacing it with his own.

Apart from his influence on the design of SLAMP, Allen's principal contribution was to provide us with an understanding of how the comb works, and how combs need to be prepared for shearing.

There are two aspects—comb preparation and sharpening. The former involves reshaping the comb teeth by hand to reduce shearing effort. The shape of the tips of the comb teeth is crucial and some aspects are shown by figure 12.2. The side profile of each tooth has to be adjusted for the type of sheep being shorn. Figure 12.3 shows the difference between a comb for Merino

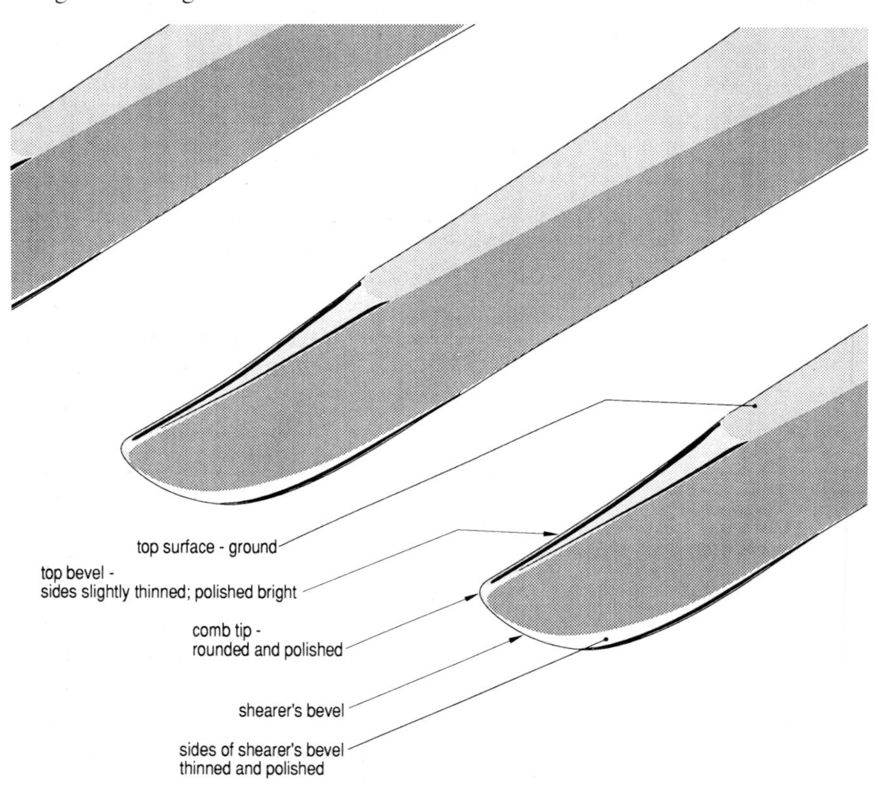

Figure 12.2 Close up view of comb tips showing finishing operations needed for shearing Merino sheep.

sheep with fine wool, and crossbred sheep with tougher skin and coarse wool. The combs are used differently too. The crossbred comb is angled with the points down about 5° to 10° and firm pressure is applied. This style relies on the tougher skin and a crossbred sheep can be shorn very quickly this way. The Merino comb is used flat on the skin, but is also pressed down firmly. In contrast to the crossbred, it is better to shear quite slowly, making sure that the full width of the comb is used whenever possible.

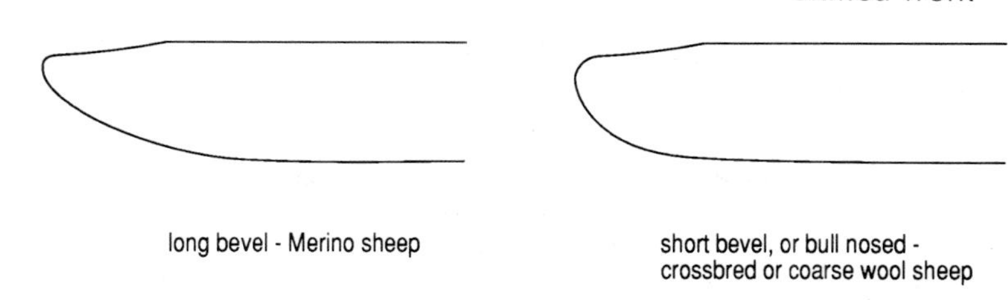

Figure 12.3 Side profiles of different shearing comb teeth.

To avoid cutting the delicate skin of the Merino, the comb tips are smoothed and do not feel sharp even when pressed on to the palm of the hand. The top bevel (figure 12.2) needs to be thinned and polished because this is where the combing and untangling of the densely packed wool fibres occur before they are severed against the sharp edges further back (figure 5.5). The shearer's bevel runs on the skin; it is thinned and polished to reduce penetration drag. The long bevel rides more easily on the skin wave ahead of the comb.

The tooth profiles are obtained by hand using a small oil stone and a great deal of care, particularly with the point on a Merino comb. The bevels are thinned with a small 'Bright Boy' wheel (a rubber disc containing abrasive grit) and the comb is finished with a polishing buff.

Once the teeth have been prepared, the comb can be sharpened. To a mechanical engineer, the sharpening process looks unimaginably crude and slapdash. The comb, located by a magnetic clip on a short pendulum, is pressed firmly against the outer face of a spinning disc coated in coarse abrasive paper (figure 12.4). For just a few seconds, a shower of sparks flies up as the shearer swings the pendulum slightly for an even finish. There is none of the precision or finesse normally needed for a surface profile which has to be accurate to within a few microns.

The finished ground top surface is hollow to a depth of about 40 μm from tip to base, and about 20 μm across the comb. The hollow is obtained partly by pressure from the magnetic clip—the force is applied by a bar across the middle of the comb—and partly by thermal effects. The rapid heating at the ground face distorts the comb during the few seconds of grinding. When the comb cools, it has the right degree of hollow needed to obtain the point contact scissoring action with the cutter. If too little pressure is applied early in the grind, the lack of mechanical and thermal distortion will cause material to be removed from the sides and tips of the comb instead.

The cutter has its own grinding requirements, but the technique is similar.

Until recently, comb manufacturers have been unable or unwilling to supply combs which are ready to use. Even though the combs were supplied ground, they were rarely even sharp enough to cut. Only a complete novice would start shearing with a comb taken from its packet. In 1989, comb manufacturers released new Merino combs which need only minor finishing by hand in response to pressure from shearers.

We helped Allen write a training manual setting out the details of preparation

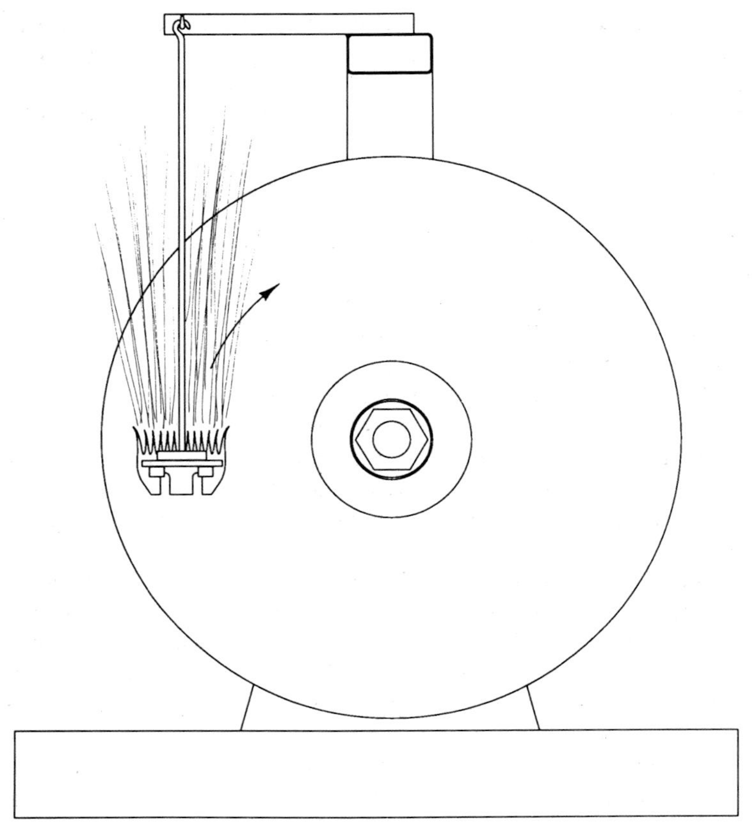

Figure 12.4 Comb grinding—refer to text.

and sharpening techniques. The manual has since been widely distributed; a note on the front explains how it was written during his stay with us—a tangible result for just a little of our work. Naturally, Allen's skills were sought after elsewhere; within a few months he was appointed as chief shearing instructor for the Corporation in Melbourne. We pressed for more of his time but he was only able to visit for a day or two. Fortunately Darryl Cole, a close associate of his, was able to join us 80% full time and has been with us since early 1988.

Expertise is only useful if it can be transferred and leads to improved performance. In the case of comb and cutter preparation Allen's skills were learned by our technicians who were soon able to produce good combs. Other skill improvements are built into the robot through software alterations. For this to occur, each item of skill must be first acquired and understood by the robot programmer. Then the programmer alters the robot control software to incorporate the skill. If the robot then exhibits the expected improvement in performance, one can assume that the skill has been effectively transferred. There are several transfer steps in this process:

(1) expert expresses skill in words and demonstrates skill by hand shearing;
(2) robot programmer learns skill and practises by hand shearing until expert is satisfied that programmer has understood the skill;
(3) robot programmer devises skill representation and encodes the skill in robot controller;
(4) expert and programmer test robot to assess robot performance of skill.

If the robot fails to show the expected performance improvement, either the skill transfer process has failed or the differences between the robot and human shearer are too great for the skill to work. If the latter turns out to be the case, then it can be concluded that the skills already built into the robot are inadequate, and need refinement. This has often been the case and the pattern of development of robot shearing has been a wandering improvement of skill. As one skill area is improved, a weakness in another skill is exposed sufficiently to require improvement. Attention then shifts to the next skill.

While Allen focussed on comb preparation, Darryl concentrated on shearing patterns and the use of the comb. By the time he joined us, the SLAMP design was completed and the SLAMP 1.75 experimental model was available for shearing tests.

The top view of the comb in figure 12.1 shows how the outer teeth are splayed out—the shearing term is 'pulled teeth'. Comb teeth were originally pulled to increase the width of the comb enabling the shearer to cut more wool along each blow. By carefully watching Darryl, and by practice on my own part, I was able to discover that the splayed teeth help in at least two ways. They do gather more wool. They also allow the comb to be run with a yaw offset which tensions the skin sideways. I have tried to illustrate how this works in figure 12.6, but I believe that it is a subtle effect which is hard to appreciate without trying it.

Failure

Often there have been failures in the skill transfer process. One example will serve to illustrate an important lesson we learned.

As I have explained, the shape of the comb tips is important for good combing. Darryl decided to modify the robot's combs to reduce skin cuts which seemed to occur too often. He did this by rounding the shearer's bevel and tips of the teeth. The more he did this, the more often skin cuts occurred. It must be emphasized that the cuts were small, and sometimes difficult to see because the robot stops and withdraws the comb teeth from the cut very quickly before any serious damage is done. However, this interrupts shearing and slows the robot.

Examination of the resistance sensor signal showed erratic variations which suggested that the comb was periodically out of contact with the skin, and then pressing too hard with little time in between. For a while I was puzzled as I was unaware of Darryl's changes. The problems occurred with the first shearing tests of the recursive motion control system and I naturally suspected some software faults were the cause. After a frustrating series of trials I decided to repeat some early sensing experiments; we measured resistance signals during hand shearing using an electrically insulated handpiece. To my great surprise, the results indicated an almost continuous skin cut, yet the skin was not being cut by the comb. It seemed that the comb was being pressed too hard against the skin. Darryl explained that heavy pressure was normal (and needed) for this type of comb. The extra rounding allowed high contact forces without cutting the skin. Darryl had assumed that the robot would press hard against the skin so his new comb would be safer for the robot to use.

The modified comb clearly could not be used because resistance sensing could not tell the difference between normal shearing and a skin cut. I repeated the hand shearing tests with thinner bevels. I asked Darryl to shear with less and less pressure until the resistance sensor signals appeared to be more usable by the robot. To Darryl's surprise, the required pressure was extremely light—he described the touch as 'featherlight'. He was even more surprised to find that the robot was capable of shearing with such a light touch, amazed in fact. He watched carefully and noticed, now, how little the skin was deflected by the comb.

Darryl realized that the comb teeth could be much sharper, reducing the force required to comb through the wool, allowing faster shearing. The tendency for the teeth to ride out of the wool would also be reduced. We realized that the more rounded teeth would need to be pressed very firmly against the skin to obtain electrical contact and once skin contact was obtained, high conductivity resulted from the large contact area and forces.

Finally, we used a series of apparently 'lethal' combs in shearing experiments with the robot. The combs had fewer teeth than normal, and almost razor-thin teeth. In the hands of any but the most expert shearers, these combs would result in long skin cuts. Yet the robot was able to shear fast without cuts, almost a 'superhuman' feat.

What did we learn from this?

The expert's advice will reflect his or her own understanding of the process. If there is a misunderstanding the advice may be ineffective or counterproductive. In this case, we had not explained how the robot works, seeking and maintaining electrical contact with the skin. Nor had Darryl actually felt the pressure of the cutter on his own skin. If he had had a better understanding, we might have had less frustration.

This problem is compounded by differences in control behaviour. The robot was (at the time) a position controlled device whereas the shearer's arm is force controlled to a large extent.

For his first year with us, Darryl felt more than a little out of place in a university laboratory. He had come in from the crisp dry dusty paddocks and the basic lifestyle of a shearer. Although well travelled with experience in the USA, New Zealand (his birthplace) and Europe, the intellectual environment of the university was hard to adjust to. He was, for a while, in awe of of our apparent ease in writing long technical reports, computer software and mathematical hieroglyphics. Yet this soon became a valuable asset. Once he understood what we were trying to do, he was able to talk about our work to farmers and shearers at a level they could understand and appreciate, and thereby access their ideas and suggestions in a way we never could.

Further developments

Even though we had the comb problem solved, I continued to think about the hand shearing tests Darryl had performed. Good shearers use firm pressure on the comb to flatten the skin—it is easier to shear efficiently that way, and it is hard to maintain a light touch with a typical mechanically driven handpiece. Could there be a fundamental problem with 'featherlight' shearing? Would a shearer work that way if it were possible?

While we were in the final stages of preparing for our major demonstration in February 1989, I began to realize that a light shearing touch had a major weakness. As long as the comb is in contact with the skin, there is no problem. But even on moderately curved parts of the sheep, this is only feasible for a few of the comb teeth. Once the teeth penetrate wool some distance above the skin, the fibres lean over, first forwards and then sideways away from the cutter as well. The leaning of the fibres increases the fibre density ahead of the comb increasing the penetration drag and causing more leaning or clogging of the gullets. Longer stubble is left behind, less wool is shorn, and the drag force rapidly increases. If the comb is pressed down firmly so the teeth run on the skin, the surface friction drag is increased but the penetration drag from the fibres remains low.

As explained in chapter 4, this observation helped to lead me to the concept of the hybrid follower—a force and position control scheme for the follower axis. Yet more was needed. We had to shear with the comb angled at about 10° to the skin because of the mechanical arrangement of the sensors and the SM cutter mechanism (figure 12.5). When the hybrid follower was first tested, I experienced a frustrating series of trials from which I learned that the comb has to be kept flat on the skin to be able to use firm pressure for shearing. Otherwise the downward combing and contact force on the comb tips pushes the tips too deep behind the skin wave and the comb 'trips up'—an effect which Merino Wool Harvesting had encountered with their force control scheme in 1985.

At the time of writing, a new cutter has just been fitted to the robot (p. 142) and flat comb shearing will be feasible for the first time. I am looking forward to being able to confirm my guesses. I expect a few more surprises yet.

Skill representation

Artificial intelligence researchers often talk about skill (and knowledge) in purely intellectual terms, as if it is something which should be embodied in software rather than hardware. Yet the skill of our robot—shearing—is represented by the entire shearing system (including the robot, sheep restraint device and controller), not just by the software in the robot controller. This may seem to be

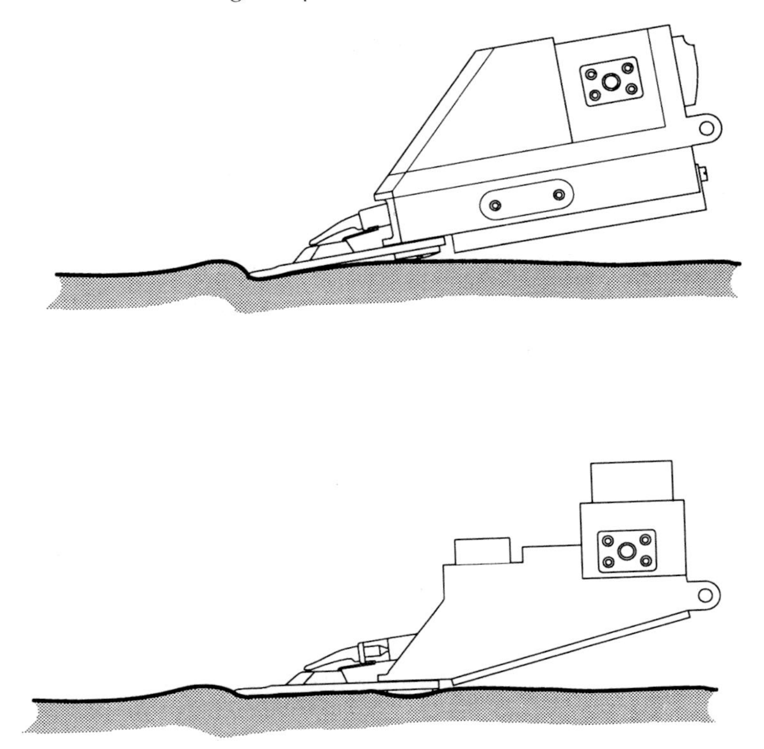

Figure 12.5 Skin wave effects—refer to text.

too broad a generalization, but the assertion can be supported by a simple series of tests.

What this means is that the skill and performance of the robot depends as much on the kinematics and dynamics of the robot arm as it does on the software.

To test this assertion, let us divide the system into elements at an arbitrary level of detail, and replace each element with a typical equivalent unit in widespread use in other industrial robots or similar devices. The effect on performance will tell us something about the extent to which each element has been specifically tailored for the task.

Take the wrist mechanism, for example. In our kinematic experiments to test different arm and wrist configurations, we found that a singularity-free wrist had major performance benefits for our task. If we substitute a conventional robot wrist mechanism, we will encounter kinematic, dynamic, control, and singularity problems which significantly reduce the performance of the robot. We can therefore deduce that the wrist mechanism is an important element of the overall skill of the robot.

The specialized features of the robot hardware provide advantages for the

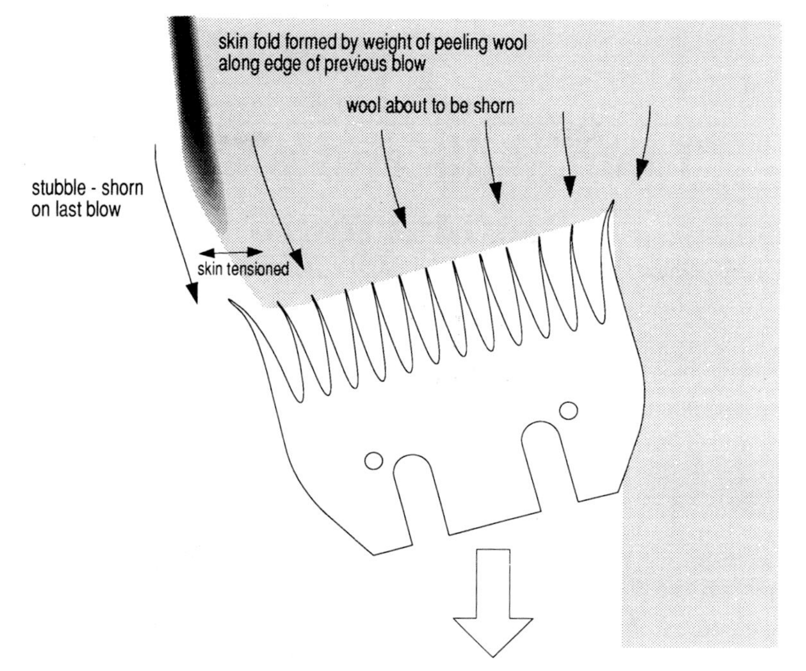

Figure 12.6 Using a pulled comb with yaw offset—refer to text.

robot which compensate partly for the huge deficiencies in information processing capacity compared to a human. A dishwasher is a good example of a specialized robot with very little information processing capacity. As robot designs improve, and electronic information processing methods improve, one might expect that less specialization of the robot hardware will be necessary for an equivalent robot performance.

When we think about other manual skills which may need to be replicated by robots, we need to recognize that we will most likely end up with specialized hardware and software. There would still be elements common to other robot applications, of course. Until there are substantial advances in information processing capacities, and techniques, this situation is likely to continue. A generalized robot capable of a wide variety of skills is not yet feasible.

13
Team skills and project management

Most engineers are accustomed to working in teams, and depending on teams to realize their dreams. Engineers need to work closely with people who have the drafting and manufacturing skills to turn sketches into working products. In turn, they also have to work closely with the client or customer who supplies the money, and the accountants and lawyers who look after the money, the client or both.

In contrast to this, researchers, particularly university researchers, like to work on their own, often in a narrow specialized field of interest. If they work in a group, it is likely to be with graduate students who are also focussed on still narrower aspects of the field. Particularly in technological research groups students are often constrained to particular approaches by the equipment or software which has been chosen for them by their mentor.

Robotics research crosses traditional boundaries and needs combinations of expertise in mechanical engineering, computer science, electrical and electronic engineering, just to get started. And the specialists who come together for robotics research can find themselves in the classic 'Tower of Babel'—unable to understand each other without realizing it. One of the challenges of the sheep-shearing project has been to bring a diverse range of interests together and to build a highly motivated team of very talented individuals. Without some team skills, we would not have been able to succeed.

On demonstrations—communicating with the client

I must now fill in the hazy space I have referred to as 'the Corporation' since the beginning of the book. The Australian Wool Corporation was set up in 1972 to operate the reserve price scheme for wool, conduct industry-wide marketing and promotion and to coordinate research and development. The Corporation inherited the long standing Wool Research Trust Fund to which all wool growers contribute a small proportion of their income (approx 1%); grower contributions are supplemented by matching contributions from the Australian government. Recently reconstituted as the Wool Research and Development Fund (WRDF) it is managed by the Wool Research and Development Council which acts on advice from several specialist committees.

The Wool Research and Development Fund has passed through several different periods of administration with progressive improvements in accounting and management practices. Yet for the duration of our project, decisions on research spending have effectively rested with the Wool Harvesting Research Advisory Committee (WHRAC) which has consisted of wool growers (farmers), economists, senior scientists and technologists assisted by Corporation staff. Usually, the wool growers have been Corporation board members or have achieved wide recognition in the wool industry. They work on the committee in an honorary capacity and are usually people with busy diaries—meetings have to be arranged 9 to 12 months in advance to make sure they can come. For most of the time span of this book the committee chairman was John Silcock, deputy chairman of the Corporation and a wool grower from the conservative western districts of Victoria. Malcolm Fraser, wool grower and Australian Prime Minister between 1975 and 1983, is a near neighbour.

Communication with this committee has never been easy. After 1986, there was a great improvement when they appointed a technical subcommittee to assume responsibility for all automated mechanical shearing research—this decision resulted in the vigorous push to complete the SLAMP manipulator. Yet there have since been communication difficulties between this group and the main committee.

I suspect these problems are common to nearly all research committees. For even the scientific and engineering members have only a distant understanding of technical detail and little time to probe deeply. In the early years they relied heavily on assessment by Corporation scientific staff; later these people had responsibility for so many research projects that they could not be expected to contribute a deep understanding either.

As a result, our progress has been judged by shearing demonstrations at 18 to 24 month intervals. And I have come to believe that successful demonstrations have justified previous, sometimes difficult, committee decisions to continue research rather than justifying new initiatives. For most of the time I directed the project, we held the opposite view—that is a successful demonstration would create the confidence needed to move the project ahead. As a result of this, we were always disappointed when the committee referred our suggestions for 'further consideration' and merely maintained the status quo after a successful demonstration. On the other hand, poor demonstrations which failed to meet expectations often resulted in changes in direction which had long-term benefits. Our two largest increases in funding, in 1981 and 1987, followed unimpressive demonstrations and months of indifferent research progress.

A time lag in the decision making process might offer an alternative explanation. Both major funding increases (in 1981 and 1987) followed outstanding technical results by 12 to 18 months. In 1980 we demonstrated repeatable shearing of most parts of the sheep for the first time and in 1985 we put on a faultless shearing performance with the new SM robot, on time. Most recently, we might have received a major funding increase a little more than 2 years after the first fully automatic shearing demonstration in February 1989. The technical subcommittee recommended doubling the research budget (for 1991/92 year) towards the end of 1990. However, the wool market collapsed and with it the confidence needed to support that decision.[1] If you subscribe to this alternative view, then the time lag is increasing—from about 12 months in the early 80's to 24 months today.

While demonstrations are useful, I believe that the real cost has yet to be seen. I have often stated that demonstrations interfere with research progress, a claim which has been vigorously disputed both within the team and by WHRAC. I have shown through this book how our research programme has been structured around four major equipment developments:

1977–1980: ORACLE
1981–1983: ARAMP
1983–1985: SM
1986–1989: SLAMP

Between 1980 and 1982 we refined the core software for robot control: surface models, cutter steering and repositioning. For most of the time, however, the equipment needed for shearing 95% of the sheep under automatic control was the main priority. Only in 1989 was the focus shifted from equipment to improved shearing time, and later, reliability. I thought we had achieved a real breakthrough in 1989 when the committee asked us to prepare a five year plan for finishing the research. At last, it seemed, the push for demonstrations had been set aside. Yet within 9 months priorities returned to demonstrations once again.

There is a distinct contrast between designing and building a new machine, and researching the software needed to operate the machine. Once the machine has been built, it is apparent to all who come to see it. It has physical form, makes noises, moves and usually impresses. Yet the software is invisible, intangible, and often for the research adviser, beyond comprehension. The ultimate test of months of research on motion control software may be just two or three shearing blows on just one part of the sheep, and to most visitors that is singularly unimpressive. So the software then has to be extended to provide all the blows to shear a sheep. In the process there is some refinement: fine tuning , a few bugs removed. At the same time months of shearing trials to make all the adjustments and refinements to shearing procedures *to guarantee as far as possible that it will work on the demonstration day*. The team members who might have been thoroughly testing the basic techniques have been diverted to what is in essence a theatrical presentation.

Apart from the equipment and software, attention to detail is crucial. Rehearsals are essential, preferably in front of a friendly audience. With experimental equipment, Murphy's law and its corollaries apply strictly: something will go wrong. An audience which cannot share the deep technical knowledge of the players on stage learns more by perhaps unconscious observation of the players' reactions to problems. If the staff presenting the demonstration handle problems with an air of routine confident familiarity, and can immediately

[1] Since writing, the Corporation and its R&D activities have been restructured and may be organized quite differently in the future.

explain the problem in simple language, and different staff members provide *consistent* responses later, the audience will be reassured and will carry some of the confidence away with them. Confidence and familiarity can only be instilled by practice and careful attention to all the small details.

Theatre it may be, but there is still a great deal of pride in a successful demonstration which has taken months of work to prepare. It is an invaluable opportunity to show farmers, ordinary farmers far removed from Corporation status, the possibilities. Videotapes have been popular and sought after.

Yet the final cost is that the invisible software work has been deferred, again and again. In the most recent review,[2] Victor Scheinman pointed to the ultimate problem which needs to be solved—biological variability. We can now shear the average size Merino sheep in average to good condition, with reasonable reliability, and that is a significant achievement. Yet we must be able to shear all the other sheep too, with reasonable reliability. There are still research questions to answer and years of trial work to achieve that.

I have to accept a share of the responsibility for deferring work on sheep variability since I have always been keen to avoid rigorous testing and evaluation when possible. And it is more satisfying to be able to work on biological variability knowing that complete sheep can be shorn in a reasonable time. I hope that relating our experience can be helpful to others.

We have now adopted the practice of planning at least one on-going research activity in which complete shearing tests are regularly repeated, typically weekly. This way, we can maintain all the practice, attention to detail and confidence which is needed to present successful demonstrations.

Team structure

This is an unusual project for an Australian university. Team members are supported by the WRDF—none are university academic staff (except for the officer in charge—a nominated senior member of the academic staff who represents the interests of the university in project administration). The university provides accommodation and basic facilities needed for research: libraries, workshops and skilled technicians, laboratories, computing support and the university community. The WRDF grant covers major overhead charges. Staff have no security of tenure beyond each twelve month funding period. With the WRDF unable to commit funds over a longer period, it has not been easy to attract and retain research staff. Good team skills have been essential and special benefits have been arranged to compensate for the wide disparity in job security and other benefits enjoyed by tenured academic staff.

Currently, the team has 17 full time and part time members (not including students) from nine different countries with six different native languages. This diversity of background has been a significant advantage in fostering creativity.

[2] A formal technical review by V. Schienman, A. J. Holzer and B. Hendy was commissioned by the Corporation in 1990. The detailed results of the review will help the Corporation choose options for future commercial development.

Team members represent the following technical skills:

mechanical design, mechanism kinematics, dynamics,
control systems, mechanical vibrations,
tribology and surface engineering, hydraulics, pneumatics,
RF and microwave electronics, instrumentation electronics,
fibre optics, microprocessor systems, electric motor controls,
real-time software, operating systems, AI software,
Fortran, Pascal and C programming, machine vision,
wool fibre technology and severance, shearing, wool growing, animal behaviour, physiology,
mechanical toolmaking, fitting, turning, electronics assembly, printed circuits,
design drafting, CAD, hydraulics and pneumatics fitting.

There are also a range of non-technical skills which I consider to be of equal importance to technical skills:

project management and control, interpersonal skills,
presentation, public relations, printed communication,
video recording and editing,
accounting, budgeting, economic modelling and forecasting,
sub-contracting, documentation control and indexing,
secretarial, parts and equipment procurement.

Nearly all of these skills are available to the team through individual team members or collaborating members of the university staff in one or two instances. How then does one focus this diversity towards a single objective?

Objective and support

The wool industry has provided a clear goal: to provide an economic alternative to manual shearing. This clear goal has helped to set priorities for solving problems.

During those difficult times when there has been no easy solution for a technical problem, a clear goal has helped us to choose one. Often, we have been able to bypass difficult problems without feeling that 'we ought to have researched that more deeply'. I contrast this with the vague aims of many robotics research groups. Take, for an example, intelligent mobile robots. The typical aim is to reproduce the behaviour of a (supposed) intelligent human operator driving a vehicle through 'complex unstructured environments' or 'the real world'. First, the goal is a distant one, and the number of alternate paths seems vast in comparison. Second, it is as hard to perceive progress as it is to know which direction to choose. It is not so easy to bypass research issues as every problem can assume equivalent significance. Finally, a wide range of applications needs to be considered—rescue operations, warfare and natural disasters are frequently cited and none is an easy situation even for human drivers.

Who needs robots in unstructured environments anyway? It may be facile to ask, but this question raises another important issue. Shearers and woolgrowers both need robots, though for different reasons. Team members respond to their needs and the personal contact enhances motivation. It helps to know that real people are interested in the results of your work, not just a Corporation. Local growers have provided strong support for our work, right from the start. Apart from personal contact and political support they have provided practical help with sheep and wool issues. They have also helped to explain our work to the wider farming community and have often called for patience and understanding when we needed it.

The support of the Corporation is best appreciated in terms of funding which has been continuous since 1976. More than once we underestimated the determination of the Corporation to push ahead with research in the face of difficulties. The funding has been adequate without being lavish, and there are few organizations anywhere with as much persistence.

Foundations of teamwork

We have found that a robotics research team is similar to any other team. The team spirit and performance is based on a triangle of respect, trust and confidence, linked by sound communication skills.

Ideally, each team member respects other members of the team for the skills which they contribute, and for their understanding of the goal. Each team member trusts the others to use their skills to the best of his or her ability. Finally, confidence, both in oneself and the team, is fostered by the first two.

Trust and confidence must also embrace the relationship with the client—the Corporation. Communication with the wool industry has also been a most important factor in maintaining this.

Communication

Respect, trust and confidence can only exist where there are adequate communication skills to build and maintain them. Communication is a basic element of teamwork, yet it is so often ignored or undermined, particularly in technically specialized fields.

Listening is the hardest communication skill to learn for many of us, yet it is essential. Many people are surprised to find that listening is a skill which can be improved through training. Bolton (1979) and Fisher and Urey (1981) provide valuable background and practical information in communication and negotiating skills.

Technical terms often undermine communication between team members. Yet they are usually invented in the first place to do just the opposite. The intention of technical terms is to find words or phrases which have unambiguous meanings so everyone (familiar with them) can understand exactly what is meant. Yet they are often used unintentionally when communicating with people who have neither the familiarity nor technical background to understand them.

Technical terms are often used to define territory—the area for which a team member feels responsible. An electronics expert may refer to his own work in terms of 'biassing resistors balancing parasitic loads'. A moment later, the same person might be heard referring to 'the hole behind the red box' which the mechanical designer calls 'the main carriage support bearing access port'.

Robotics theory can be equally obscure. A 'Non-holonomic, high lateral mobility vertical load bearing device' can be called a 'wheel'. The longer phrase is meaningful to a mathematician specializing in robotics because it defines a much wider class of objects in a special way, but to most of us, even technically oriented people, it is baffling. Technical specialists like to preserve territory, and even cover uncertainty, by using their own technical jargon. And in any case, no research organization wants to be accused of reinventing wheels.

Trust and respect can be reinforced when team members feel confident enough to challenge specialized terms—'Can you please explain what you mean by that'. But it is not easy for anyone to do that. Most people believe that they will seem to be ignorant if they do not know what a term means, particularly researchers who think they are expected to know. So they carry on as if they understand the term, hoping that they will pick up enough clues to understand what is meant. Fertile ground for misunderstandings indeed. When the barriers created by technical jargon are pierced, there are great benefits. Respect for one another is enhanced, further reinforcing trust and confidence.

I have already referred to instances of communication problems, particularly between technical experts in different fields, but it is worth repeating some of the lessons learned.

First, to transfer skills from shearers to the robot, we learned how important it is for the shearer to understand how the robot works. Thus, the communication and understanding has to be a two-way process: the researcher must come to understand the shearer's skill, and the shearer must come to understand how the researcher proposes to mechanize the skill.

Second, the only safe assumption is that technical terms, both commonplace and obscure, will be misunderstood just as any words in human communication can be. Human language is context dependent. Pretending that technical terms overcome this problem is likely to cause problems. One way to avoid this is to communicate in plain and simple language. A further measure which helps is for the listener to echo the message using different words until the speaker is satisfied that it has been understood.

Since then, I have become aware of a further more cogent advantage of describing technical concepts in simple, everyday terms. It is hard work—more difficult sometimes than inventing original ideas. Yet it leads to a simplified understanding, which allows simplification of the original ideas. And simple ideas catch on.

To describe a technical idea in simple terms, one often has to eliminate detail. Then, one can ask whether the detail is needed in the first place. The value of this process is that the essence of the idea can be rescued from the clutter of detail, much of which may be neither necessary nor desirable.

As long as robots are seen to be complex machines, they will be regarded with suspicion by most people. They will be adopted much more quickly if they are seen to be simple machines, preferably understandable by six-year-olds. Describing them in simple terms is a first step towards this.

Other factors

Communication skills linking respect, trust and confidence provide a firm foundation for a team. We have also found that other factors help to maintain a good team:

(1) a group identity, expressed by a logo (T shirts and sweaters carrying the Shear Magic logo are always in strong demand from team members, friends, relatives and visitors);
(2) publicity, in both learned and popular press;
(3) regular, informal sporting activities with other nearby groups . . . e.g. 5-a-side soccer, squash, cricket, etc;
(4) external competition;
(5) freedom from fear (job security) and strict numerical goals;
(6) regular tea breaks at which all staff meet on an equal basis;
(7) birthday and other special occasion celebrations, occasional social events;
(8) subsidized parking fees;
(9) superannuation (subsidized staff pension scheme);
(10) pay levels equivalent to equivalent industrial research groups (in lieu of longer term security);
(11) exclusive contracts ... no out-of-hours consulting or other commitments;
(12) access, in groups, to training courses;
(13) participatory training workshops and seminars.

Team building workshop

One of the most successful steps taken to improve team skills was a team building workshop presented by a professional facilitator. Training of this kind is increasingly common among management groups in industry and the public service, but it is almost unheard of in a university research department.

Most of the team members were both curious and apprehensive to begin with. By the end of the two day (full time) workshop, most were exhausted! One person used to hard physical labour said that it was the hardest day's work he remembered. By the end of the workshop, team members had a better understanding of the attitudes of their colleagues, and appreciated that communication skills could be a great help. Team members had the opportunity to express personal frustrations in a safe atmosphere, knowing that respect and privacy would be protected by the facilitator.

It was a very beneficial experience which enhanced respect among team members, and communication skills, particularly listening skills. It was followed about a month later by a shorter workshop on conflict resolution.

Project management

In the contemporary climate of accountability, planning and forecasting, research plans and budgets have been subjected to meticulous examination and our progress has been closely watched. With the Corporation, we have tried several approaches to project management and control and several of these have proved to be worthwhile, though not always for the expected reasons. Chapters 7, 8 and 9 contain practical observations.

Critical path methods

We used critical path methods to plan and monitor progress during the construction of our major new equipment items—SM and SLAMP. They were useful, though I believe that the discipline of planning and monitoring progress is as important as the choice of technique. There are several computer packages which can be used, though the two we have used are tricky to apply where human resources are strictly limited and each person has several divergent daily responsibilities. *Timeline* (which runs on IBM-PC computers) and *MicroPlanner* (Macintosh) provide basically similar facilities; both are difficult to set up to provide realistic estimates. I prefer my own programs which Michael Ong and I wrote for the SM robot construction phase, but the results from a 'commercial' package seem to carry more credibility.

Some other robotics researchers have been amazed to hear me talk about using critical path planning methods. How can one plan research?

The answer is that it is possible to plan the work needed to do the research, while accepting that the outcomes are unknown and hence the future needs to be replanned when necessary. Equipment construction, trials, maintenance, meetings, report and paper preparation, even reading, can be allowed for in plans. And the discipline of thinking about plans is the most valuable aspect of the process because the numbers calculated by the computer package reflect that thinking.

As in all project management, progress is often hard to measure. Particularly in research. After a heated discussion at one meeting, I exclaimed: 'Listen, the time we make the most progress with our research is when nothing seems to be happening!' It is often hard for project administrators to know whether to believe engineers' claims on progress. A good project controller will come to know when the engineers are deceiving; not by being dishonest, but by unconsciously ignoring difficulties which they prefer to face another day. The controller is often perceived as a 'bogeyman' who will be cross or disappointed when progress lags behind schedule. Yet the controller can be a great help. Engineers, like anyone else, like to talk about their work to anyone who will listen and understand; this is often the project controller's most valuable contribution.

I believe that a useful way to avoid problems is by focusing on 'observable outcomes': results of activities which can be observed by all. In design and construction this is usually not difficult—the outcomes can be sets of drawings, reports, parts, assemblies and working machinery. Even software can be monitored by counting lines of code completed. Research outcomes can be reports, but I prefer experiments and trials which can be watched.

We have claimed, as I am sure others would, that plans impose too much rigidity in research. In planned research, one has to negotiate a change in the plan to follow up a new idea. Why impose such a restriction? Surely opportunities can be missed by following a strict plan.

Now I believe that this is a misconception. A good research plan will allow enough time to follow up new ideas, for reading, conferences, seminars, discussions. The process of planning is the important part—the time spent thinking about time allocation for all the 'indirect' activities will repay handsome benefits by avoiding unrealistic expectations. Our greatest administrative difficulties have mostly come from differences in expectations between ourselves and the Corporation, and good planning helps to create realistic expectations. Life can be much more relaxed by planning a schedule which allows spare time for the unexpected, and time to enjoy what we do.

Documentation

Many administrative and support jobs in a research team result in few lasting achievements. The cleaning staff return every week to find messy floors, neatly typed memos are soon filed in circular filing bins and carefully maintained budgets soon become irrelevant when the winds of politics change.

One routine and mundane job which has created a lasting achievement is the indexing of our documentation. We have maintained on-line indexes of most of our documentation: reports, technical notes, commercial catalogues, papers, software notes, photographs—even the contents of working files of notes and papers. Naturally, we do not have 100% coverage and there is a lag in keeping it up to date. But it has saved countless hours searching for reports and notes, and provides a sense of security in knowing that all the information gathered in the project can be accessed easily. Once a technical reference has been found in our index, we know it is available; we do not have to wait for photocopy requests or inter-library loans to be arranged.

At one stage, we indexed the contents of journals we received—titles, authors and abstracts. Now we subscribe to an on-line database of robotics publications.

Michael Ong and I refined a database program based on a student exercise, tidied it up and made it more robust. Partly because the commercial database software for our computer seemed too expensive and partly because I wanted a program which would provide links to the robot operating software code so that the reasons for software changes could be conveniently recorded and retrieved. I also wanted to use a free format for data entries, and a retrieval system tolerant of minor spelling and interpretation errors.

The program retrieves index entries on the basis of a search key which allows for Boolean combinations of words. I could request a search of software problem reports:

TOM mem* protect error* or software crash* and not repos*

* at the end of a word allows for any number of other letters—e.g. crash, crashes etc.. Combinations of words without operators between them imply **and**, thus 'software crash*' would be considered found within the following text: ' . . . the robot crashed against a pile of wool during software tests'

The program allows for minor spelling errors, at the expense of retrieving some irrelevant entries. Searches can be confined to authors, titles, keywords, etc. Each entry has a loose format, a typical one for a work file being:

$n JPT1304-WF-02/91 (green, 1 cm)
$t C3R capacitance sensor trial notes
$w JPT's filing cabinet
$a J. Trevelyan
Trial reports t348, t349, t351, t352
Listings capcal.ftn, sensor.ftn, csens.ftn
chart recordings
spare copy of sensor documentation

Each line starting with a $ sign provides loosely formatted data of a known type—classification number (first line), title, whereabouts, and author. The remaining text is assumed to be comments or abstract information. Entries can be as long or short as desired.

At the time of writing our indexes comprise about 3 MBytes of text which has been accumulated in about 8 years. The brief description of the physical appearance of the file is a recent innovation, and helps when the file is missing from its designated resting place. The data base is maintained with a text editor or word processor by our loyal clerical support staff.

A personal note of thanks

The experiences related in this chapter were obtained by leading an exceptionally talented team of colleagues to whom I am most grateful. They suffered my mistakes with patience and great loyalty.

The team (April 1990)

Edward Tabb—Project Director (1989–)
Responsible for direction of the project, financial control, project management, public relations. Appointed to lead the project into a commercial development phase. Background: 25 years experience in industrial research management and commercialization.

Professor Brian Stone—Officer in Charge (1981–)
Represents the university interest in the project and asks difficult questions when most needed. A source of encouragement and constructive criticism.

Research staff

James Trevelyan—Technical Consultant (40% full time) (1977–)
Technical director (full time) until 1989 with responsibility for research direc-

tion, financial control and project management. Research interests include control systems, shearing software, robot dynamics, kinematics, and machine vision. Background: mechanical engineering, 20 years experience in control systems, digital electronics, computer software, human factors, aerospace, mechanical system modelling and simulation.

David Elford—Mechanical Designer (and Deputy Director) (1981–)
Responsible for construction of robots and associated equipment, research and development on sheep restraint and manipulation, cutter design. Supervision of mechanical research engineer, draftsperson, workshop technicians. Background: 38 years of mechanical and machine design work, with particular focus on automation plant, handling of food items varying in size and consistency, fruit canning equipment, etc.

Peter Kovesi—Research Engineer (1981–)
Responsible for software development and machine vision research, and supervision of research engineers. Responsible for numerous original developments including SM robot (co-inventor of ET wrist), robot control software, surface measurement software, etc. Background: mechanical engineering degree, 10 years experience in robotics research.

Jan Baranski—Electronics Engineer (1982–)
Responsible for all electronics and sensor hardware, and for supervising electronics technician. Background: 15 years experience in digital and semiconductor electronics design and manufacture. Interests include microprocessor development and interfacing, RF electronics and sensor design.

Darryl Cole—Shearing Consultant (60% time) (1988–)
Responsible for expert advice and knowledge on shearing, combs and cutters, sheep and wool. Present during shearing trials, analysis of videotape records. Background: 15 years professional shearing and shearing instruction in New Zealand, Australia, USA and Europe.

Lotta Sundqvist—Research Engineer (1990–)
Responsible for design and construction of mechanical (strain gauge) sensing devices, pressure transducers, and analysis of hydraulics system. Background: graduate mechanical engineer.

Steve Ridout—Research Engineer (1987–)
Responsible for detailed software design, development, testing and documentation. Background: computer science, contract programming.

Graham Walker—Research Engineer (1986)
Responsible for software developments, research in machine vision. Also other tasks such as stress and stiffness analysis of mechanical structures, statistical analysis of data. Background: mechanical engineering, 4 years robotics research.

Support staff

Wai Chee Yao—Project Administration Controller (80% time) (1982–)
Responsible for project administration, including financial control, budgeting, accounts management, salaries, operation of project management computer system, collection of technical progress details, maintenance of technical index and records system. Supervision of administration clerk, and casual assistant. Background: accountancy and administration, 15 years experience.

Lynette Lynn—Project Secretary (80% time) (1982–)
Responsible for secretarial duties for technical director and team members. Record keeping, technical and administrative typing and correspondence, public relations, requests for information, videotape and camera operator for trials. Background: secretarial.

Virginia Skipworth—Administration Clerk (50% time) (1987–)
Responsible for accounts, invoices, purchasing and expediting ordered items. Also maintenance of technical index and library.

Andrew Whitehead—Mechanical Design Drafting (1986–)
Background: 15 years of mechanical design drafting experience

Ian Hamilton (Senior technician, 1979–), Dennis Brown (1985–) Mechanical Technicians
Background: 20 years experience of mechanical toolmaking in aircraft and specialized machine manufacturing industry.

Bela Luik—Electronics Technician (1987–)
15 years experience in electronics maintenance and repair, especially TV and VTR equipment, design and specialized test equipment construction.

Brunhilde Prince—Shepherd and Casual Assistant (50% time) (1990–)
Feeding, transport of experimental animals; collection and delivery of parts and supplies; cleaning of laboratory and animal facilities.

Principal past team members mentioned in the book

Roy Leslie—Research Engineer (1976–1981)
Responsible for initial research, then mechanical design of ORACLE robot and HOMP manipulator.

Stewart Key—Research Engineer (1979–1987)
Responsible for software development, economic modelling and forecasting, shearing trials, project accounts. Created software package to control ARAMP manipulator.

Zbigniew Lambert—Mechanical Designer (1981–1985)
Responsible with David Elford for design, construction and testing of SM robot and ARAMP mechanisms.

Michael Ong—Research Engineer (1982–1985)
Co-inventor of ET wrist and SM robot concept, design and analysis of SM robot—kinematics, dynamics, control system and performance. Originator of numerous developments in sensing, mechanisms and control.

Michael Crook—Electronics Engineer (1981–1982)
Co-inventor of vibration contact sensor, and commenced improvements to capacitance sensing devices.

Robyn Owens—Research Officer, Mathematics (1981–1983)
Developed statistical methods for predicting sheep surface models.

List of contributors

It has not been possible to mention the individual contributions of every person who has helped with our work in the last 15 years. I hope the following list includes all who have contributed, if not I apologize. Some individuals have contributed in a number of ways, but they have only been named once.

Principal Staff:
- David Elford*
- Stewart Key†
- Roy Leslie†
- André Roszko†
- Ed Tabb*
- James Trevelyan*
- Zbigniew Lambert†
- Jan Baranski*
- Peter Kovesi*
- Graham Walker*
- Steve Ridout*
- Robyn Owens†
- Michael Ong†
- Lotta Sundqvist*
- Perry Cohn†
- Michael Crooke†
- Michele Della'Aqua†

Head of Department, Officer in Charge:
- Professor David Allen-Williams† (Prof. A-W)
- Professor Bob Brown
- Professor Brian Stone*
- John Appleyard†

University Staff:
- Dr Jim Blair (Mechanical Engineering)
- Dr Duncan Steven (Electrical Engineering)
- Professor David Lindsay (Animal Science)
- Ross Armstrong (Animal Science)
- Susan Lewis (Animal Welfare)
- Peter Johnson (Research Administration)
- Eric Hawkins (Media Services) (dec'd)
- Gerald Wright (Media Services)
- Colin Murphy (Media Services)

Workshop Technicians:
- Alan Bunn
- Ian Hamilton*
- Dennis Brown*
- Renato de Pannone
- Paul Marsh
- Mike Cowell†
- Rob Greenhalgh†
- Bela Luik*
- Roberto di Biaggio†
- Derek Goad†
- Alan Rogers†
- Noel Hines
- Stewart Howard†
- Don Baker†
- Albert Ter Horst†

Support Staff
- Lynette Lynn* (Secretary, documentation)
- Virginia Skipworth* (Accounts, documentation)
- Wai Chee Yao* (Administration, project control, documentation)

Brunhilde Prince* (Animal care)
Trent Smith† (Animal care)
Alison Harding† (Animal care)
Beth Spencer (Secretary)
Morag Bennett† (Secretary)
Paula Elliott† (Secretary)
Sheena Carter† (Secretary)
Lyn Moss† (Secretary)
Krystina Roszko† (Drafting)
Andy Whitehead* (Drafting)
Dan Pitic† (Drafting)
Jeremy Price† (Accounts)
Matthew Dwyer† (Animal care)
Debbie Skipworth (Animal care)
Karla Pavlinovich (Animal care)
Steve Ramsay (Animal care)

Australian Wool Corporation:
Neil Evans (R&D Department)
Dr. Roger Thiessen (R&D Department)
Peter Silk (R&D Department)
Paul Hudson (R&D Department)
Brian Stevens (R&D Department)
Peter Booth (R&D Department)
Bruce MacKay (R&D Department)
Alan Richardson (R&D Department)
David Hindle (R&D Department)
Lyn Coyne (R&D Department)
Rodney Jones (R&D Department)
Paul Wong (R&D Department)
Russell Flack (R&D Consultant)
Tom Lodge (Fremantle office)

WA Liaison Committee:
Norman Lewis
Marjorie Lewis
Tony Gooch (AWC Board)
Pat Moore (AWC Board)
Frank Merry
Ian Campbell
Dexter Rick

Wool Harvesting Research Advisory Committee:
John Silcock (Deputy chairman of AWC)
John Allen
Ken Langley
Jan Kolm
Len Whelan
Ken Bishop
Michael McBride
Ernest Barr
Ken Whiteley

Cambridge University:
Dr. Tony Nutbourne†

University of Adelaide:
Prof. Bob Bognor
Rod Bryant
Henry Nissinck

University of Melbourne:
Bruce Field
Paul Burrow
Alastair Mackenzie
John Bryson
Bob Freeman

Shearers:
Graeme Tyers
Neville Marney
Don Munday
Allen White† (AWC Shearer Training)
Darryl Cole*

Woolgrowers:
Ian Peacock
Humphrey Park
Claude Mills
Bevan Taylor
Bill Creswell

Students:
Ralph Martin†
Piero Velletri
Reg Williamson
Pat Foale
Huan Li
Garry Junkin†
Frank Wittwer
Tim Sharrock†
Mark Nelson†
Malcolm Reeson
Ian Howard†
Harold Hochstadt
David Devenish
Greg Kirkaldy
Justin Marwick
Ralph Cammarano†

Merino Wool Harvesting:
John Baxter
Lance Lines

Chris Abell
Allan Wallace
Rod Smith
Malcolm McGuinness
Kevin Rogers
Jeff Groves
Ian Atkinson
Stewart Page

CSIRO:
David Henshaw
Roman Malaniak
Peter Lamb

Patent Attorney:
Trevor Beadle

PATSCentre:
Jim Brown

Others:
Victor Scheinman
Alex Holzer
Barry Hendy
Adrian van den Avoort

14
Prospects for robot shearers

Shearing robots work, and with a modest amount of further research could shear sheep reliably and fast enough to make a handsome operating profit. Yet there is still a great deal of scepticism. In the end, the prospects for robot shearers depend far more on politics and economics than on technical factors.

Rather than attempt to make predictions, or point out all the political obstacles to be overcome, I would prefer to provide some answers to the questions we are most commonly asked. The questions reveal a little about scepticism and uncertainty; the answers fill in the remaining parts of our vision of what the future could provide.

Will shearing pills make robots obsolete?

Biological defleecing works. For several years following the announcement of EGF in 1981 (p. 174), CSIRO pursued genetic engineering techniques for synthesizing the compound. Working in collaboration with the prominent agricultural chemical company, Coopers, enough EGF was available by 1987 for meaningful field trials. In 1989 Coopers announced their brand name 'Rybuck' at the World Sheep and Wool Congress in Hobart (Bligh 1989). Soon farmers would be able to give their sheep an injection and the wool could literally be peeled off the sheep's back six weeks later.

Early attempts to induce a partial break in the wool fibres to prevent the main fleece from falling off before harvesting were unsuccessful. The variability in sheep response to EGF was too great. Even allowing for sheep breed, age, sex and weight, and administering the compound under laboratory conditions, some fleeces fell off and others showed only patchy signs of weakening. CSIRO increased the dose to the point where a complete break in the wool fibres could be guaranteed. By the time 'Rybuck' was announced, the sheep were to be fitted with elastic nets, like hair nets, to keep the wool in place. This was essential, not only to retain the valuable fleece, but also to allow a new coat of stubble to grow underneath to protect the sheep after the fleece is removed. An exposed nude sheep will suffer fatal sunburn or chilling .

Coopers are planning to release 'Rybuck' before long. Biological shearing could be attractive, especially for special flocks of sheep which are kept in sheds all the time. For sheep in open paddocks, the cost and effort to muster the sheep for dosing and fitting the nets offsets the ease of removing the wool.

There are still suspicions of side effects and further research is needed. The compound was found to be non-selective in its effects on the sheep's skin which, in a biological sense, continues through the digestive and reproductive tracts. Here some side effects have been found. There are also problems with fitting the nets, and with keeping the fleece wool in good condition inside the nets. The wool has a tendency to collect under the belly and form a solid mat of worthless felt. The possibility of long term meat residues is a further concern.

Research is continuing; CSIRO hopes to modify the compound to make its effect on wool follicles more selective, and to reduce the risk of side effects. Biological defleecing, if successful, could eliminate skin damage due to cuts in shearing and promote closer attention to animal health and well-being.

How have Merino Wool Harvesting progressed?

MWH completed their first robot in 1984 and commenced their own shearing trials. By 1986, they had incorporated several refinements in a second robot (to shear the other side of the sheep) and were achieving reasonable shearing performance on the back and side fleece. They shared our apprehensions during the 1986 review by Aptech (p. 255); they felt vulnerable and commissioned their own competing review at the same time. In the end, they were reluctant to demonstrate their robot. Time after time they postponed shearing until they were forced to meet a final deadline and their demonstration fell short of their claims.

The Corporation ended their financial support the following June and accelerated our work on the SLAMP manipulator.

MWH were not finished. Fearing the AWC would withdraw their funding, they had already approached Elders IXL for more funds. This appeared to be a move of last resort, because it meant relinquishing financial control after so many years. Elders responded with a promise of \$5 million over $2^1/_2$ years to build a field trial unit—but there had to be a demonstration at the end of it.

The Elders decision astonished both us and the Corporation. There was some advantage to them; there were significant new tax incentives for research and development by which they could reduce company tax payments. They had already injected substantial funding, and to withhold funds would mean writing off the money they had already put in. Elders, under the chairmanship of John Elliot, were riding the crest of a wave—huge take-over deals had made them Australia's largest brewers and they were making huge profits.

John Baxter retired and Rod Smith took over as managing director, though he commenced without any technical staff. Nearly all of the early work had been carried out by consultants who were by now reluctant to rejoin. Key engineering staff were recruited and the organization grew rapidly. When I first met Rod in Sydney at the end of 1988, there were more than 40 people working for MWH. Now they were self-contained, there was little contact between us for over three years.

The magnitude of the funding worried us. We had little confidence that they could succeed, given their past history and their failure could affect our prospects. An investor might well ask how we could succeed if MWH had failed with more than \$10 million from Elders and other sources.

By early 1990, results began to appear. Two new robots had been built with hydraulic and electric actuators and a control computer based on transputer chips. Impressive film of the robots in action appeared on commercial television, with Rod Smith confidently predicting sales of transportable twin-robot units.

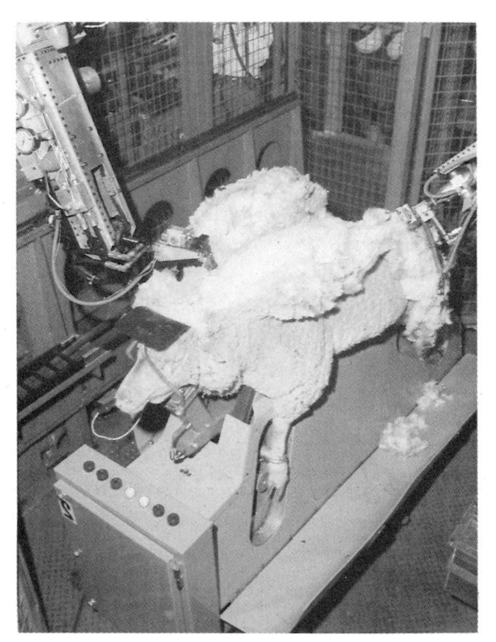

Figure 14.1 Twin Merino Wool Harvesting robots in action, 1990. The sheep was loaded on to the supporting cradle outside the robot enclosure. The back and side wool was shorn by the robots; the rest was removed by two shearer/wool classers. Three or four people were needed to load sheep, shear, class and press wool. The shearing cycle time was between 105 and 120 seconds.

We were quietly sceptical, and still for the same reasons. Basically, two robots were being used to shear wool off the back, and upper sides of the sheep. Three or four men were needed to load sheep on to the cradles, retrieve shorn wool and shear the remaining wool. At its most basic level, two robots and three men performed the work of two shearers and a shed hand. While the machinery might cost less with time and manufacturing improvements, the labour cost would always make the economics marginal.

In 1990, the Elders wave crashed and spent its energy. John Elliot organized a highly geared management buyout which left the group staggering under a huge burden of debt. Assets were sold in a collapsing market as the financial excesses of the eighties brought their inevitable consequences. When Elders decided in July 1990 that they had no more money, MWH were just months away from completing a mobile field trial unit. Elders by now were starved of cash, but in complete control and no one was interested in speculative high technology investments any more.

Some months later we visited MWH for the first time in five years to see the results of their work. With a completely new team of engineers they had created a prototype mobile shearing plant and had demonstrated a continuous shearing cycle time of less than two minutes for each sheep. Four men were needed: a sheep handler to load sheep on to cradles, two shearer/wool classers and a shed hand/press operator. The robots removed the back and side wool and the shearers removed the rest. It was an impressive achievement particularly as they had set up all the foundations for a manufacturing phase as well. The robots, based on the original design by Abell and Wallace, the computer control system, and the sheep and wool handling equipment had all been redesigned and built into a mobile semi-trailer rig (Smith 1990; Rogers 1990). With justifiable pride, they discussed their longer-term plans for improved robots and fully automated shearing.

The core group of engineers remain. They are seeking consulting and design work now, hoping to stay together while new finance can be arranged.

The two new robots built with the Elders money are mothballed. They live on in television film, used as 'high tech' film clips by a commercial science and technology television show.

We were all sad at their passing. They were never short of enthusiasm. They helped us as a competitor; we had to perform to meet their threat. The loss of accumulated expertise which could have helped us is also sad. But their demise came at a time of financial stringency for everyone, particularly the wool industry, and the Corporation was in no position to offer them work with us.

How much will it cost?

I cannot give an accurate answer because no decision has been taken to build machines for commercial use. Further, the Corporation has not provided us with their cost estimates. Until about 1986, Stewart Key estimated equipment costs for us and maintained an economic model of a shearing system which would predict operating costs, profits and annual investment return over the life of the machinery. By then, it became clear that our estimates were given little credibility and the Corporation has been responsible for all economic evaluations since.

What we do know is that the shearing time and machinery cost are the most critical performance factors—the more sheep shorn per day, the better the economics. And our earlier experience with economics revealed that factors such as interest rates, depreciation, and allowable manufacturing profit margins were highly significant and hard to predict. For example, the effective interest rate depends on the perceived risk, and the manufacturer's price depends on political decisions as well as the actual manufacturing cost. The price which can be charged for shearing may depend on the labour rates at the time and they have been remarkably difficult to predict. Typically, shearing rates are static, or even slightly declining in real terms, except for certain periods when marked rapid increases occur. One such period between 1973 and 1975 stimulated this research programme, and another occurred between 1987 and 1989.

Shearing quality is also an important factor. A reduction in skin cuts could improve animal health and increase the value of the skins. A reduction in second cuts (long stubble shorn twice) could increase the yield of fleece wool which sells for the highest prices. So far we have not been able to estimate the likely gains, but the impact on economics could be considerable.

The cost of injuries to shearers is an unknown factor. Few shearers claim compensation yet anecdotal evidence suggests high rates of occupational injuries. In terms of animal handling alone, a shearer is regularly heaving 50 to 60 kg of meat several metres every four minutes on average—equivalent to at least 6 tonnes per day. For crutching that figure rises to 25 tonnes per day. Recently governments imposed lifting limits to reduce the incidence of costly occupational strain injuries—shearers probably exceed the limits regularly. Both shearers and wool growers agree that the physically stressful part of shearing needs to be eliminated if possible—and robots could provide a solution.

Whatever the extent to which economic considerations influence decision making, the Corporation has continued to support the project with high priority for funding. I have presumed that the economics must seem favourable in the long term.

How fast can it shear?

At the time of writing the typical shearing cycle takes about 17 minutes with our laboratory equipment of which only about 4 minutes is spent moving the cutter forwards on the sheep. We need to reduce this shearing component to about 2.5 minutes, and eliminate as much of the remaining wasted time as possible. We are happy that improvements now being made will allow the shearing time to be reduced sufficiently, which is why the main research emphasis is being changed to reliability.

Can it shear all sheep—what about rams?

In principle, yes. Contrary to the popular view that Australian sheep stations have hundreds of thousands of sheep spread out on endless plains, most farms have flocks of between 2000 and 4000 sheep. Robot shearing equipment needs to be operating for as much of the time as possible, and we have designed machinery to suit farms with more than 3000 sheep. Hand shearing is likely to be more economic for stragglers, rams and smaller flocks. We can shear around the horns and pizzles on wethers but rams may require strengthened equipment for mechanized handling. However this may be worthwhile, even for hand shearing, as rams are the largest sheep for shearers to lift.

What will the unions say?

I mentioned earlier that there was a bitter dispute over the allowable width of the shearing comb in the early 1980's (p. 142). With robots and automation being a far more significant issue, one might expect some industrial action.

The shearing industry is less unionized than it has been. The principal union is the Australian Workers Union which shearers formed in the 1890's and have dominated since then. Though they form a tiny minority of the 120,000 members, their influence on the union executive far outweighs their numerical strength. Membership among shearers has been declining and a perceived need to raise the union profile may have motivated the wide comb dispute in 1982 and 1983. Yet the petty nature of the dispute and the overt intention to exclude West Australians and New Zealanders (non-union) with more efficient work practices put many shearers offside.

Until recently, the union has taken a sceptical attitude to shearing robots. In a surprising article in the Australian,[1] the union secretary Errol Hodder accepted that the introduction of robots and biological shearing was inevitable and that the union had to accept and adopt new technology. Like many unions in Australia, they favour new technology which improves working conditions and productivity, provided that the interests of the present and future work force are taken into account.

Certainly, we think that shearers are needed to operate the machinery effectively. The shearer will load and prepare the sheep for shearing and will provide subjective assessment of the animals for the robots to select appropriate shearing programs. He (or she) will probably shear two or three blows, opening up tricky sections of the fleece and thus saving the robot far more time. The shorn animals are returned to the operator for inspection and any finishing which may be needed, providing feedback on the effectiveness of preparation. Human shearing experience will be essential for keeping the machinery working efficiently.

Some sheep will still have to be shorn by hand: sheep with injuries, diseased sheep, and animals in very poor condition, and the oddly proportioned animals which just will not fit.

At the same time, working conditions will be much better. The hard, continuous, physical strain will be gone and the workspace will be air-conditioned. Our earlier cost estimates included comfortable mobile on-site accommodation for areas too remote for town commuting. (Most shearers are now based in country towns and commute to work daily.) With the change in work patterns, diets will also change, and cooking for shearers may be less of a full time job than it is at the moment.

In the end, though, the union attitude will depend more on the attitude of wool growers than on technological change. Few shearers, even now, fit the traditional mould—tough men who express themselves with physical strength and stamina, with a huge capacity for food and drink. Most shearers are quiet, hard working and articulate; intelligence, concentration, fitness and skill contribute far more to high tallies than brute force. Shearers responsible for operating robot shearing plants will need the same qualities and will demand to be treated with respect, as working partners.

[1] *The Weekend Australian*, October 15th 1988, p. 21.

What about complex machinery? How will shearers be able to maintain it without degrees in computer science and engineering?

I usually reply to this by asking: 'How many homes equipped with video cassette recorders or personal computers have resident electronics engineers to maintain them?' Not many. The resident experts are usually children between 6 and 16.

Video cassette recorders are complex machines with moving parts built to far greater precision than shearing robots need. Naturally, enormous design effort has been expended to produce such remarkable reliability and operational simplicity. Computers which are built into the machines provide much of the apparent simplicity by making automatic adjustments and guarding against operator mistakes. In the same way, computers built into robots can make automatic adjustments and carry out regular maintenance checks automatically—the operator is alerted when repairs or replacements are needed.

The concept we have proposed for a mobile shearing plant allows for robot shearing stations to be replaced easily if they need maintenance, and a failure of one part of the machinery will not impede the rest of the plant (figure 9.18, p. 271). The operators will be able to replace some parts, such as cutters, on a routine basis. Other repairs and maintenance checks will be performed at fully equipped support bases, just as with modern aircraft. This way, repairs last much longer, reducing operating costs in the long run.

Careful design and attention to detail provides machines with remarkable endurance, and we expect shearing robots to be durable and reliable.

How will mobile units survive being hauled over rough and dusty corrugated country roads?

Merino Wool Harvesting gave far more consideration to this question than we have because they were about to mobilize their equipment at the time they were closed down. Again, it is a matter of design. The most vulnerable components are moving mechanical parts—the manipulators and robots. Electronics units can be provided with vibration isolating mountings, and clean cooling air, and can be shut down for transport. Mechanical parts which move need to be secured. Some rolling bearings are particularly vulnerable because small dents appear in the races when there is severe vibration without other movement to distribute lubricants and spread out the wear along the rolling surfaces. Problems like this, though, are commonplace in many industries and there are many different solutions for them. Mechanical technology improvements in bearings, seals, lubricants, materials and design during the last thirty years have resulted in spectacular reliability gains, though these changes are rather less obvious than miniaturization of electronics components.

How much has been spent on research so far?

Nearly $7 million has been spent on research at The University of Western Australia since 1976 and further amounts have been spent with other research groups—Adelaide University, Melbourne University, University of New South Wales, CSIRO, Monash University and others. Approximately $10.5 million was invested with Merino Wool Harvesting.

How does sheep shearing research relate to other robotics research and development?

Had we been immersed in the robotics research mainstream of the mid 1970's, we might never have started. On the other hand, we were certainly not naïve; it was just that we were able to start independently with a fresh point of view, and we remained safely isolated for the first few years. We looked at the Unimate and other robots and decided to design our own machine. When we submitted our first paper to an international robotics conference it was initially dismissed as a practical joke. Even in the early 1980's, the idea of shearing sheep with robots seemed so difficult, so far removed from what was considered possible, that a leap of faith was needed to accept it.

From our starting point, we have continued on a path at the fringe of robotics research. We have built our techniques around a single application, incorporating adaptation throughout; the more traditional concerns for accuracy and versatility have taken second place. The precision of most factory robots is of little value when the workpieces vary unpredictably.

I tend to view a computer as a fast calculator with severely limited thinking abilities. The mainstream of artificial intelligence (AI) and robotics research sees the computer as the source of 'intelligence' for a robot. I have seen researchers at respected symposia define an intelligent robot as a machine connected to a computer which runs LISP—a programming language used for AI research. There are many research teams working hard to extend the thinking abilities of computers, but this has been much more difficult than most people expected two decades ago. Computers can perform some surprising intellectual feats which 'intelligent' people find hard. Yet the results in other less obvious aspects of intelligence—common sense, visual perception—have been scarce and unimpressive. The thought processes we take for granted, with which all humans, and probably nearly all animals are endowed, are yet to be understood and are not easy to replicate with computers. Therefore, I see the programmer and designer as the sources of intelligence for robots and I think this viewpoint has helped to bypass problems which traditional research regards as 'mountains to be climbed'.

We have adopted the results of mainstream research when they seem to be appropriate. Our motion control techniques are original, and possibly unique, yet they are based upon the same concepts as more common approaches. These concepts have stemmed from the differential and analytical geometry developed by the French mathematicians of the 19th century, the computational geometry developed for computer-aided design and manufacturing, and more recent developments in software structure from computer science.

We have avoided expert systems and artificial neural networks so far because we think it is unlikely that they are useful solutions for our current problems. Expert systems may be useful for guiding maintenance and repair operations of course, and there may be future applications for neural networks. However, we focus our resources on problems which will not be solved by anyone else.

There is another dimension to our work for which there are few parallels in robotics research. Mechanical design has been as important for us as software, and many software problems have been avoided because of this. Unfortunately, mechanical design skills are more scarce than programming skills. To students, computers seem new and exciting, and mechanisms seem to be old-fashioned and obsolete because so many elaborate mechanisms have been replaced by silicon chips. Robotics researchers have often tried to solve problems with software, when good mechanical design would provide a simpler solution. We have been fortunate to have some of the best in these and many other skills and this has helped us find simple solutions.

The value in relating our work to other research projects emerges in unexpected ways. When we look at our own work in a more general robotics research framework, we receive a different view which stimulates new ideas. And good ideas are essential in our type of work.

Is this the end of the story?

Of course not. We can now shear average-sized Merino sheep in good condition with reasonable reliability and improving speed. The remaining research challenge is to broaden the classes of sheep which can be shorn to cover the vast majority of sheep in Australia. To do this, a mobile field trial research unit will be needed, with a robot and manipulator.

I have been with the project since the start of research at the University of Western Australia. By 1981, I had satisfied myself by experiments with ORACLE that robotic shearing was definitely possible and could be an economic proposition. For the next 8 years, my colleagues and I worked towards a practical embodiment of our ideas, through a series of major reviews and equipment demonstrations. We were sustained by the belief that if we demonstrated enough of the components of a shearing system, reliably enough, the Corporation would embark on a development phase and make robotic shearing a commercial reality. Time after time, we attempted to focus attention towards finding 'how it could be done' rather than 'if it could be done'. At the beginning of 1989 I realized that this was an unrealistic expectation. Instead I decided that fully automatic shearing of the whole sheep was, for myself, a very satisfying achievement.

I reduced my involvement to part-time consulting and, freed of the responsibilities of directing the project, I turned my attention to the hybrid follower, shearing skills and machine vision. Here, I felt, I still had useful contributions to make. I had time to write this book and broaden my interests in other directions besides. For this I have to thank Ed Tabb who assumed the difficult role of project director with his predecessor still around, and my colleagues for assuming my technical responsibilities.

A decade may pass before shearing robots earn their keep on farms and that is too long to withhold this part of the story for the sake of completeness.

How long will it be before we see robots in shearing sheds?

. . . Farmers usually ask this question **first**.

Our next step is to build a field research unit—a single mobile shearing robot and manipulator system to obtain shearing experience with a much larger population of sheep. It could take about two years to design, build and test this machine, but the plans have been deferred while the wool industry recovers from its current financial problems.

Once a field research unit is operating, a further two to three years will be needed to develop and manufacture a commercial model. In 1981, the Corporation estimated that commercial development would cost $20 million,[2] though this cost would be recovered through profits on manufacture and operating royalties. Recent estimates are said to range between $20 million and $35 million.

These figures seem to cause great concern in the Corporation, particularly at a time when the debt on the stockpile of unsold wool exceeds $3 billion.

In relative terms, the cost is almost negligible. Shearing costs the wool industry at least $400 million every year. All that is needed to finish the project is the vision, drive and enthusiasm which started it.

[2] P. Hudson quoted in *The Countryman*, August 20th 1981, p 4.

Chronology

The diagram illustrates the chronology of the various major phases of the project and indicates the approximate timespan covered by chapters which describe historical events. The darkness of the bands represents the intensity of work on each major robot development and the thickness represents the number of people involved. Research on software, sensors, cutters and other supporting technology continued between the major robot construction phases, and is described in the other chapters.

Major dates

1972: Farmers Norman Lewis, Jim Shepherd and Lance Lines independently conceive idea of shearing robots.

June 1974: Shearers receive 39% increase—work on shearing robots commenced.

1975: First automated shearing blows by CSIRO machine.

August 1976: Roy Leslie commences work on a study of the control systems aspects of automated shearing at the University of Western Australia.

January 1977: James Trevelyan, Alan Richardson and Lionel Stern review PATS and CSIRO shearing machines.

November 1977: James Trevelyan, Roy Leslie commence work on ORACLE robot.

October 1978: CSIRO withdraw from automated sheep shearing research.

October 1978: ORACLE moves under hydraulic power—long delays due to industrial disputes affect delivery of several key components.

July 1979: First ORACLE shearing test.

April 1980: Software sheep invented.

December 1980: Report 'Automated Shearing—Development Costs, Timescale and Capital Costs' by Roy Leslie, Stewark Key & James Trevelyan predicts $10M development cost, 4 robot system cost $380,000. CAD requested, new robot seen as essential.

March 1981: The first hand-operated sheep manipulator (HOMP) completed.

March 1981: First major review of UWA work by AWC. Review recommends construction of automated manipulator, a target of shearing 95% of sheep under automatic control, and increasing the team from 4 to 13 people.

July 1981: David Elford (and others) join team and start work on an automatic sheep manipulator.

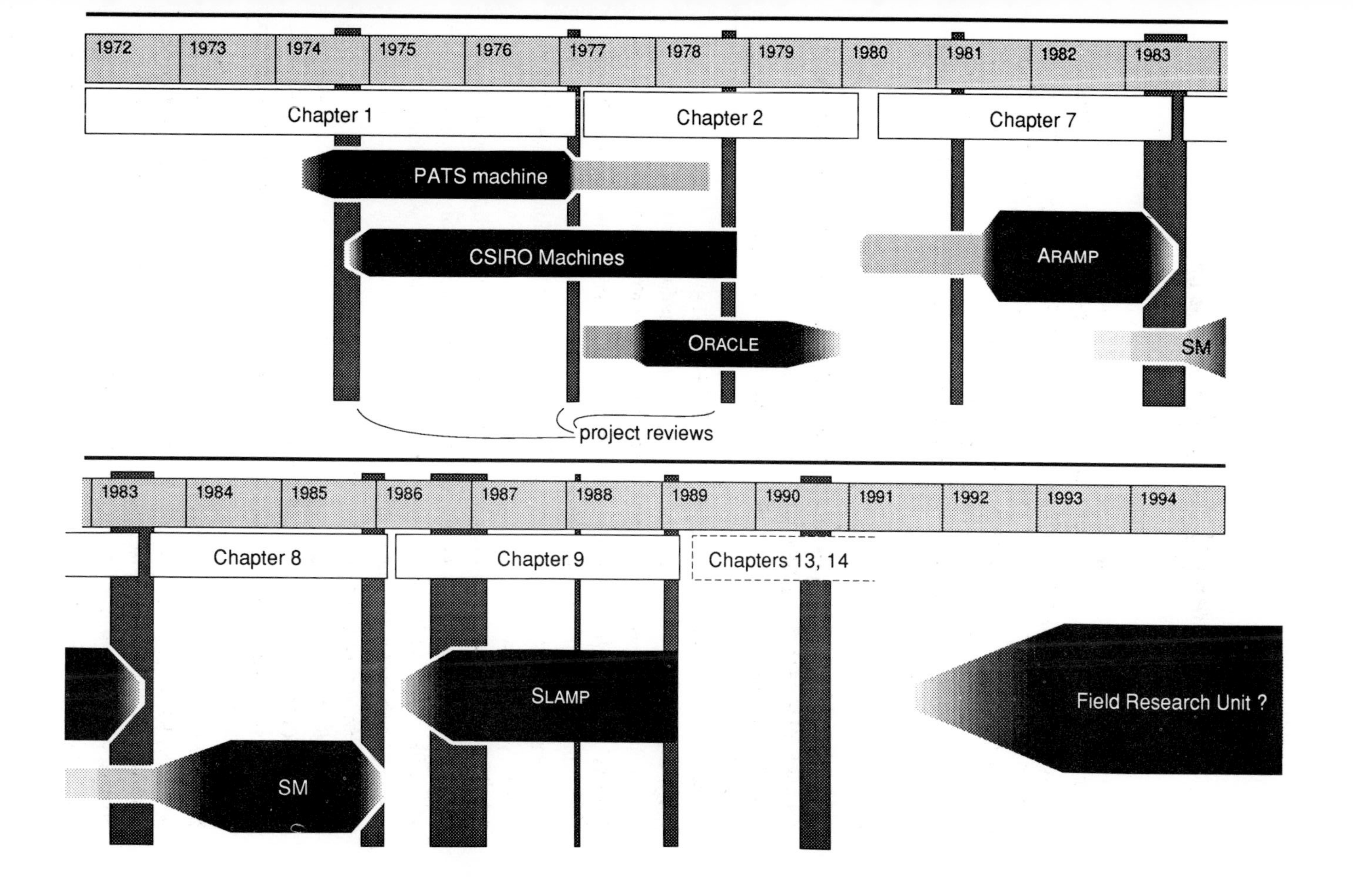

1972
1973
1974
1975
1976
1977
1978
1979
1980
1981
1982
1983
Chapter 1
Chapter 2
Chapter 7
PATS machine
CSIRO Machines
ARAMP
ORACLE
SM
project reviews
1983
1984
1985
1986
1987
1988
1989
1990
1991
1992
1993
1994
Chapter 8
Chapter 9
Chapters 13, 14
SLAMP
Field Research Unit ?
SM

August 1981: Paul Hudson (of AWC R&D Dept) predicts $20M cost to develop automated shearing.

December 1982: First test of ARAMP cradle with a sheep.

December 1982: Michael Ong commences studies on a new robot design. The wrist design is seen as critical.

March 1983: Another AWC review. ORACLE shears a sheep in mostly completed ARAMP manipulator. Leg manipulation still performed manually.

July 1983: ARAMP manipulator completed—reliably manipulates sheep into shearing positions for ORACLE robot.

October 1983: Serious concept evaluation work on SM robot commenced.

January 1984: ET wrist mechanism invented.

May 1984: AWC authorize contruction of new SM robot—completion expected by June 1985.

April 1985: ORACLE decommissioned after about 400 trials.

September 1985: First shearing test of SM robot.

November 1985: AWC review SM robot and see faultless demonstration of robot shearing 80% of a sheep in ARAMP manipulator, only 7 weeks after first shearing test. 14 sheep similarly shorn before demonstration in practice trials.

March 1986: 97% Wool shorn by SM robot on ARAMP manipulator but blows are set up manually.

April 1986: David Elford adopts roller cradle as basis for SLAMP development.

July 1986: Jan Baranski and Peter Kovesi commence work on vision system.

January 1987: Review by technical subcommittee recommend increased funding and purchase of CAD system for mechanical design.

Late October 1987: SLAMP 1.75 prototype completed. Sheep loading tests reveal serious weaknesses in concept.

November 1987: David Elford conceives a fresh approach for sheep loading, solving problems and simplifying design. New design called SLAMP 3.

February 1988: AWC technical subcommittee disappointed with progress but decide to provide more funds, and offer a bonus if team is succesful in demonstrating SLAMP in Feb 1989 with 12 minute shearing time.

March 1988: SLAMP 3 design completed—construction underway. Completion forecast for September 1988.

December 1988: SLAMP 3 sufficiently complete by Christmas Eve for first shearing test.

February 1989: SLAMP 3 demonstrated to Corporation review committee—20 sheep shorn automatically. Time is longer than expected—25 mins—and this causes some disenchantment.

June 1989: James Trevelyan steps aside from directing the project to focus on technical issues.

August 1989: Ed Tabb appointed as Project Director.

July 1990: AWC commission an external technical review of project by V. Scheinman, A. Holzer and B. Hendy. Shearing cycle reduced to 17 minutes.

August 1990: MWH stop work.

November 1990: Funding crisis in wool industry causes R&D budget to be reduced. Work on field research unit deferred.

Glossary of shearing terms

Bale—pack containing up to 180 kg of pressed wool.

Bare belly —a sheep which has naturally shed its belly wool—a shearer's delight.

Belly—wool shorn from the belly.

Bins—space partitioned to hold different grades or types of wool before it is pressed into **bales**.

Blades—earliest shears, before machines. Also known as tongs, scissors, swords, shears, daggers.

Blow—a single shearing stroke performed by a shearer.

Blue tongue—condition of a sheep with heart problems—puff and pant, froth at the mouth, unless released will die.

Board—section of shearing shed where shearers work.

Bog-eye—a handpiece.

Boot lace —when a comb tooth snags and removes a long strip of skin, also a rouseabout.

Break—a short thin section of wool, usually caused by illness or a hot day off water, at which the fibre predictably breaks when pulled. Can severely discount value of fleece if located in middle region of fibre.

Brisket—the boney section of the sheep on the underside between the front legs.

Broken-mouthed—an 8-toothed sheep losing its teeth. A sheep is full-mouthed as long as the teeth remain perfect, but when they decline the sheep becomes failing-mouthed, broken-mouthed, and finally gummy. The age of a sheep is judged by the number and condition of the teeth.

Burry—wool containing clover seeds (burrs) which are prickly and hard to remove.

Catching pen—situated inside shed adjacent to the shearing board, from which shearers catch sheep to shear. Usually two shearers share each pen.

Chicken feet—Also called 'forks'—pivoted V-shaped prongs which drive the cutter across the comb and distribute the tension force evenly to each tooth of the cutter.

Classer—a senior wool shed worker who determines the grade (type or class) of each fleece.

Clout—a trade name for a systemic chemical **lice** treatment. The chemical is squirted from an applicator in a line along the backbone of a newly shorn sheep.

Cobbler—a very rough shearing sheep.

Comb— the lower part of the severance mechanism used for cutting wool.

Condition—generally refers to how well fed the sheep are.

Cots—matted locks of wool.

Counting-out pen—situated outside shed to receive shorn sheep.

Crimp—regular waviness of wool fibres. Also a rough measure of fibre diameter; fine crimp can imply fine wool.

Crossbred—the progeny of two distinct pure breeds of sheep. Also refers particularly to sheep which are not pure Merino.

Crutching—shearing the wool between the rear legs and around the backside of the sheep. This is usually carried out twice a year to reduce problems caused by blowflies and contamination by excreta.

Crutchings—wool removed during crutching.

Cull—uneconomic sheep classed out to retain an even flock.

Cutter— a metal plate with three or four V-shaped teeth, which forms the upper moving half of the severance mechanism for cutting wool.

Dag—consolidated excreta in region of sheep's crutch.

Dewlap—fold of skin up to 100 mm wide along the bottom of the neck of a Merino sheep.

Drag force—the force required to push a shearing **handpiece** through the wool while shearing.

Dump—wool baling term. Also big, strong gun shearers who 'dump' their sheep to gain speed i.e. putting sheep, into its smallest position.

Ewe—female sheep.

Flat to the boards—shearing at top speed, like a blue-arse fly; like a jew-lizard up a gum tree; flat out like a lizard drinking.

Fleece— wool shorn from the main body areas of the sheep, hanging together in one piece. Wool from the belly and legs is shorn separately and is usually of lower quality than the fleece wool.

Fly-strike—blow-fly maggots eating the skin and tissues of the living sheep.

Flyer—very good shearing sheep.

Frib—fribs, fribby, refer to small, short locks, sometimes left on sheep. Also second cuts.

Gullets—gaps between teeth of comb.

Gun—a skilled and fast shearer.

Gun Crutcha—a trade name for a machine for moving sheep to a shearer for **crutching** which incorporates a means of turning the sheep upside down and restraining it during the crutching process.

Handpiece—a powered tool used for manual shearing.

Hogget—sheep 12 months old, as yet unshorn.

Jumbuck—sheep. In 1896 A.S. Meston (Qld) wrote: 'The word jumbuck for sheep appears in Aboriginal speech as jimba, jimbock, dombock and dumbog. In each case it meant the white mist preceding a shower, to which a flock of sheep bore a strong resemblance. It seemed the only thing to which the Aboriginal mind could compare it.

Lice—a small insect which eats dead skin and lives in the fleece wool. As in other animals these insects cause severe irritation to the sheep, breed rapidly and are easily transferred from one sheep to another.

Line—a type of sheep (breeding line) or classification of wool sorted after classing and evaluation by Corporation.

Locks—(second cuts) any staple of wool which is cut twice to remove it or portions of wool from the lower parts of the legs and edges of fleece, greasy staples from under forearm, flank and crutch. Lower quality trimming. Frib wool.

Manipulation—the process of repositioning the sheep (and the parts of the sheep relative to each other) to present the sheep in the right position for shearing a given part of the sheep, with the skin suitably stretched.

Merino—a breed of sheep originating in Spain which has been refined by Australian sheep farmers over the last 200 years. The wool produced by Merino sheep is normally used for clothing as it is the finest of all the different wools produced by different breeds of sheep.

Mob (of sheep)—a flock.

Mulesing—operation performed on the wrinkled back-side of a sheep to remove skin to reduce the risk of **fly-strike**.

Muster—bring sheep together in a mob (or flock) to be brought in for shearing, dipping, etc.

Noils—short, chewed-up, wasty portions of wool.

Pendulum—a fixture which holds the comb or cutter for grinding.

Pieces—odd pieces of wool shorn from legs, or parts of fleece which have dropped on-to the floor.

Pink—to shear so cleanly that the pink skin of a healthy hide shows. Top quality shearing, 'Pinking 'em'. The pink skin can reflect in various breed types, usually plain-bodied sheep.

Press—a machine for pressing wool into bales for transport to wool processing plants.

Pulled comb—comb with outer teeth splayed apart to increase amount of wool shorn each blow.

Race—passageway to guide sheep.

Ram—male sheep which has not been castrated.

Reactive wool—project term for wool which interferes with sensing. Also known as conductive wool when resistance sensing affected. Associated with high **suint** content and **yolky** wool, particularly in crossbreds.

Ringer—best shearer in the shed.

Rouseabout—a person who is the general runabout—picking up wool, doing odd jobs.

Scouring—washing of wool—the first step in wool processing. Also a form of dysentery among sheep pastured on soft, lush feed—hated by shearers.

Second cuts—short wool which has been shorn off by a second **blow**. This wool is of low value and it is best to leave it on the sheep until the next shearing.

Shed, wool shed, shearing shed—specially designed building used for shearing sheep.

Skin conditioning—the combined process of positioning the different parts of the animal and stretching the skin to make it easier to shear the wool.

Skirting—the process of removing dirty and muddy bits of wool from around the edge of the **fleece**. This is currently done by hand.

Skirtings—the wool removed during **skirting**.

Stag—imperfectly castrated male sheep, usually with some of the aggressive characteristics of a ram.

Stained pieces—urine-stained wool and small portions of yolk-stained wool from the crutch.

Staple—number of wool fibres which naturally form themselves into a clump or cluster.

Staple strength —a standardized wool strength measurement.

Stragglers—sheep missed in general muster. Later they are shorn at the straggler-shearing.

Stubble—short wool remaining on the sheep after shearing

Suint—contamination of greasy wool associated with high levels of inorganic salts, particularly potassium salts. Often associated with discolouration of wool.

Tally—number of sheep shorn by shearer daily, weekly, or for the duration of shearing at a particular shed.

Tender—wool weakened by a break or poor conditions.

Tension—actually compression! The contact force between comb and cutter.

Tooth — number of teeth in sheep's mouth is used to indicate age; 2 tooth—1 year old; 4 tooth—2 years old; full mouth—3 years or more; broken mouthed—too old.

Top—wool after scouring and combing processing—resembles an endless sausage of soft white cotton wool.

Top notch, Top knot—wool on top of head.

Vegetable Content—vegetable matter (grass, seeds, burrs, etc.) in wool. Quite small values can heavily discount price of wool because of acid process needed to remove vegetable matter after scouring.

Weaner—6-9 month old sheep.

Wether—castrated ram.

Whipping side—last side of the sheep when being shorn by hand.

Wig—wool on the top of the head of the sheep, often of low commercial value.

Wigging—shearing the wool on the top and sides of the head. This is generally carried out at the same time as **crutching**.

Wool blind—sheep with long wool growing round the eyes, interfering with sight.

Wool table—a table in a shearing shed, about 2 m by 3 m, made from wire mesh or wooden slats over which a shorn fleece is thrown for classing and skirting, outside uppermost. Second cuts and small pieces fall through to the floor.

Yolk—yellow discolouration of wool caused by combined secretions of the several glands of sheep. Often associated with reactive wool (see chapter 6).

Acknowledgement is made for many of the terms in this Glossary to *The Shearers*, Patsy Adam-Smith (1982). (By permission)

References

Note: Copies of references marked 'private communication' are available from the authors cited, or have since been published. See end of list for patent literature.

Adam-Smith, P. (1982). *The shearers*. Thomas Nelson, Melbourne.

Anderson, B. D. O. and Moore, J. B. (1979). *Optimal filtering*. Prentice Hall Information and System Science Series.

Atkinson, K. R. and Parnell, B. G. (1979). A review of experiments on wool severance. In *Wool harvesting research and development: 1st National Conference*, (ed. P. R. W. Hudson), pp. 117–28, Melbourne.

Australian Wool Corporation (AWC) (1990). Annual Report of Wool Research and Development Council, 369 Royal Parade, Parkville, Melbourne.

Baranski, J. (1983). Inductive Sensor. Technical Report 805/9 – IR – 3/83, Automated Sheep Shearing Project, Department of Mechanical Engineering, The University of Western Australia.

Baranski, J. (1986*a*). Inductive Sensor. Technical Report 805/5 – IR – 2/86, Automated Sheep Shearing Project, Department of Mechanical Engineering, The University of Western Australia.

Baranski, J. (1986*b*). J3 Capacitance Sensor. Technical Report 805/6 – IR – 2/86, Automated Sheep Shearing Project, Department of Mechanical Engineering, The University of Western Australia.

Baxter, J. R. (1981). Development of a small mobile transportable automated wool harvesting system. In *Wool harvesting research and development: 2nd National Conference*, (ed. P. R. W. Hudson), pp. 209–14, Sydney.

Beadle, T. N. (1987). Survey of patents relating to wool harvesting. Technical Report 808/1-IR-11/87, Automated Sheep Shearing Project, Department of Mechanical Engineering, The University of Western Australia.

Bligh, R. (1989). Shearing without shears—fact or fantasy? A report on research into biological harvesting of wool. *World Sheep and Wool Congress*, Vol. 1, paper 10, Hobart.

Bognor, R. and Nissinck, H. (1981). Ultrasonic sensing through the fleece. In *Wool harvesting research and development: 2nd National Conference*, (ed. P. R. W. Hudson), pp. 167–76, Sydney.

Bolles, R. C., Kremers, J. H., and Cain, R. A. (1981). A simple sensor to gather three-dimensional data. Technical Note 249, Industrial Automation Department Computer Science and Technology Division, Stanford Research Institute.

Bolton, R. (1979). *People skills—how to assert yourself, listen to others and resolve conflicts*. Spectrum Books, Prentice Hall Inc, USA.

Bookstein, F. L. (1978). *The measurement of biological shape and shape change*. Lecture Notes in Biomathematics, Springer Verlag.

Bowditch, H. G. and Dennis, R. A. (1979). The development of a rotary cutter-type shearing handpiece. In *Wool harvesting research and development: 1st National Conference*, (ed. P. R. W. Hudson), pp. 107–16, Melbourne.

Bowditch, H. G. and Elphick, G. R. (1981). Investigation of rotary severance mechanisms. In *Wool harvesting research and development: 2nd National Conference*, (ed. P. R. W. Hudson), pp. 265–66, Melbourne.

Bowman, M. E. and Forrest, A. K. (1987). Transformation calibration of a camera mounted on a robot. *Image and Vision Computing*, **5**, (4), pp. 261–66.

Brooks, R. A., Flynn, A. M., and Marill, T. (1988). Self-calibration of motion and stereo vision for mobile robots. *Robotics Research: The 4th International Symposium*, Santa Cruz, USA, pp. 267–76, MIT Press.

Bryant, R. C., Nissink, H. F., and Piccirrillo, D. (1983). Ultrasonic sensing for control of a shearing robot—final report of a feasibility study for the AWC. Technical Report, Department of Electrical Engineering, University of Adelaide.

Bryson, J.A. and Field, B. W. (1981). Automatic wool removal and handling devices. In *Wool harvesting research and development: 2nd National Conference*, (ed. P. R. W. Hudson), pp. 183–90. Sydney.

Burrow, P. (1981). Catching sheep automatically. In *Wool harvesting research and development: 2nd National Conference*, (ed. P. R. W. Hudson), pp. 143–50, Sydney.

Canny, J. F. (1983). Finding edges and lines in images. M.Sc. thesis, Massachusetts Institute of Technology, Cambridge, USA.

De Boor, C. (1972). On calculating with B-splines. *Journal of Approximation Theory*, **6**, 50–62.

Dennis, R. A. (1979). Alternative drives for the shearing handpiece. In *Wool harvesting research and development: 1st National Conference.*, (ed. P. R. W. Hudson), pp. 163–86, Melbourne.

Dickmanns, E. D. (1988). Dynamic computer vision for mobile robot control. *International Symposium and Exposition on Robots*, (19th ISIR), pp. 314–27, Sydney.

Ewbank, R. (1968). The behaviour of animals in restraint. In *Abnormal behaviour in animals*, (ed. M. W. Fox), Saunders Co., Philadelphia, USA.

Faux, I. D. and Pratt, J. (1979). *Computational geometry for design and manufacture*. Ellis Horwood, London.

Field, B. W. (1981). Severance requirements for automated shearing. In *Wool harvesting research and development: 2nd National Conference*, (ed. P. R. W. Hudson), pp. 177–82, Sydney.

Field, B. W., Mackenzie, A. J., and Amery, M. I. (1981). Physical characteristics of wool shearing actions. In *Wool harvesting research and development: 2nd National Conference*, (ed. P. R. W. Hudson), pp. 267–84, Sydney.

Fisher, R. and Urey, W. (1981). *Getting to Yes*. Century Hutchison, London.

Gabel, R. A., and Roberts, R. A. (1980). *Signals and linear systems*, (2nd edn.). Wiley.

Gosling, A. B., Marsh, J. D., and Smith, M. J. (1979). Investigation of wool severance techniques. *In Wool harvesting and research development: 1st National Conference*, (ed. P. R. W. Hudson), pp. 95–106, Melbourne.

Harmon, S. Y. (1986). Practical implementation of autonomous systems: problems and solutions. *1st. International Conference on Intelligent Autonomous Systems*, pp. 47–59, Amsterdam, Elsevier.

Hayward, V. and Paul, R. P. (1986). Robot manipulator control under Unix RCCL: a robot control 'C' library. *International Journal of Robotics Research*, **5**, (4), 94–111.

Hochstadt, H. (1986). Performance of a hydraulic gear motor. B. Eng. thesis, Department of Mechanical Engineering, The University of Western Australia.

Howard, I. M. (1987). The development of a transfer function technique for the modelling of the vibration characteristics of any beam-like structure. Ph.D. thesis, Department of Mechanical Engineering, The University of Western Australia.

Hunt, K. H. (1978). *Kinematic geometry of mechanisms*. Clarendon Press, Oxford.

Jarvis, R. A. (1986). Eye in hand robotic vision. In *Robots in Australia's Future, National Conference of Australian Robot Association*, pp. 149–61, Perth.

Kass, M., Witkin, A., and Terzopoulos, D. (1988). Snakes: active contour models. *International Journal of Computer Vision*, **1**, 321–31.

Key, S. J. (1981). Possible automated shearing system productivities. In *Wool harvesting research and development: 2nd National Conference*, (ed. P.R.W. Hudson), pp. 191–96, Sydney.

Key, S. J. (1983). Automated restraint and manipulation platform: report on driver software status. Technical Report 221–R–6/83, Automated Sheep Shearing Project, Department of Mechanical Engineering, The University of Western Australia.

Key, S. J. (1985). Productivity modelling and forecasting for automated shearing machinery. *Agri-Mation 1 Conference*, pp. 200–209, Chicago, ASAE.

Key, S. J. and Elford, D. (1983). Animal positioning, manipulation and restraint for a sheep shearing robot. *International Conference on Intelligent Machines in Agriculture*, pp. 45–51, University of Florida, Tampa, USA. (Edited version in chapter 7)

Kovesi, P. D. (1981). Sheep profile modelling for automated shearing. B. Eng. thesis, Department of Mechanical Engineering, The University of Western Australia.

Kovesi, P. D. (1985). Obstacle avoidance. *2nd International Conference on Advanced Robotics* (ICAR), pp. 51–58, Tokyo.

Kovesi, P. D. (1987). Machine vision system for SLAMP and its lighting requirements. Technical Report, Automated Sheep Shearing Project, Department of Mechanical Engineering, The University of Western Australia.

Kovesi, P. D. and Walker, G. J. (1988). Vision generation of surface models for guiding a sheep shearing robot. *International Symposium and Exposition on Robots*, (19th ISIR), pp. 417–26, Sydney.

Kovesi, P. D. and Walker, G. J. (1989). The vision system user's manual. Technical Report, Automated Sheep Shearing Project, Department of Mechancial Engineering, The University of Western Australia.

Lee, M. H., Barnes, D. P., and Hardy, N. W. (1985). Research into error recovery for sensory robots. *Sensor Review*, **5**, (4), pp. 194–7.

Leslie, R. A. (1980). Preliminary report on animal manipulation. Technical report 200-R-80, Automated Sheep Shearing Project, Department of Mechanical Engineering, The University of Western Australia.

Leslie, R. A. and Trevelyan J. P. (1979). Mechanical development of an automated shearing rig. In *Wool harvesting research and development: 1st National Conference*, (ed. P. R. W. Hudson), pp. 411–20, Melbourne.

Mackenzie, A. J., Freeman, R. B., and Burrow, R. P. (1979). The MUSHEEP programme in mechanical wool severance. In *Wool harvesting research and development: 1st National Conference*, (ed. P. R. W. Hudson), pp. 129–40, Melbourne.

Marr, D. C. (1983). *Vision*. Freeman, San Francisco.

Meriam, J. L. (1975). *Dynamics.* (2nd edn.). Wiley, New York.

Milenkovic, V. (1984). Effects of robot wrist singularity on path control. *14th International Symposium on Industrial Robots* (ISIR), Gothenburg, Sweden, pp. 279–86, IFS Conferences Ltd.

Milenkovic, V. and Huang, B. (1983). Kinematics of a major robot linkage. *13th*

International Symposium on Industrial Robots (ISIR), Chicago. USA., pp. 31–47, Society of Manufacturing Engineers.
Milenkovic, V., Milenkovic, V. J., and Milenkovic, P. H. (1990). Inverse kinematics of not fully serial robot linkages with nonsingular wrists. Private communication.
Morrone M. C. and Owens R. A. (1987). Feature detection from local energy. *Pattern Recognition Letters*, **6**, pp. 303–13, North Holland.
Nelson, M. J. (1986). Automatic error recovery for adaptive robots. Technical Report 86/3, Department of Computer Science, University of Western Australia.
Nelson, M. J. (1990). An adaptive system for mobile robot navigation. *National Conference of Australian Robot Association*, pp. 186–97, Melbourne.
Ng, K. and Alexander, R. (1986). 3D shape measurement by active triangulation using a projected array of coded dots or stripes of light. In *Robots in Australia's Future, National Conference of Australian Robot Association*, pp. 242–50, Perth.
Ong, M. (1981). An approach to the mechanical design of a sheep shearing robot. In *Wool harvesting research and development: 2nd National Conference*, (ed. P. R. W. Hudson), pp. 161–6, Sydney.
Ong, M. (1985). Technical reports and working papers on force sensing. Technical Reports 1598-WF-7/85 through to 1674. Automated Sheep Shearing Project, Department of Mechanical Engineering, The University of Western Australia.
Ong, M., Howard, I. M., and Trevelyan, J. P. (1989). Manipulator arm design for fast response with simple controls. Technical Report, Automated Sheep Shearing Project, Department of Mechanical Engineering, The University of Western Australia.
Owens, R. A. (1982). Report on model testing. Technical Report 445/501-R-82, Automated Sheep Shearing Project, Department of Mechanical Engineering, The University of Western Australia.
Owens, R. A. (1983). Documentation of statistical procedures used in automated shearing of sheep. Technical Report 445/502-R-83, Automated Sheep Shearing Project, Department of Mechanical Engineering, The University of Western Australia.
Owens, R. A. (1985). Surface prediction and adaptation in a robot's workspace. *Mechanical Engineering Transactions*, Institution of Engineers, Australia, **ME10**, (3), 203–7.
Paul, R. P. and Stephenson, C. N. (1983). Kinematics of robot wrists. *International Journal of Robotics Research*, **2**, (1), 31–8.
Penrose, R. (1990). *The Emperor's new mind.* Vintage (Oxford University Press).
Phillips, C. L. and Nagel, H. T. (1984). *Digital control systems analysis and design.* Prentice Hall, Englewood Cliffs, USA.
Randell, P., Lee, P. A., and Treleaven, P. C.(1978). Reliability issues in computing system design. *Computing Surveys*, **10**, (2), 123–65.
Rogers, D. F. and Adams, J. A. (1985). *Mathematical elements for computer graphics.* McGraw-Hill.
Rogers, K. J. (1990). Wool harvesting with robots. *Electrical and Electronic Engineering Transactions*, Institution of Engineers, Australia, **EE10**, (3), 207–15.
Rosheim, E. I. (1988). *Robot wrist actuators.* Wiley Interscience Publications, New York.
Sharkey, I. M., Daniel, R. W., and Elosegui, P. (1989). Transputer based real time robot control. Private communication, Robotics Research Group, Department of Engineering Science, Oxford University.
Sharrock, T. J. (1983). Biquadratic local patches on sheep surfaces. Technical Report

440/503-R-12/83, Automated Sheep Shearing Project, Department of Mechanical Engineering, The University of Western Australia.

Siciliano, B. (1990). Closed loop inverse kinematic scheme for on-line joint based robot control. *Robotica*, **8**, (3) 231–7.

Smith, R. (1990). Automated wool harvesting. *International Advanced Robotics Programme—Workshop on Robotics in Agriculture and the Food Industry*, Avignon, France.

Smith, R. E. and Gini, M. (1986). Reliable real-time robot operation employing intelligent forward recovery. *Journal of Robotic Systems*, **3**, (3), 281–300.

Stackhouse, E. (1979). A new concept in wrist flexibility. *9th International Symposium on Industrial Robots* (ISIR), pp. 589–600, Washington, DC.

Syme, L. A. and Durham, I. H. (1982). Microprocessor control of sheep movement. In *Wool harvesting research and development: 2nd National Conference*, (ed. P. R. W. Hudson), pp. 237–46, Sydney.

Taylor, R. H., Hollis, R. L., and Lavin, M. A. (1985). Precise manipulation with endpoint sensing, *Robotics Research: The 2nd International Symposium*, pp. 59–69, MIT Press.

Terzopoulos, D., Witkin, A., and Kass, M. (1987). Snakes: active contour models. 1st *International Conference on Computer Vision*, pp. 269–76.

Texas Instruments, (1985). Control systems compensation and implementation with the TMS32010. Digital Signal Processing Application Reports SPRA009.

Trevelyan, J. P. (1981a). Software for automated sheep shearing. *51st ANZAAS Congress*, Brisbane, Queensland.

Trevelyan, J. P. (1981b). Automated shearing experiments. In *Wool harvesting research and development: 2nd National Conference*, (ed. P. R. W. Hudson), pp. 151–60. Melbourne.

Trevelyan, J. P. (1982). Report on deviation functions. Technical Report 440-R-08/82, Automated Sheep Shearing Project, Department of Mechanical Engineering, The University of Western Australia.

Trevelyan, J. P. (1985). Skills for a shearing robot: dexterity and sensing. *Robotics research: 2nd International Symposium*, Kyoto, Japan, pp. 273–80, MIT Press.

Trevelyan, J. P. (1989a). Sensing and control for sheep shearing robots. *IEEE Transactions on Robotics and Automation*, Special issue on sensing and control, **5**, (6), 716–27.

Trevelyan, J. P. (1990*a*). Replicating manual skills. *Robotics research: 5th International Symposium*, Tokyo, Japan, pp. 333–40, MIT Press.

Trevelyan, J. P. (1990*b*). Team skills for mechatronics. *SERC Conference on Mechatronics*, St Albans, I. Mech E., London.

Trevelyan, J. P. and Nelson, M. (1987). Adaptive robot control incorporating automatic error recovery. *3rd International Conference on Advanced Robotics (ICAR)*, pp. 385–98, Versailles, France.

Trevelyan, J. P. and Leslie, R. A. (1979). A control system for automated shearing. In *Wool harvesting and research development: 1st National Conference*, (ed. P. R. W. Hudson), pp. 443–60. Melbourne.

Trevelyan, J. P. and Elford, D. (1988). Sheep handling and manipulation for automated shearing, *International Symposium and Exposition on Robots (19th ISIR)*, pp. 397–416, Sydney.Trevelyan, J. P. Key, S. J., and Owens, R. A. (1982). Techniques for surface representation and adaptation in automated shearing. *12th International Symposium on Industrial Robots* (ISIR), pp. 163–74, Paris.

Trevelyan, J. P., Kovesi, P., and Ong, M. (1984). Motion control for a sheep shearing robot. *Robotics research: 1st International Symposium*, pp. 175–90, MIT Press.

Trevelyan, J. P., Kovesi, P., Ong, M., and Elford, D. (1986). A wrist mechanism without singular positions. *International Journal of Robotics Research*, **4**, (4), 71–85.

Trevelyan, J. P., Kovesi, P. D., and Nelson, M. (1988). Adaptive motion sequencing for process robots. *Robotics research: 4th International Symposium*, Santa Cruz, USA, pp. 445–54, MIT Press.

Tsai, R. Y. (1986). An efficient and accurate camera calibration technique for 3D machine vision. *Conference on Computer Vision and Pattern Recognition*, pp 364–74.

Velletri, P. (1989). Controlling robot manipulators at motion limits. B. Eng. thesis, Department of Mechanical Engineering, The University of Western Australia.

Walker, G. J. (1987). Preliminary evaluation of surface prediction using vision measurements. Technical Report 870/6-N-08/87, Automated Sheep Shearing Project, Department of Mechanical Engineering, The University of Western Australia.

Weatherburn, C. E. (1927). *Differential geometry of three dimensions*, Vol. 1. Cambridge University Press.

Whitney, D. E. (1969). Resolved motion and control of manipulators and human arms and prostheses. *IEEE Transactions on Man Machine Systems*, **MMS-10**, (2), 47–53.

Whitney, D. E. (1986). Real robots don't need jigs. *IEEE Conference on Robotics and Automation*, **Vol. 2**, pp. 746–52, San Francisco.

Whitney, D. E. and Brown, M. L. (1987). Metal removal models and process planning for robot grinding. *17th International Symposium on Industrial Robots* (ISIR). pp. 19–29 – 44, Chicago.

Williams, R. A. and Phillips, R. N. (1981). Wool harvesting systems—an operations research study. In *Wool harvesting and research development: 2nd National Conference*, (ed. P. R. W. Hudson), pp. 197–208, Sydney.

Yoshikawa, T. (1985). Manipulability of robotic mechanisms. Robotics Research: 2nd International Symposium, pp. 439–46. Kyoto, Japan, MIT Press.

Patents

Anderson, C. C. (1976). Shearing cradle. Australian patent 501,826(15901/76).

Anderson, C. C. (1987). Shearing cradle. Australian patent application 53,982/79.

Baranski, J. (1987*a*). Inductive sensor. Australian patent 604059.

Baranski, J. (1987*b*). Capacitance sensor arrangement. Australian patent 602178.

Baranski, J., Greenhalgh, R., Steven, D. H., and Trevelyan, J. P. (1982). Capacitive sensing. Australian patent 582543.

Brooker, R. E. and Brooker, D. L. (1983). Sheep shearing table. Australian patent application 29867/84.

Drew, G. V. (1976). Animal support platform. Australian patent 501,613 (15738/76).

Elford, D. (1989). Sheep handling and manipulation for automated shearing. Australian patent application 22059/88, United States patent 4892062.

Elford, D. and Leslie, R. A. (1984). Manipulation and restraining of animals during shearing—ARAMP. Australian patents 567764, 574295, 574614.

Elford, D., Kovesi, P., Ong, M., and Trevelyan, J. P. (1988). Wrist mechanism for robotic manipulators. Australian patent 591834, United States patent 4862759.

Greenhalgh, R., Leslie, R. A., and Trevelyan, J. P. (1981). Resistance sensing. New Zealand patent 210154.

Milenkovic, V. (1988). Hollow non singular robot wrist, U.S. patent 4,744,264.

Milenkovic, V. (1990). Non singular industrial robot wrist, U.S. patent 4,907,937.

Moffatt, G. C. (1981). Animal handling device. Australian patent 550,883 (80310/82).

Susnjara, K. J. and Fleck, M. A. (1982). Wrist construction for industrial robots. United States patent 4,353,677.

Trevelyan, J. P. (1985). Sensor shielding device. Australian patent 591838.

Trevelyan, J. P. and Crooke, M. J. (1982). Contact sensing device. Australian patent 563897.

Trevelyan, J. P. and Whitehead, A. (1988). Animal head restraint. Australian patent 595129, United States patent 4887553.

Tunsey, J. P. (1974). Handling animals. Australian patent 492,434 (application 85,841/75).

Van der Heyden, E. (1977). Sheep handling cradle. Australian patent application 32,292/78.

Wytkin, A. J. (1981). Shearing table restraint. Australian patent 556,242 (application 83647/82).

Young, H. S. C. (1979). Crutching machine for sheep (Gun Crutcha). Australian patent 536,338 (application 60834/80).

Index

Page references in bold are boxed explanations; references in italics are illustrations. Names in bold refer to software commands or functions; names in italics refer to variable names used in text.